Addressing New Therapeutic Strategies Using Models

Addressing New Therapeutic Strategies Using Models

Editors

Ana Faustino
Paula A. Oliveira

MDPI • Basel • Beijing • Wuhan • Barcelona • Belgrade • Manchester • Tokyo • Cluj • Tianjin

Editors
Ana Faustino
Department of Zootechnics
University of Évora
Évora
Portugal

Paula A. Oliveira
Departamento de Ciências Veterinárias
Universidade de Trás-os-Montes e Alto Douro
Vila Real
Portugal

Editorial Office
MDPI
St. Alban-Anlage 66
4052 Basel, Switzerland

This is a reprint of articles from the Special Issue published online in the open access journal *Veterinary Sciences* (ISSN 2306-7381) (available at: www.mdpi.com/journal/vetsci/special_issues/Therapeutic_Models).

For citation purposes, cite each article independently as indicated on the article page online and as indicated below:

LastName, A.A.; LastName, B.B.; LastName, C.C. Article Title. *Journal Name* **Year**, *Volume Number*, Page Range.

ISBN 978-3-0365-7083-9 (Hbk)
ISBN 978-3-0365-7082-2 (PDF)

Contents

Hee-Ryung Lee, Gareeballah Osman Adam and Shang-Jin Kim
Application of Patient-Specific Instrumentation in a Dog Model with Antebrachial Growth Deformity Using a 3-D Phantom Bone Model
Reprinted from: *Vet. Sci.* **2022**, *9*, 157, doi:10.3390/vetsci9040157 . . . . **1**

Ioannis Lyros, Efstratios Ferdianakis, Demetrios Halazonetis, Theodoros Lykogeorgos, Antigoni Alexiou and Konstantina-Eleni Alexiou et al.
Three-Dimensional Analysis of Posterior Mandibular Displacement in Rats
Reprinted from: *Vet. Sci.* **2022**, *9*, 144, doi:10.3390/vetsci9030144 . . . . **13**

Ana Mafalda Rasteiro, Eva Sá e Lemos, Paula A. Oliveira and Rui M. Gil da Costa
Molecular Markers in Urinary Bladder Cancer: Applications for Diagnosis, Prognosis and Therapy
Reprinted from: *Vet. Sci.* **2022**, *9*, 107, doi:10.3390/vetsci9030107 . . . . **29**

Debra Hickman, Anjali Prakash and Richard Bell
Predictive Value of Grooming Behavior for Development of Dermatitis in Selectively Bred P Rats as a Model of Trichotillomania Hair Pulling Disorder
Reprinted from: *Vet. Sci.* **2022**, *9*, 89, doi:10.3390/vetsci9020089 . . . . **65**

Tae-Kyeong Lee, Junkee Hong, Ji-Won Lee, Sung-Su Kim, Hyejin Sim and Jae-Chul Lee et al.
Ischemia-Induced Cognitive Impairment Is Improved via Remyelination and Restoration of Synaptic Density in the Hippocampus after Treatment with COG-Up® in a Gerbil Model of Ischemic Stroke
Reprinted from: *Vet. Sci.* **2021**, *8*, 321, doi:10.3390/vetsci8120321 . . . . **75**

Anand Paramasivam, Suresh Mickymaray, Saikarthik Jayakumar, Mathew Jeraud, Periasamy Perumal and Abdullah Alassaf et al.
Locomotor Behavior Analysis in Spinal Cord Injured *Macaca radiata* after Predegenerated Peripheral Nerve Grafting—A Preliminary Evidence
Reprinted from: *Vet. Sci.* **2021**, *8*, 288, doi:10.3390/vetsci8120288 . . . . **91**

Ekaterina Nefedova, Vyacheslav Koptev, Anna S. Bobikova, Viktoria Cherepushkina, Tatyana Mironova and Vasily Afonyushkin et al.
The Infectious Bronchitis Coronavirus Pneumonia Model Presenting a Novel Insight for the SARS-CoV-2
Dissemination Route
Reprinted from: *Vet. Sci.* **2021**, *8*, 239, doi:10.3390/vetsci8100239 . . . . **111**

Tae-Kyeong Lee, Jae-Chul Lee, Hyun-Jin Tae, Hyung-Il Kim, Myoung Cheol Shin and Ji Hyeon Ahn et al.
Therapeutic Effects of Risperidone against Spinal Cord Injury in a Rat Model of Asphyxial Cardiac Arrest: A Focus on Body Temperature, Paraplegia, Motor Neuron Damage, and Neuroinflammation
Reprinted from: *Vet. Sci.* **2021**, *8*, 230, doi:10.3390/vetsci8100230 . . . . **125**

Sara Caceres, Angela Alonso-Diez, Belén Crespo, Laura Peña, Maria J. Illera and Gema Silvan et al.
Tumor Growth Progression in Ectopic and Orthotopic Xenografts from Inflammatory Breast Cancer Cell Lines
Reprinted from: *Vet. Sci.* **2021**, *8*, 194, doi:10.3390/vetsci8090194 . . . . **141**

Geng-Ruei Chang, Po-Hsun Hou, Chao-Min Wang, Jen-Wei Lin, Wei-Li Lin and Tzu-Chun Lin et al.
Imipramine Accelerates Nonalcoholic Fatty Liver Disease, Renal Impairment, Diabetic Retinopathy, Insulin Resistance, and Urinary Chromium Loss in Obese Mice
Reprinted from: *Vet. Sci.* **2021**, *8*, 189, doi:10.3390/vetsci8090189 . . . . **153**

Ambre Baillou, Nathalie Kasal-Hoc, Céline Barc, Juliette Cognié, Anne Pinard and Jérémy Pezant et al.
Establishment of a Newborn Lamb Gut-Loop Model to Evaluate New Methods of Enteric Disease Control and Reduce Experimental Animal Use
Reprinted from: *Vet. Sci.* **2021**, *8*, 170, doi:10.3390/vetsci8090170 . . . . **177**

Jennifer Davis, Anthea L. Raisis, Claire R. Sharp, Rachel E. Cianciolo, Steven C. Wallis and Kwok M. Ho
Improved Cardiovascular Tolerance to Hemorrhage after Oral Resveratrol Pretreatment in Dogs
Reprinted from: *Vet. Sci.* **2021**, *8*, 129, doi:10.3390/vetsci8070129 . . . . **193**

Elisabete Nascimento-Gonçalves, Bruno A.L. Mendes, Rita Silva-Reis, Ana I. Faustino-Rocha, Adelina Gama and Paula A. Oliveira
Animal Models of Colorectal Cancer: From Spontaneous to Genetically Engineered Models and Their Applications
Reprinted from: *Vet. Sci.* **2021**, *8*, 59, doi:10.3390/vetsci8040059 . . . . **209**

veterinary sciences

MDPI

Article

Application of Patient-Specific Instrumentation in a Dog Model with Antebrachial Growth Deformity Using a 3-D Phantom Bone Model

Hee-Ryung Lee [1,†], Gareeballah Osman Adam [2,3,†] and Shang-Jin Kim [4,*]

1 Hansarang Animal Hospital, Seoul 02880, Korea; hr-lee@hanmail.net
2 Department of Veterinary Medicine and Surgery, College of Veterinary Medicine, Sudan University of Science and Technology, Khartoum P.O. Box 204, Sudan; gorba000@gmail.com
3 R&D Division, HUVET Co., Ltd., Iksan 54531, Korea
4 College of Veterinary Medicine, Jeonbuk National University, Specialized Campus, Iksan 54596, Korea
* Correspondence: abbasj@jbnu.ac.kr; Tel.: +82-63-850-0963
† These authors contributed equally to this work.

Abstract: One of the most frequent bone deformities in dogs is antebrachial growth deformity (AGD), which results from malunion of the distal growth plates. The objective of the present study was to re-align the limbs, which can correct the length mismatch and reset the coherence of the joint with the aid of a 3-D phantom model for surgical preplanning. A 14-month-old, intact female Golden Retriever with an angular deformity of the left radius and ulna was selected for the study. The diagnosis was confirmed by orthogonal radiographs. Moreover, computed tomography (CT) scans revealed a multiplane deformity with valgus, procurator, and external rotation of the left radius. The pre-surgical planning started with the quantification of the angular deformity, followed by a simulated virtual osteotomy, and concluded with an in vitro rehearsal surgery on 3-D printed phantom bone models. In the operating room, prefabricated patient-specific instrumentation (PSI) was attached at the planned site of the radial bone surface for a precise closing wedge osteotomy. Then two locking plates were fixed routinely. Post-operative radiographs showed accurate correction of the deformity as we had planned. At 12 weeks post-operatively, the follow-up surveys revealed improved gait, weight-bearing, and progression of bone healing. Our PSI design, based on novel surgical planning, was steady yet straightforward during the osteotomy. The osteotomy was performed without difficulty since the PSI that pre-determined the sites and angles let the surgeon perform the antebrachial malformation surgery. This method of operation reduces stress on the operator and helps to improve accuracy, repeatability, and surgery time.

Keywords: antebrachial growth deformity; dog; osteotomy; patient-specific instrumentation; 3-D

Citation: Lee, H.-R.; Adam, G.O.; Kim, S.-J. Application of Patient-Specific Instrumentation in a Dog Model with Antebrachial Growth Deformity Using a 3-D Phantom Bone Model. *Vet. Sci.* **2022**, *9*, 157. https://doi.org/10.3390/vetsci9040157

Academic Editors: Ana Faustino and Paula A. Oliveira

Received: 7 February 2022
Accepted: 18 March 2022
Published: 25 March 2022

1. Introduction

The disparity in growth between the radius and the ulna is associated with multiplanar contortions, which result in elbow joint deformity or inappropriate bodyweight stress on the carpal joint, as well as osteoarthritis in the elbow, carpal joints, or both [1]. In dogs, antebrachial growth deformities (AGDs) are the most common deformities, typically attributable to the extremes of chondrodystrophic formation or damaging premature closure of the distal growth plates [2–7].

Corrective osteotomy is advised if mobility and motion intolerance have changed, decreased, or secondary osteoarthritis of the carpal and elbow joints has progressed [3–5]. The purpose of corrective angular deformity osteotomy is to re-align the limbs, which can fix the length mismatch and reset the coherence of the joint. Corrective osteotomy requires a precise description of the features and range of deformation to plan open wedge osteotomy, closed wedge osteotomy, or dome osteotomy [4,6–9]. Correction can be immediate at the time of surgery or gradual with the use of external fixation incorporating hinged motors. Limb alignment using the external skeletal fixator has many advantages [10,11]. However, external skeletal fixation has the risk of pin-tract infection that is adversely related to morbidity compared to internal fixation [12,13].

Critical iatrogenic translational defects can result from applications of osteotomies that are inappropriate because of inadequately quantified bone deformities [14]. In the case of angular limb deformity corrective osteotomy, many studies have been conducted using 3-D techniques and rapid prototyping (RP) bone models. In particular, the improvement of the operation time, surgical invasion, surgical accuracy, patient pain, and patient-specific instrumentation (PSI) usefulness, such as risk, pre-operative planning, and surgical error reduction, have recently been demonstrated [15].

In general, programs that convert computed tomography (CT) data, which are used in many studies, into a 3-D printable file format (Stereo Lithography; STL; .stl) after bone segmentation and a 3-D surgical planning simulation program are expensive and require a professional workforce for the operation. Therefore, attempting to request 3-D print production of an affected bone model and patient-specific osteotomy guides from an external company may result in more than two weeks, several modifications, and relatively high costs [16,17]. Therefore, in this experiment, the PSI was designed for osteotomy saw blade guide and the bone plate and screw fixation with closing wedge corrective osteotomy.

In our previous report in the journal of animals [18], we demonstrated that the limitations of the freehand method could be overcome protocol in an in vitro environment using the PSI. Therefore, the goal of this study is to use the PSI protocol in the operating room on a dog with AGD. We reported the results and proposed novel methods based on successful findings.

2. Materials and Methods

2.1. Case Selection

At Hansarang animal clinic, a 14-month-old intact female Golden Retriever with an angular deformity of the left radius and ulna was treated with corrective osteotomy. Left forelimb lameness, restriction in flexion, radial shortening, and valgus of the left distal radius and ulna due to the partial premature closure of the epiphyseal growth plates were observed. A conventional orthogonal radiographic evaluation was performed. The anatomical angles of the proximal and distal radius were measured on frontal and sagittal view radiographs of the affected and unaffected contralateral radii. On the frontal view, the initial anatomical medial proximal radius angle (aMPRA; ref 82–83°; mean value 83°) and the lateral distal radius angle (aLDRA; ref 85–87°; mean value 86°) were 75° and 57°, respectively. On the sagittal view, the anatomical caudal proximal radial angle (aCdPRA; ref 84–86°; mean value 85°) measured value was 92°, while the caudal distal radial angle (aCdDRA; ref 76–78°; mean value 77°) measured value was 62° (Figure 1). The unaffected radius's joint orientational angles, aMPRA, aLDRA, aCdPRA, and aCdDRA were 84°, 85°, 87°, and 77°, respectively (Figure 2). The magnitude of CORA for the case was 25° valgus and 30° procurvatum. The reference values were obtained from a previous study [19].

2.2. CT and Bone Image Segmentation

Stereolithographic 3-D triangular images of the radius and ulna of the affected and contralateral unaffected sides were generated by segmenting the bones from CT images (scanned with a 16-detector helical CT scanner (slice thickness, 0.7 mm; 120 kV; helical CT Alexion, TOSHIBA, Japan)) using 3-D slicer freeware (3-D Slicer, https://www.slicer.org/; version 4.8.0, accessed on 12 March 2021) [20–22]. The CT protocol for corrective osteotomies of the forearms included contiguous scanning of the whole forearm from the elbow to the radiocarpal joint. Manual thresholding and region growing were used for the bone image segmentation.

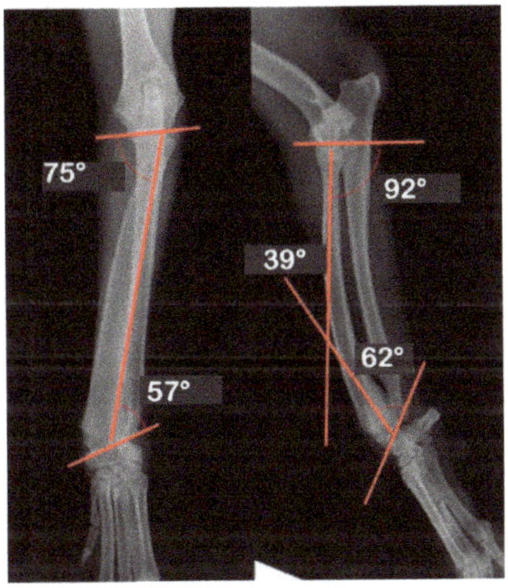

Figure 1. Orthogonal radiographs of the affected antebrachium. The anatomical angles of the proximal and distal radii were measured. The anatomic medial proximal radius angle (aMPRA) and the lateral distal radius angle (aLDRA) were initially 75° and 57°, respectively (Left). The anatomic caudal proximal radial angle (aCdPRA) measured 92°, while the caudal distal radial angles (aCdDRA) were 62°.

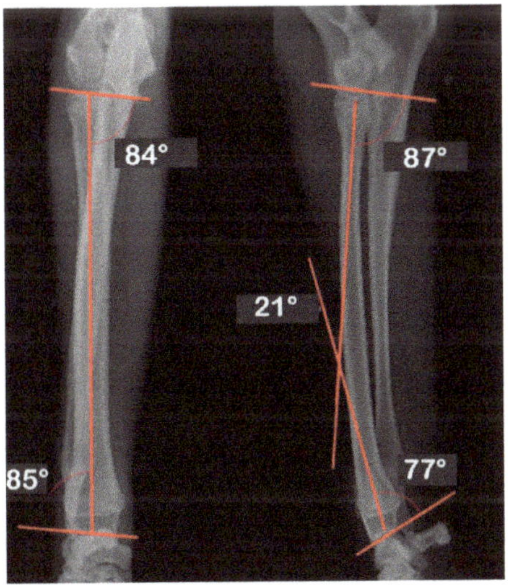

Figure 2. Orthogonal radiographs of the unaffected contralateral antebrachium. The anatomical angles of the proximal and distal radii were measured. The unaffected radius's joint orientational angles, aMPRA, aLDRA, aCdPRA, and aCdDRA were 84°, 85°, 87°, and 77°, respectively.

2.3. Preoperative Surgical Planning

The computer planning was performed on a desktop personal computer using the Windows 10 software application, 3-D builder (Microsoft Windows free application program, Microsoft Corporation, Redmond, WA, USA). The contralateral virtual bone model was mirrored and aligned with the pathological bone using the Iterative Closest Plane surface registration algorithm, to quantify the AGD in 3-D, as described in our published study [18]. A closing wedge osteotomy was planned [23]. To achieve a unique fit, PSI was designed to contain irregular convex and concave surfaces covering the bone from other directions, and for the corresponding surface of the guide to be placed on the bone as an exact replication of the surface of the bone model. A drill sleeve hole was designed for the most distal screw hole on the proximal fragment of the radius that can be drilled before the pre-contoured y-universal locking plate (YULP) is allowed to settle on the precise site. The PSI was fabricated in-house using the PMMA (Polymethyl methacrylate) material, based on the PLA model that was previously manufactured with the desktop 3-D printer device.

The authors chose 2.7 mm, a YULP, a compression locking plate (CLP), and locking screws (BS.COREM, Jeonbuk, Korea). After osteotomy simulation, the distal part was reduced, followed by stable plates that were pre-contoured based on the RP bone model surface in the planned correction. Cancellous bone or scaffold material for bone tissue engineering was not planned for this operation.

2.4. Pre-Medication and Anesthesia

Fluid therapy, 5% DW; antibiotics, Cefovecin 8 mg/kg SQ; analgesics, Butorphanol 0.3 mg/kg; Lidocaine, 0.5 mg/kg; and Ketamine, 0.5 mg/kg were mixed in one syringe. Meloxicam 0.2 mg/kg intravenous bolus injection was followed by Butorphanol 0.2 mg/kg/h (3.3 g/kg/min), Lidocaine 1.5 mg/kg/h (25 g/kg/min), and Ketamine 1.2 mg/kg/h (20 g/kg/min) added to 5% DW 100 mL bag; 50 mL/h constant rate of infusion. Ranitidine, 1 mg/kg, was subcutaneously injected as an antacid. General anesthesia was induced by Propofol 5 mg/kg IV followed by maintenance with 1–3 Vol% Isoflurane-oxygen. Lidocaine, 1 mL, was used to perform radial, ulnar, median, and musculocutaneous (RUMM) blocking anesthesia.

2.5. Surgery and Evaluation

As described in the previously published work [18], corrective osteotomies were preoperatively planned in 3-D, and patient-specific drills and cutting guides were designed based on the 3-D virtual simulation followed by fabrication. The final model of the PSI was applied intraoperatively to perform the closing wedge corrective osteotomy.

First, a lateral approach to the ulna was conducted for an ulnar osteotomy, with the approach made through a skin incision from the ulnar styloid to the midshaft. The radius was then reached cranially, with the skin incision extending from the radial diaphysis to the carpal joint.

Intraoperatively, the PSI was placed on the planned position with navigation pin and calipers, under c-arm fluoroscopy images that had contributed to the precise positioning of the PSI (Figure 3). Each PSI was temporarily fixed with two or three K-wires. The saw blade was aligned with the PSI slope, and a radial closed wedge osteotomy was performed with a Micro E oscillating saw (Rancho Cucamonga, CA, USA).

The osteotomies were performed, resulting in proximal and distal fragments. Next, a bone screw hole was predrilled through the SK ESF Drill Sleeves drill sleeve (IMEX Veterinary, Inc., Rancho Cucamonga, CA, USA) in the proximal PSI unit (Figure 4). The PSI was removed, and the carpus was aligned intraoperatively to the elbow to correct the external rotation. Each fragment had five locking screws and was connected by two plates. At least three locking screws were placed on each side of the bone fragment. The pre-contoured locking plate (YULP) was first fixed to the proximal bone fragment on the cranial bone surface using a pre-drilled hole, then the distal fragment was reduced directly by plate fixation. The second fixation was performed with a pre-contoured compression locking plate (CLP) on the medial bone surface as planned without any difficulty (Figure 5). In this surgery, we do not use or promote bone healing methods such as a cancellous bone graft. The surgical wound was closed in a routine manner. Orthogonal postoperative radiographs showed accurate implant positioning and antebrachial alignment (Figure 6).

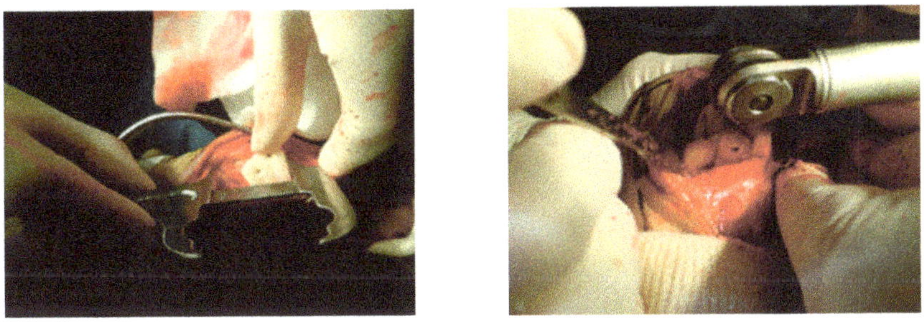

(a) (b)

Figure 3. Distal patient-specific instrumentation (PSI) unit setting in situ using a caliper (**a**). The osteotomy was performed after temporary fixation with K-wire (**b**). Intraoperatively, the PSI was placed on the planned position with navigation pin and calipers contributed to precise aid positioning of the PSI (**a**). Each PSI was temporarily fixed with two or three K-wires. The saw blade was aligned with the PSI slope and a radial closed wedge osteotomy was performed with an oscillating saw (**b**).

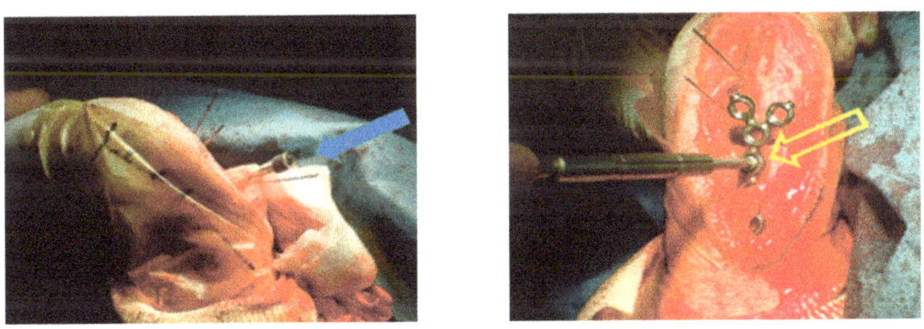

(a) (b)

Figure 4. The proximal patient-specific instrumentation (PSI) unit has a design element for the drill sleeve hole (**a**). The pre-contoured locking plate (YULP) was first fixed to the proximal bone fragment on the cranial bone surface using a pre-drilled hole (**b**).

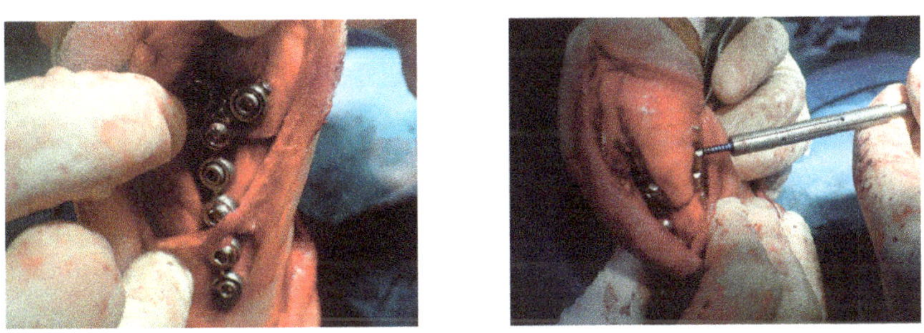

(a) (b)

Figure 5. Locking plate (YULP) and locking screws on the cranial surface of the distal radius (**a**). The second locking plate and screws were placed on the caudomedial surface as planned (**b**).

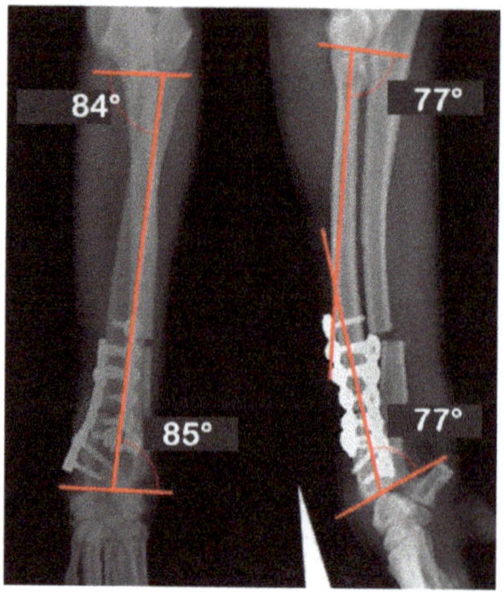

Figure 6. The immediate postoperative radiographs with the joint orientation line precisely meet the target angles as planned.

2.6. Post-Operative Care

The dog recovered from anesthesia without complications, and carprofen (1.5 mg/kg, orally every 24 h for seven days) was administered for analgesia. Cage rest was directed for four weeks. The activity was restricted to a 10-min walk with a short lead walking for toileting twice per day and gradually increased until the fracture line disappeared.

3. Results

The dog was in excellent physical condition preoperatively. A general physical examination, thoracic radiographic evaluation, and laboratory investigations, including hematological, serum biochemical analysis, and urinalysis, were performed before the anesthesia, and all the results were within the normal limit. C-arm fluoroscopic intraoperative imaging was performed during the operation to confirm PSI positioning, K-wire pinning, and pre-drilling in situ. The time consumed for registration of the distal unit and proximal unit was 9 min 6 s and 1 min 20 s, respectively; for osteotomy of the distal and proximal cutting planes was 1 min 45 s and 51 s, respectively. No complications occurred during the anesthesia or immediately after surgery. The osteotomy results were confirmed as pre-operatively planned by orthogonal radiographic images. As shown in the preoperative surgical planning and simulation, the bone fixation implants could be mounted without interference between the screws of the two plates. The summary of the post-operative joint orientation angles evaluation for the corrective osteotomy is presented in Tables 1 and 2. The reference values in Tables 1 and 2 were collected from a previous report [24].

Table 1. Joint orientation angles with joint orientation lines.

Golden Retrievers		Mean Value	Reference Value
Frontal	aMPRA	83	82~83
	aLDRA	86	85~87
Sagittal	aCdPRA	85	84~86
	aCdDRA	77	76~78
	Θ	27	21~32

aMPRA, anatomical medial proximal radial angle; aLDRA, anatomical lateral distal radial angle; aCdPRA, anatomical caudal proximal radial angle; aCdDRA, anatomical caudal distal radial angle; Θ, the angular intersection (procurvature).

Table 2. Joint orientation angles before and after surgery.

JOA	Pre-OP	Post-OP	Target	Mean Values	Reference Values
aLDRA	57	85	85	86	85~87
aCdPRA	92	87	87	85	84~86
aMPRA	75	84	84	83	82~83
aCdDRA	62	77	77	77	76~78
θ	38	21	21	27	21~32

Joint Orientation Angles (JOA) of pre-operative (Pre-OP), post-operative (Post-OP), Target, and mean values aMPRA, anatomical medial proximal radial angle; aLDRA, anatomical lateral distal radial angle; aCdPRA, anatomical caudal proximal radial angle; aCdDRA, anatomical caudal distal radial angle; θ, Angular intersection of two segmental joint reference axes.

The surgical site edema reduced significantly on the second day after the operation, so the gentle pressure bandage for the surgical site edema was removed on the third day after the operation.

The patient improved significantly after two weeks. A lameness and mild painful carpal palpation with reduced carpal flexion range of motion (ROM) signified were noticed. Radiographs were rechecked immediately post-operatively and 12 weeks later. The radiographic findings showed excellent healing of the osteotomy. The 12-week radiographs showed osteogenesis with enhanced radiopacity in the radial callus formation (Figure 7). After 8 weeks, weight-bearing had improved, and the radiographs showed a progressive new bone formation at the osteotomy sites. At the final 12-week postoperative check, the patient presented no lameness with improved, pain-free carpal ROM, and increased antebrachial muscular tone on palpation.

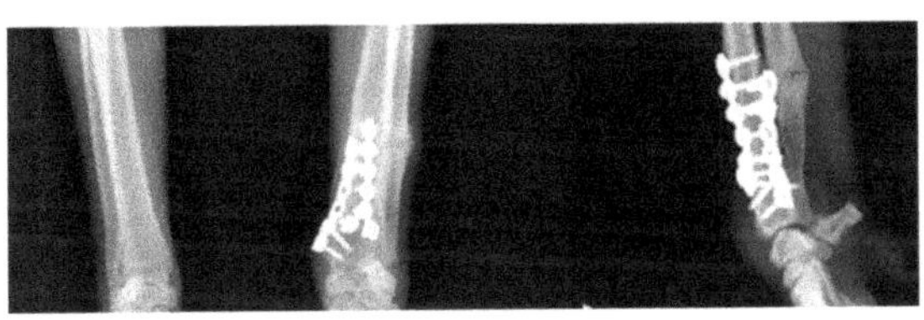

Figure 7. At the final 12-week postoperative radiography. The fracture-line on the distal radius had disappeared.

4. Discussion

We describe a novel method for precise planning of a closing wedge osteotomy in 3-D using CT and CAD applications. RP modeling was used to create models for saw guide fabrication, bone plate pre-contouring, and rehearsal surgery. The patient had a successful outcome with significantly improved limb use and cosmetic appearance. Carpal degeneration may be reduced following the reform of the loading through the antebrachial bones.

Ideally, the veterinary surgeon will have access to the normal axis, as determined in 3-D, on the opposite limbs obtained from measurements of several individuals unaffected animals or with the same breed description. Currently, however, it is difficult to search for antebrachial joint orientational angle data on other breeds of dogs. Hence, for the work, we selected the Golden Retriever dog as the animal model, which had the same skeletal structure and normal value research data of joint orientation angles and lines [19].

Due to the deformity, there was a superimposition of the radiocarpal and accessory carpal bones with the distal radius and carpal rotation, which obscured the normal radiographic landmarks. It may have been possible to use the segmental radiography method outlined by Dismukes et al. [24] to avoid the superimposition of the carpal bones and the radius caused by the rotation. However, as this method is still two-dimensional, it was decided to use 3-D CT to augment and improve our understanding of the deformity within the radius.

The decision to use CT in the planning of this case stemmed from the need for more precise and accurate joint orientational angles to perform close wedge corrective osteotomy.

The CT image gave us not only the 3-D virtual visualization to try a range of potential osteotomies and assess the optimal site and angle of correction, but also real-sized antebrachial bone models and patient-specific cutting guides that can be successfully used for ALD. However, pre-operative labor does increase, especially computing work, which was significantly more when the 3-D printer was used [25,26]. CT allowed us to handle the deformity in 3-D bone models. In this way, CT enables rehearsal surgery to prevent potential mistakes in the operating room, and in the process, it can improve the protocol [27,28]. In our case, we were able to improve the protocol by finding out the problem with the osteotomy plan during the rehearsal surgery and assessing the risks of over- or under-correction as it is not possible to adjust during the postoperative period and revise the surgical plan.

Computer simulation has been described in human surgery and has been used for corrective osteotomies for the treatment of angular deformities, including malunions [20,29–33]. After reading reports about human surgery, we decided to create our simple cutting guides by molding PMMA onto the plastic model abutting our outlined cut planes. It improved the accuracy of the rehearsal and subsequent surgery. Rather than visually translating measurements from the CT data, the location and angle of the osteotomy were determined by the saw guides that were held rigidly in position on the patient's radius.

It is reported in human surgery that the creation of a cutting guide template from a model and surgical simulation overcomes the inaccuracy associated with the translation of the rehearsal surgery to the patient. The accuracy of the osteotomy is determined by the resolution of the CT data (1 mm) and the accuracy of the model reproduction (0.18 mm), rather than the surgeon's intraoperative measurements and judgment [32,33].

Our simple guides aided the use of the oscillating saw and illustrated how a specifically designed template or jig would be greatly advantageous. Stereo-lithographically produced jigs can be fabricated from plastic or metal as shells and used in the same manner as the PMMA guides used in this case. With greater familiarity with the technique, the surgery could be rehearsed electronically, and the saw guides themselves could be printed without the direct requirement for the bone models. The use of specifically manufactured surgical templates is an area that requires further study and may have application in veterinary surgery.

The major limitation of this study was that it was the first attempt at the proposed method. Both unfamiliarity with the capabilities of the software systems and the lack of normal data contributed to the flaws in the method that we reported. However, with further work, it will be easier to determine whether this method is primarily valid and how it relates to previously described methods of radiographic measurements.

The PSI design, in this case, did not require the modification of the angle to reduce the correction or leave a degree of cranial bowing. Although this adjustment was empirical as it was based on the already defined wedge, it was performed with confidence and the changes could be rehearsed.

The use of CT data to create stereolithographic models to aid in surgery is widely reported in the human literature, and there are early reports in the veterinary literature as well [24,34–37]. It has seen the most use in craniomaxillofacial surgery, but it is also used in many areas of orthopedics and, more recently, human cardiovascular surgery [38–40]. Models can be used to rehearse surgery, contour implants, improve diagnostic abilities, and aid in both the clinician's and the involved client's understanding. Human studies have concluded that the use of stereolithographic bio-models in combination with standard imaging has greater utility in surgical management than standard imaging data alone [38]. It is also reported that surgeons feel the use of bio-models reduces the time for surgical procedures; although, due to the inability to repeat surgery with the two methods, this observation is purely an anecdotal finding [36,38].

The positioning of the bone plate in situ is one of the vital issues for the outcome of computer-assisted osteotomies. The plate rigidly defined the displacement of the distal radius relative to the radial shaft. Thus, the surgeon had to ensure that the screw holes were drilled in the right locations and that the plate lay flat and perpendicular to the bone fragments. If these conditions were not met, errors in the placement of the plate and, hence, in the final position of the distal radius could occur. If the holes for the fixation screws were not drilled in the planned location, errors in translations and rotations could be induced in the location of the distal radius.

Of particular concern is the location of the most proximal hole of the proximal radial bone fragment: a literal translation of that hole by 1.0 mm can cause the second proximal hole to be a pivot point that rotates the plates and, hence, the centroid of the distal radius by approximately 1.9 mm. Since the bone surface is not planar, angular errors could be introduced. Traditionally, to attach the screws to the distal radius, the bone surface to which the fixation plate was to be attached needed to be shaved flat [41]. Failure to properly shape the bone could introduce translational errors in the y- and z-axes and rotational errors about x- and z-axes. However, using the locking screw plate in this method does not require the shaving procedure, which potentially creates an error [42]. The radial shaft is approximately cylindrical. This geometry allows for small rotational errors about the shaft's axis to pass undetected. In addition, the screw holes in the fixation plate allowed the screw to lay anywhere within a cortical region. This reduced the effectiveness of the screw as a spatial constraint of the plate–bone system [43]. However, locking screw and plate usage can solve this problem. For the screws, pilot holes were created using K-wires of 1.2 mm diameter. For each such wire, the drill had to be calibrated, introducing a possible source of error. Due to the length of the wire protruding approximately 50 mm from the drill chuck and the weight of the drill, bending of the wire during the procedure was possible. If such bending occurred during the calibration registration or alignment of the wire to the drill locations, an error could be introduced in positioning the drill location and/or orientation. Careful calibration and use of the drill can reduce such errors. Though this was not strictly part of the planning, it is vital to mention its importance for the overall outcome of a 3-D computer-aid procedure. Registration was used to find the transformation from the proximal target coordinates to the CT coordinates [29]. A poor registration could have led to a misplaced drill location and, hence, a misplaced fixation plate. Robust algorithms, including the contouring on the real-size RP bone model and validation of the registration, can reduce the likelihood of poor registration.

Although this study, a 3-D-based assessment, can be more useful for complex deformities [44,45], the post-operative evaluation relied on 2-D radiographs.

We found that having the models was very valuable in the planning of the case, not only for the procedure itself but also for explaining the procedure to the owner. This case report showed the application of this new technology in veterinary surgery and demonstrated that it is possible to improve the accuracy of surgical procedures involving complex geometry. It does, however, highlight the need for further study. Furthermore, an assessment of the extra cost of CT and 3-D printing work, and a comparison of the analysis of merits and demerits achieved using traditional radiography versus 3-D-aided surgical protocol is needed.

5. Conclusions

From this in-house surgical procedure, we concluded that the advanced data gained with digital imaging, stereolithographic models, 3-D printed real-sized bone models, and fabricating PSCGs may benefit the patient, client, and surgical team. This protocol enriched the morphological experience of bone deformities, aided client communication, gained the surgeon's confidence, maximized surgical precision, minimized surgical wound exposure and anesthetic times, and was intuitive. We propose to veterinary surgeons that computer-aided orthopedic surgery affords many advantages and advocates further research.

Author Contributions: Conceptualization, H.-R.L. and S.-J.K.; Methodology, H.-R.L. and G.O.A.; Formal analysis, H.-R.L. and G.O.A.; Data curation, H.-R.L. and G.O.A.; Writing—original draft preparation, H.-R.L., G.O.A. and S.-J.K.; Project administration, H.-R.L. and S.-J.K. Funding acquisition, H.-R.L. and S.-J.K. All authors have read and agreed to the published version of the manuscript.

Funding: This research was supported by the Ministry of Science and Technology (NRF-2018R1A2B6003332 and NRF-2018R1D1A3B07049191), South Korea. This study was also partially financed by the Korea Institute of Planning and Evaluation for Technology in Food, Agriculture, and Forestry (IPET) via Agriculture, Food, and Rural Affairs Convergence Technologies Program for Educating Creative Global Leader, Ministry of Agriculture, Food, and Rural Affairs (MAFRA; 320005-4).

Institutional Review Board Statement: The animal procedures in this study were approved by the Committee on the Care of Laboratory Animal Resources of Jeonbuk National University (CBNU2016-67) and were performed following the Guide for the Care and Use of Laboratory Animals published by the U.S. National Institutes of Health (Bethesda, MA, USA; NIH Publication no. 85–23, revised 1996).

Informed Consent Statement: Written informed consent has been obtained from the owner to publish this paper.

Data Availability Statement: Not applicable.

Acknowledgments: The authors would like to thank the staff of Hansarang Veterinary Clinic, Seoul, Korea for their technical and clinical assistance.

Conflicts of Interest: The authors declare no conflict of interest.

References

1. Knapp, J.L.; Tomlinson, J.L.; Fox, D.B. Classification of Angular Limb Deformities Affecting the Canine Radius and Ulna Using the Center of Rotation of Angulation Method. *Vet. Surg.* **2016**, *45*, 295–302. [CrossRef] [PubMed]
2. Johnson, J.A.; Austin, C.; Breur, G.J. Incidence of canine appendicular musculoskeletal disorders in 16 veterinary teaching hospitals from 1980 through 1989. *Vet. Comp. Orthop. Traumatol.* **1994**, *7*, 56–69. [CrossRef]
3. Fox, D.B. Physeal injuries and angular limb deformities. *Vet. Clin. N. Am. Small Anim. Pract.* **2021**, *51*, 305–322. [CrossRef]
4. Ramadan, R.; Vaughan, L. Premature closure of the distal ulnar growth plate in dogs—A review of 58 cases. *J. Small Anim. Pract.* **1978**, *19*, 647–667. [CrossRef]
5. O'brien, T.R.; Morgan, J.P.; Suter, P. Epiphyseal plate injury in the dog: A radiographic study of growth disturbance in the forelimb. *J. Small Anim. Pract.* **1971**, *12*, 19–36. [CrossRef]
6. Morgan, P.; Miller, C. Osteotomy for correction of premature growth plate closure in 24 dogs. *Vet. Comp. Orthop. Traumatol.* **1994**, *7*, 129–135.
7. Henney, L.H.S.; Gambardella, P.C. Premature closure of the ulnar physis in the dog: A retrospective clinical study. *J. Am. Anim. Hosp. Assoc.* **1989**, *25*, 573–581.
8. Fox, D.B.; Tomlinson, J.L.; Cook, J.L.; Breshears, L.M. Principles of uniapical and biapical radial deformity correction using dome osteotomies and the center of rotation of angulation methodology in dogs. *Vet. Surg.* **2006**, *35*, 67–77. [CrossRef]
9. Nikolaou, C.; Black, C.; Ochoa, J.J.; Fitzpatrick, N. Guidelines for the Execution of True Spherical Osteotomies Using a Modified Dome Blade Design. *Vet. Comp. Orthop. Traumatol.* **2019**, *33*, 71–81. [CrossRef]
10. Quinn, M.K.; Ehrhart, N.; Johnson, A.L.; Schaeffer, D.J. Realignment of the radius in canine antebrachial growth deformities treated with corrective osteotomy and bilateral (type II) external fixation. *Vet. Surg.* **2000**, *29*, 558–563. [CrossRef]
11. Fox, S.; Bray, J.; Guerin, S.; Burbridge, H. Antebrachial deformities in the dog: Treatment with external fixation. *J. Small Anim. Pract.* **1995**, *36*, 315–320. [CrossRef] [PubMed]
12. Harari, J. Complications of external skeletal fixation. The veterinary clinics of North America. *Small Anim. Pract.* **1992**, *22*, 99–107. [CrossRef]
13. Green, S.A. Complications of external skeletal fixation. *Clin. Orthop. Relat. Res.* **1983**, *180*, 109–116. [CrossRef]
14. Paley, D. *Principles of Deformity Correction*; Springer Science & Business Media: Berlin/Heidelberg, Germany, 2002.
15. Arzi, B.; Cissell, D.D.; Pollard, R.E.; Verstraete, F.J. Regenerative approach to bilateral rostral mandibular reconstruction in a case series of dogs. *Front. Vet. Sci.* **2015**, *2*, 4. [CrossRef]
16. Hespel, A.M.; Wilhite, R.; Hudson, J. Invited review-applications for 3D printers in veterinary medicine. *Vet. Radiol. Ultrasound* **2014**, *55*, 347–358. [CrossRef]

17. Castilho, M.; Rodrigues, J.; Vorndran, E.; Gbureck, U.; Quental, C.; Folgado, J.; Fernandes, P.R. Computational design and fabrication of a novel bioresorbable cage for tibial tuberosity advancement application. *J. Mech. Behav. Biomed. Mater.* **2017**, *65*, 344–355. [CrossRef]
18. Lee, H.-R.; Adam, G.O.; Yang, D.K.; Tungalag, T.; Lee, S.-J.; Kim, J.-S.; Kang, H.-S.; Kim, S.-J.; Kim, N.S. An Easy and Economical Way to Produce a Three-Dimensional Bone Phantom in a Dog with Antebrachial Deformities. *Animals* **2020**, *10*, 1445. [CrossRef]
19. Fasanella, F.; Tomlinson, J.; Welihozkiy, A. Radiographic measurements of the axes and joint angles of the canine radius and ulna. *Vet. Comp. Orthop. Traumatol.* **2010**, *23*, A11.
20. Jastifer, J.R.; Gustafson, P.A. Three-dimensional printing and surgical simulation for preoperative planning of deformity correction in foot and ankle surgery. *J. Foot Ankle Surg.* **2017**, *56*, 191–195. [CrossRef]
21. Auricchio, F.; Marconi, S. 3D printing: Clinical applications in orthopaedics and traumatology. *EFORT Open Rev.* **2016**, *1*, 121–127. [CrossRef]
22. Kwun, J.-D.; Kim, H.-J.; Park, J.; Park, I.-H.; Kyung, H.-S. Open wedge high tibial osteotomy using three-dimensional printed models: Experimental analysis using porcine bone. *Knee* **2017**, *24*, 16–22. [CrossRef] [PubMed]
23. Balfour, R.J.; Boudrieau, R.J.; Gores, B.R. T-plate fixation of distal radial closing wedge osteotomies for the treatment of angular limb deformities in 18 dogs. *Vet. Surg.* **2000**, *29*, 207–217. [CrossRef] [PubMed]
24. Dismukes, D.I.; Fox, D.B.; Tomlinson, J.L.; Essman, S.C. Use of radiographic measures and three-dimensional computed tomographic imaging in surgical correction of an antebrachial deformity in a dog. *J. Am. Vet. Med. Assoc.* **2008**, *232*, 68–73. [CrossRef]
25. Morris, J.B. CAD/CAM options in dental implant treatment planning. *J. Calif. Dent. Assoc.* **2010**, *38*, 333–336. [PubMed]
26. Martelli, N.; Serrano, C.; van den Brink, H.; Pineau, J.; Prognon, P.; Borget, I.; El Batti, S. Advantages and disadvantages of 3-dimensional printing in surgery: A systematic review. *Surgery* **2016**, *159*, 1485–1500. [CrossRef] [PubMed]
27. Potamianos, P.; Amis, A.; Forester, A.; McGurk, M.; Bircher, M. Rapid prototyping for orthopaedic surgery. *Proc. Inst. Mech. Eng. Part H J. Eng. Med.* **1998**, *212*, 383–393. [CrossRef]
28. Frame, M.; Huntley, J.S. Rapid prototyping in orthopaedic surgery: A user's guide. *Sci. World J.* **2012**, *2012*, 838575. [CrossRef]
29. Cartiaux, O.; Paul, L.; Francq, B.G.; Banse, X.; Docquier, P.L. Improved accuracy with 3D planning and patient-specific instruments during simulated pelvic bone tumor surgery. *Ann. Biomed. Eng.* **2014**, *42*, 205–213. [CrossRef]
30. Meyer, D.C.; Siebenrock, K.A.; Schiele, B.; Gerber, C. A new methodology for the planning of single-cut corrective osteotomies of mal-aligned long bones. *Clin. Biomech.* **2005**, *20*, 223–227. [CrossRef]
31. Michielsen, M.; Van Haver, A.; Bertrand, V.; Vanhees, M.; Verstreken, F. Corrective osteotomy of distal radius malunions using three-dimensional computer simulation and patient-specific guides to achieve anatomic reduction. *Eur. J. Orthop. Surg. Traumatol.* **2018**, *28*, 1531–1535. [CrossRef]
32. Omori, S.; Murase, T.; Kataoka, T.; Kawanishi, Y.; Oura, K.; Miyake, J.; Tanaka, H.; Yoshikawa, H. Three-dimensional corrective osteotomy using a patient-specific osteotomy guide and bone plate based on a computer simulation system: Accuracy analysis in a cadaver study. *Int. J. Med. Robot. Comput. Assist. Surg.* **2014**, *10*, 196–202. [CrossRef] [PubMed]
33. Omori, S.; Murase, T.; Oka, K.; Kawanishi, Y.; Oura, K.; Tanaka, H.; Yoshikawa, H. Postoperative accuracy analysis of three-dimensional corrective osteotomy for cubitus varus deformity with a custom-made surgical guide based on computer simulation. *J. Shoulder Elb. Surg.* **2015**, *24*, 242–249. [CrossRef] [PubMed]
34. Murphy, S.B.; Kijewski, P.; Simon, S.; Chandler, H.; Griffin, P.; Reilly, D.; Penenberg, B.; Landy, M. Computer-aided simulation, analysis, and design in orthopedic surgery. *Orthop. Clin. N. Am.* **1986**, *17*, 637–649. [CrossRef]
35. Webb, P. A review of rapid prototyping (RP) techniques in the medical and biomedical sector. *J. Med. Eng. Technol.* **2000**, *24*, 149–153. [CrossRef] [PubMed]
36. Harrysson, O.L.; Cormier, D.R.; Marcellin-Little, D.J.; Jajal, K. Rapid prototyping for treatment of canine limb deformities. *Rapid Prototyp. J.* **2003**, *9*, 37–42. [CrossRef]
37. Faber, J.; Berto, P.M.; Quaresma, M. Rapid prototyping as a tool for diagnosis and treatment planning for maxillary canine impaction. *Am. J. Orthod. Dentofac. Orthop.* **2006**, *129*, 583–589. [CrossRef]
38. D'Urso, P.S.; Effeney, D.J.; Earwaker, W.J.; Barker, T.M.; Redmond, M.J.; Thompson, R.G.; Tomlinson, F.H. Custom cranioplasty using stereolithography and acrylic. *Br. J. Plast. Surg.* **2000**, *53*, 200–204. [CrossRef]
39. Minns, R.; Bibb, R.; Banks, R.; Sutton, R. The use of a reconstructed three-dimensional solid model from CT to aid the surgical management of a total knee arthroplasty: A case study. *Med. Eng. Phys.* **2003**, *25*, 523–526. [CrossRef]
40. Kim, M.S.; Hansgen, A.R.; Wink, O.; Quaife, R.A.; Carroll, J.D. Rapid prototyping: A new tool in understanding and treating structural heart disease. *Circulation* **2008**, *117*, 2388–2394. [CrossRef]
41. Grujicic, A.; LaBerge, M.; Xie, X.; Arakere, G.; Pandurangan, B.; Grujicic, M.; Jeray, K.; Tanner, S. Computational investigation of the relative efficacies of nail-and plate-type proximal femoral-fracture fixation implants. *Multidiscipline Model. Mater. Struct.* **2011**, *7*, 212–244. [CrossRef]
42. Ma, B.; Kunz, M.; Gammon, B.; Ellis, R.E.; Pichora, D.R. A laboratory comparison of computer navigation and individualized guides for distal radius osteotomy. *Int. J. Comput. Assist. Radiol. Surg.* **2014**, *9*, 713–724. [CrossRef] [PubMed]
43. Soubeyrand, M.; Wassermann, V.; Hirsch, C.; Oberlin, C.; Gagey, O.; Dumontier, C. The middle radioulnar joint and triarticular forearm complex. *J. Hand Surg. (Eur. Vol.)* **2011**, *36*, 447–454. [CrossRef] [PubMed]

44. Miyake, J.; Oka, K.; Kataoka, T.; Moritomo, H.; Sugamoto, K.; Murase, T. 3-dimensional deformity analysis of malunited forearm diaphyseal fractures. *J. Hand Surg.* **2013**, *38*, 1356–1365. [CrossRef] [PubMed]
45. Tobias, K.M.; Johnston, S.A. *Veterinary Surgery: Small Animal-E-BOOK: 2-Volume Set*; Elsevier Health Sciences: Amsterdam, The Netherlands, 2013.

veterinary sciences

MDPI

Article

Three-Dimensional Analysis of Posterior Mandibular Displacement in Rats

Ioannis Lyros [1,*], Efstratios Ferdianakis [1], Demetrios Halazonetis [1], Theodoros Lykogeorgos [2], Antigoni Alexiou [1], Konstantina-Eleni Alexiou [3], Maria Georgaki [4], Emmanouil Vardas [4], Zafeiroula Yfanti [3] and Apostolos I. Tsolakis [1,5]

1 Department of Orthodontics, School of Dentistry, National and Kapodistrian University of Athens, 11527 Athens, Greece; stratis-fer@hotmail.com (E.F.); dhal@dhal.com (D.H.); alexiou.antigoni@gmail.com (A.A.); apostso@otenet.gr (A.I.T.)
2 "Hatzikosta" General Hospital of Messolonghi, 30200 Messolonghi, Greece; theolyk@gmail.com
3 Department of Oral Diagnosis & Radiology, School of Dentistry, National and Kapodistrian University of Athens, 10679 Athens, Greece; kalexiou20@gmail.com (K.-E.A.); zafeiroula86@gmail.com (Z.Y.)
4 Department of Oral Medicine & Pathology and Hospital Dentistry, School of Dentistry, National and Kapodistrian University of Athens, 10679 Athens, Greece; mar1georgaki@gmail.com (M.G.); mbardas@gmail.com (E.V.)
5 Department of Orthodontics, Case Western Reserve University, Cleveland, OH 44106, USA
* Correspondence: yiannislyros@hotmail.com

Abstract: Mandibular protrusion and its treatment is challenging for the orthodontist. The aim of the present research was to identify macroscopic changes in the mandible, based on three-dimensional Cone Beam Computed Tomography analysis. Seventy-two male Wistar rats were divided into two equal groups, experimental (group A) and control (group B). Each consisted of three equal subgroups of 12 rats (A1, A2, A3, B1, B2, B3). Full-cast orthodontic intraoral devices were attached to the maxillary incisors of the experimental animals, and effected functional posterior mandibular displacement. Throughout the experimental period, all animals were fed with mashed food. Animals were sacrificed at 30 days (A1, B1), 60 days (A2, B2) and 90 days (A3, B3). At the 60th day of the experiment, the orthodontic devices were removed from the remaining experimental subgroup A3. Measurements revealed significant differences in the anteroposterior dimensions between experimental and control subgroups. However, the observed changes in the vertical dimensions, Condylion/Go'–Menton and the Intercondylar distance proved insignificant. Posterior mandibular displacement of the mandible in growing rats affects the morphology of the mandible and culminates in the development of a smaller mandible at a grown age.

Keywords: mandibular growth; mandibular posterior displacement; mandibular length; condylar growth; rat; class III malocclusion; orthodontic treatment

Citation: Lyros, I.; Ferdianakis, E.; Halazonetis, D.; Lykogeorgos, T.; Alexiou, A.; Alexiou, K.-E.; Georgaki, M.; Vardas, E.; Yfanti, Z.; Tsolakis, A.I. Three-Dimensional Analysis of Posterior Mandibular Displacement in Rats. *Vet. Sci.* **2022**, *9*, 144. https://doi.org/10.3390/vetsci9030144

Academic Editors: Ana Faustino and Paula A. Oliveira

Received: 6 February 2022
Accepted: 18 March 2022
Published: 20 March 2022

1. Introduction

Orthodontists frequently face the challenge of treating class III skeletal malocclusion during the period of growth [1–4]. Their intervention may comprise functional removable and fixed appliances [5,6], depending on the severity of the disorder and patient characteristics such as age, gender or the ability to cooperate [7]. It would be interesting to assess the osseous alterations effected by an intraoral device causing distal mandibular displacement and, potentially, growth restriction on the molecular or cellular level [8–11]. Despite the fact that such appliances were introduced early in the 20th century [12], the exact skeletal effect of the loading is still elusive, and it has been the subject of a recent systematic review [13]. Understanding the pathway of bone remodeling may have repercussions on treatment planning and the stability of the intervention.

Mandibular and condylar growth have been repeatedly studied [14,15], and reportedly they are affected by heredity [16,17], hormones [18–20], environment [21], dental occlusion [22], and by systemic disease [23]. The mandible makes a major contribution to the shape and function of the dentofacial complex, and thus, it is regarded as a significant determinant of self-esteem [24–26].

Lateral cephalometric radiography remains important in orthodontic diagnostics [27], although it is sometimes omitted [28]. However, traditional two-dimensional (2D) imaging suffers from drawbacks, like overlapping structures, magnification, and linear distortion [29–31], which may lead to erroneous interpretation [32,33]. The above-mentioned inherent shortcomings may be overcome with the implementation of three-dimensional (3D) imaging, Cone Beam Computed Tomography (CBCT) [34–37], which is gradually gaining popularity in Orthodontics [38–40], in particular due to the reduced radiation dose compared to traditional computed tomography [41,42], despite its higher dose compared to the traditional (2D) lateral radiograph [43,44], a fact that merits prudence when prescribing such an examination in younger patients [45]. Comparisons involving patients and human skulls have demonstrated the reliability [46] and validity of the CBCT in estimating the actual anatomical distance between assigned cephalometric points [47] and similar [48–50] or better [51–54] estimations of cephalometric points in comparison to (2D) lateral cephalometry.

The rat is likely the most preferred lab animal for conducting experiments on cranial growth despite existing anatomical and physiological differences with humans [55,56]. There is a long history and experience of using rats to study mandibular and condylar growth [57–59]. However, growth evaluation has been-based mainly on 2D lateral radiographs [60,61].

In studies with rats, appliances are used to retract the mandible with the aim to inhibit growth and also to enhance additional orthopedic effects [57]. The present original research aspires to identify macroscopic changes in the mandibular bone as appear in CBCT-based three-dimensional analysis.

2. Materials and Methods

The study experimental protocol was approved by the Veterinary Directorate and received protocol number 598742/04-10-2019, registered as EL 25 BIO 05, according to Greek national legislation (P.D 56/2013), conforming to European Directive 2010/63/EE and that of the European Council (276/33/20.10.2010) related to the protection of vertebrate animals used in experiments and for other scientific purposes.

2.1. Experimental Design

In the present experimental study, seventy-two (72) four-week-old male Wistar rats were used. After their initial four-week breeding in the Hellenic Pasteur Institute, all the animals were transferred and housed at the Laboratory for Experimental Surgery and Surgical Research "N. S. Christeas" at the University School of Medicine in Athens. Standardization following national and European legislation determined cage selection (Tecniplast S.P.A., Italy) and stable centrally ventilated (15 air changes/h) environmental conditions at 55% relative humidity, temperature at 20 °C ± 2 °C, and artificial 12 h span of alternating light–dark cycles. Access to food and water was ad libitum.

The animals were randomly allocated to equal groups, namely groups A (experimental) and B (control), each been divided into three equally sized subgroups featuring twelve rats (A1, A2, A3, B1, B2, B3). The online Random Team Generator tool was used for the grouping.

Modified orthodontic intraoral devices that have been previously described [62] were placed in the experimental animals and led to posterior mandibular displacement. The full-cast metal orthodontic devices were constructed in the laboratory, following a digital intraoral scanning (TRIOS 3, 3Shape intraoral scanner) of an animal selected at random. The modified guiding appliances were cemented to the maxillary incisors with zinc phosphate cement (Harvard Cement Normal Setting; Harvard Dental International GmbH, 15366 Hoppegarten, Germany) (Figure 1). During the whole experimental period, all animals (experimental and control) were fed with mashed food, produced by blending pellets with water in standardized proportions to achieve a porridge-like consistency.

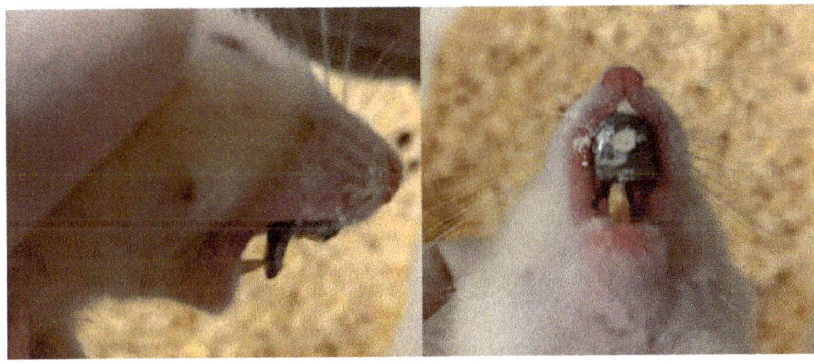

Figure 1. The modified orthodontic intraoral device cemented to the maxillary incisors.

In total, the experimental period lasted for 90 days. Animals were sacrificed at 30 days (subgroups A1, B1), 60 days (subgroups A2, B2) and 90 days (subgroups A3, B3). At the 60th day of the experiment, orthodontic devices were removed from the subjects still remaining in the experimental subgroup A3. Throughout the entire experimental period, all animals were kept closely monitored for normal growth and development.

2.2. Three-Dimensional Analysis

To determine the three-dimensional morphology of the mandible, initial (day 1 of the experiment) and final (day of sacrifice) CBCTs were performed in every rat. The rats were injected intramuscularly for anesthesia with ketamine-xylazine combination at a dosage of 0.2 mL/kg. Once the rats were adequately sedated, they were positioned in the head-resting cushion. All rats were scanned with the same CBCT unit (New Tom VGi, Cefla SC, Imola, Italy) using the same field of view (8 × 8 cm, high-resolution, denture scan) with exposure settings 110 kV. Each scan was performed by an Oral and Dentomaxillofacial Radiologist, who assessed the presence or absence of obvious motion artefacts. In cases of obvious motion artefacts, the scans were performed again and the volumetric data of all scans were exported as Dicom 3 datasets. Three-dimensional reconstruction and analysis were conducted by using Viewbox software (Viewbox© version 4.1.0.10, dHAL Software, Kifissia, Greece). Table 1 and Figure 2 present the detected mandibular anatomic landmarks, while Figure 3a,b show the performed linear measurements.

Table 1. Description of anatomic landmarks detected in Cone Beam CT reconstructed images.

Anatomic Landmarks	Description
Go′	the lowest point of the gonial process
Go	the most posterior point of the gonial process
Menton	the lowest point of the mental process
Coronoid	the tip of the coronoid process
Condylion	the most posterior and highest point of the condylar process
I′	the most anterior point of the alveolar process at the side of the concavity of the lower incisor
Id	the most anterior point of the alveolar process at the side of the convexity of the lower incisor
Incisal	Incisal edge of the lower incisor

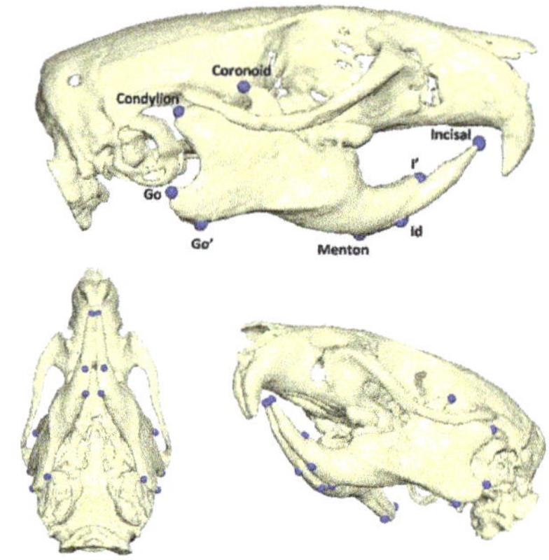

Figure 2. Anatomic landmarks detected in Cone Beam CT reconstructed images.

2.3. *Statistics*

The groups should be kept sufficiently small, for ethical reasons, while reliably detecting potentially statistical results. The number of animals was calculated using power analysis. In addition, the size of the respective samples in the study was finalized after allowing for the low probability that some experimental animals might not cope with the stress of the experimental process.

Subgroups consisting of 12 rats were calculated using standard statistical criteria (a = 0.05, b = 0.10), yielding a power of 90% for detecting a 0.5 mm difference (26.5 vs. 27.0 SD 0.37) for the primary outcome of the study, namely mandibular length (Condylion–I'). Therefore, 72 rats were used, equally divided into experimental and control group.

To calculate the intra-observer and inter-observer errors, double measurements, 4 weeks apart, were made independently by two observers that were blinded to the groups undergoing evaluation. Lin's concordance correlation coefficient and Bland and Altman analysis were used for the estimation of inter- and intra-observer agreement [63,64].

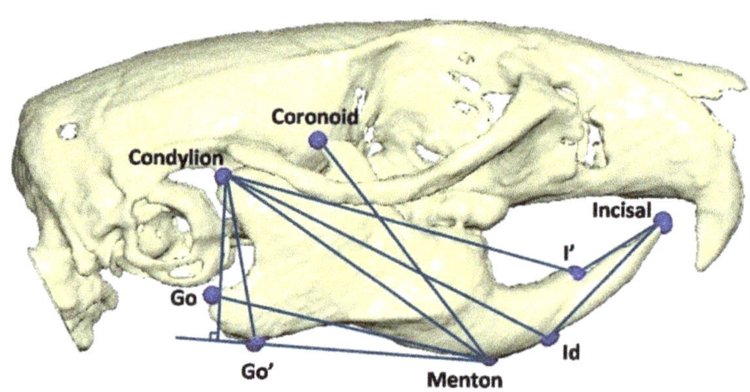

(a)

Figure 3. *Cont.*

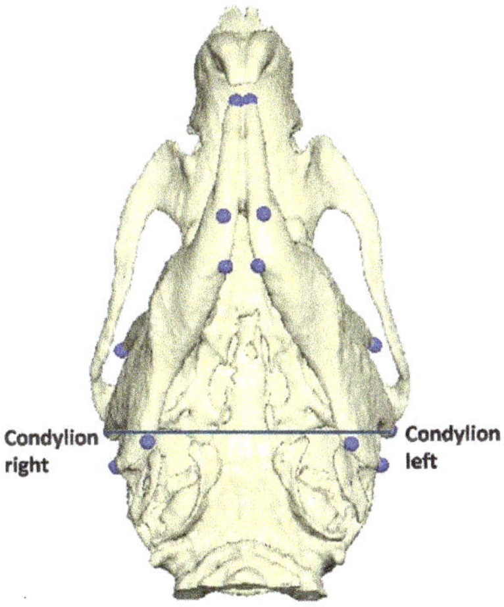

(b)

Figure 3. (**a**) Linear measurements: Go′–Menton (mandibular body length a); Go–Menton (mandibular body length b); Coronoid–Menton; Condylion/Go′–Menton (Condylion height); Condylion–Go′ (Ramus height); Condylion–Menton; Condylion–Id; Condylion–I′ (mandibular length); Incisal–Id; Incisal–I′. (**b**) Linear measurement: Condylion right–Condylion left (Intercondylar distance).

First, seeking to detect any meaningful differences, the dimensional means of the right and left mandibular sides were calculated and were subsequently used for the statistics. Next, differences in measurements related to "group" and "timing (subgroups)" were assessed using linear regression models. Each measurement was regressed on group, timing, and their interaction. When initial measurements were assessed, models were adjusted for initial weight when appropriate. Models for the final measurements were adjusted for the initial ones. Estimated changes from the initial measurements (final minus initial) were also investigated using regression models with group, timing, and their interaction as dependent variables, adjusting for initial weight when appropriate. When normality assumption for the residuals was violated, quantile regression was used. Estimates were adjusted for multiple comparisons, using the Bonferroni method.

Analysis was performed at the $\alpha = 5\%$ level of statistical significance (p-value < 0.05 indicates a statistically significant result). Data were coded and analyzed using the statistical software Stata ver.14 (Stata Statistical Software: Release 14, College Station, TX, USA: StataCorp LP.).

3. Results

Only minor deviation was observed in Lin's concordance correlation coefficient and values of 0.98 or greater were predominant. These values indicate excellent agreement in measurements. The level of agreement (LoA) of Bland and Altman analysis is displayed in Supplementary Tables S1 and S2. Descriptive statistics with estimated means and standard deviations (SD) are presented in Tables 2 and 3.

Table 2. Descriptive statistics (mean and standard deviation) for each measurement by subgroups and overall, for experimental group A.

Experimental Group A	Subgroups—Timing				
	A1—0d	A2—0d	A3—0d	Overall	
	Mean (SD)	Mean (SD)	Mean (SD)	Mean (SD)	*p*-Value *
Weight Initial (grams)	117.2 (17.2)	117.6 (17.5)	115.8 (13.4)	116.9 (15.7)	0.956
Go′–Menton Initial (mm)	13.60 (0.52)	14.80 (0.51)	13.70 (0.53)	14.03 (0.74)	**<0.001**
Go–Menton Initial (mm)	16.75 (0.52)	17.14 (0.55)	16.63 (0.46)	16.84 (0.54)	**0.018**
Coronoid–Menton Initial (mm)	16.17 (0.44)	16.48 (0.48)	16.00 (0.56)	16.22 (0.52)	**0.016**
Condylion/Go′–Menton Initial (mm)	8.29 (0.40)	8.20 (0.30)	8.36 (0.33)	8.28 (0.34)	0.137
Condylion–Go′ Initial (mm)	8.81 (0.38)	8.52 (0.31)	8.82 (0.26)	8.71 (0.34)	**0.002**
Condylion–Menton Initial (mm)	18.55 (0.42)	18.91 (0.45)	18.48 (0.42)	18.64 (0.46)	**0.005**
Condylion–Id Initial (mm)	20.65 (0.36)	20.95 (0.52)	20.71 (0.49)	20.77 (0.47)	0.140
Condylion–I′ Initial (mm)	20.85 (0.51)	21.11 (0.48)	20.75 (0.38)	20.90 (0.48)	0.076
Incisal–Id Initial (mm)	7.87 (0.45)	7.94 (0.29)	7.82 (0.28)	7.88 (0.34)	0.741
Incisal–I′ Initial (mm)	5.06 (0.24)	5.13 (0.27)	5.27 (0.26)	5.15 (0.26)	0.124
Intercondylar Initial (mm)	17.78 (0.40)	17.43 (0.33)	17.50 (0.48)	17.57 (0.42)	0.068
	A1—30d	A2—60d	A3—90d	Overall	
	Mean (SD)	Mean (SD)	Mean (SD)	Mean (SD)	*p*-Value *
Weight Final (grams)	256.1 (24.5)	320.7 (25.0)	337.0 (58.6)	304.6 (52.2)	**<0.001**
Go′–Menton Final (mm)	15.86 (0.79)	16.58 (0.69)	16.83 (0.69)	16.43 (0.82)	**0.003**
Go–Menton Final (mm)	19.48 (0.46)	20.81 (0.51)	21.38 (0.77)	20.56 (0.99)	**<0.001**
Coronoid–Menton Final (mm)	18.74 (0.40)	20.13 (0.49)	20.50 (0.98)	19.79 (1.01)	**<0.001**
Condylion/Go′–Menton Final (mm)	9.72 (0.38)	10.58 (0.26)	10.78 (0.51)	10.36 (0.60)	**<0.001**
Condylion–Go′ Final (mm)	10.12 (0.40)	11.02 (0.31)	11.32 (0.53)	10.82 (0.66)	**<0.001**
Condylion–Menton Final (mm)	20.93 (0.48)	22.24 (0.48)	22.87 (0.81)	22.01 (1.01)	**<0.001**
Condylion–Id Final (mm)	23.13 (0.54)	24.48 (0.60)	25.10 (0.95)	24.23 (1.09)	**<0.001**
Condylion–I′ Final (mm)	23.06 (0.48)	24.35 (0.49)	25.13 (0.75)	24.18 (1.04)	**<0.001**
Incisal–Id Final (mm)	7.40 (0.93)	9.22 (1.62)	12.59 (1.00)	9.73 (2.48)	**<0.001**
Incisal–I′ Final (mm)	4.25 (0.97)	5.63 (1.52)	8.90 (1.00)	6.26 (2.29)	**<0.001**
Intercondylar Final (mm)	18.02 (0.35)	18.19 (0.33)	18.08 (0.54)	18.10 (0.41)	0.142

* derived from linear regression models; pairwise comparisons are provided in Tables 4 and 5.

Table 3. Descriptive statistics (mean and standard deviation) for each measurement by subgroups and overall, for control group B.

Control Group B	Subgroups—Timing				
	B1—0d	B2—0d	B3—0d	Overall	
	Mean (SD)	Mean (SD)	Mean (SD)	Mean (SD)	*p*-Value *
Weight Initial (grams)	109.4 (12.2)	105.4 (22.1)	121.7 (14.7)	112.2 (17.8)	0.062
Go′–Menton Initial (mm)	13.18 (0.43)	13.40 (0.61)	13.74 (0.38)	13.44 (0.52)	0.075
Go–Menton Initial (mm)	16.34 (0.44)	16.28 (0.66)	16.70 (0.55)	16.44 (0.57)	0.804
Coronoid–Menton Initial (mm)	15.90 (0.39)	15.74 (0.74)	15.98 (0.59)	15.87 (0.58)	0.348
Condylion/Go′–Menton Initial (mm)	8.13 (0.34)	8.26 (0.37)	8.55 (0.37)	8.31 (0.39)	**0.040**
Condylion–Go′ Initial (mm)	8.67 (0.34)	8.72 (0.38)	8.97 (0.41)	8.79 (0.39)	0.413
Condylion–Menton Initial (mm)	18.11 (0.50)	18.18 (0.71)	18.50 (0.52)	18.26 (0.59)	0.400
Condylion–Id Initial (mm)	20.36 (0.59)	20.37 (0.81)	20.61 (0.64)	20.45 (0.68)	0.345
Condylion–I′ Initial (mm)	20.42 (0.52)	20.44 (0.78)	20.72 (0.58)	20.53 (0.63)	0.546
Incisal–Id Initial (mm)	7.64 (0.26)	7.73 (0.32)	7.95 (0.54)	7.77 (0.40)	0.381
Incisal–I′ Initial (mm)	5.17 (0.31)	5.15 (0.22)	5.33 (0.43)	5.21 (0.33)	0.865
Intercondylar Initial (mm)	17.60 (0.47)	17.43 (0.50)	17.65 (0.55)	17.56 (0.50)	0.604
	B1—30d	B2—60d	B3—90d	Overall	
	Mean (SD)	Mean (SD)	Mean (SD)	Mean (SD)	
Weight Final	282.3 (18.1)	365.3 (35.0)	430.2 (30.1)	359.2 (67.4)	**<0.001**
Go′–Menton Final (mm)	17.43 (0.43)	18.74 (0.51)	19.92 (0.80)	18.69 (1.19)	**<0.001**
Go–Menton Final (mm)	20.30 (0.60)	21.63 (0.67)	22.93 (0.56)	21.62 (1.24)	**<0.001**
Coronoid–Menton Final (mm)	19.35 (0.51)	20.63 (0.43)	21.61 (0.47)	20.53 (1.05)	**<0.001**
Condylion/Go′–Menton Final (mm)	9.66 (0.44)	10.56 (0.58)	11.23 (0.23)	10.48 (0.78)	**<0.001**
Condylion–Go′ Final (mm)	9.86 (0.46)	10.72 (0.58)	11.32 (0.25)	10.63 (0.75)	**<0.001**
Condylion–Menton Final (mm)	21.60 (0.67)	23.06 (0.33)	23.91 (0.37)	22.86 (1.08)	**<0.001**
Condylion–Id Final (mm)	23.90 (0.72)	25.79 (0.54)	26.45 (0.41)	25.38 (1.23)	**<0.001**
Condylion–I′ Final (mm)	23.93 (0.58)	25.69 (0.39)	26.44 (0.48)	25.36 (1.17)	**<0.001**
Incisal–Id Final (mm)	10.05 (0.38)	11.06 (0.37)	11.81 (0.48)	10.98 (0.83)	**<0.001**
Incisal–I′ Final (mm)	6.92 (0.23)	7.72 (0.28)	8.13 (0.32)	7.59 (0.58)	**0.001**
Intercondylar Final (mm)	18.02 (0.41)	18.07 (0.48)	18.18 (0.49)	18.09 (0.45)	0.642

* derived from linear regression models; pairwise comparisons are provided in Tables 4 and 5.

Table 4. Pairwise group per timing measurements comparisons' *p*-values derived from linear regression models, adjusted for multiple comparison (Bonferroni).

	p-Values *		
Measurement	**A1 vs. B1**	**A2 vs. B2**	**A3 vs. B3**
Weight Initial	0.764	0.222	>0.999
Go'–Menton Initial	0.262	**<0.001**	>0.999
Go–Menton Initial	0.436	**0.003**	>0.999
Coronoid–Menton Initial	>0.999	**0.016**	0.810
Condylion/Go'–Menton Initial	>0.999	**0.027**	0.953
Condylion–Go' Initial	>0.999	**0.001**	>0.999
Condylion–Menton Initial	0.177	**0.005**	0.900
Condylion–Id Initial	>0.999	0.363	0.206
Condylion–I' Initial	0.345	0.059	0.605
Incisal–Id Initial	0.775	>0.999	>0.999
Incisal–I' Initial	0.353	0.873	>0.999
Intercondylar Initial	>0.999	0.590	>0.999
Weight Final	0.202	**0.007**	**<0.001**
Go'–Menton Final	**<0.001**	**<0.001**	**<0.001**
Go–Menton Final	**<0.001**	**<0.001**	**<0.001**
Coronoid–Menton Final	**0.002**	**0.001**	**<0.001**
Condylion/Go'–Menton Final	>0.999	>0.999	0.104
Condylion–Go' Final	0.793	0.056	>0.999
Condylion–Menton Final	**<0.001**	**<0.001**	**<0.001**
Condylion–Id Final	**0.001**	**<0.001**	**<0.001**
Condylion–I' Final	**<0.001**	**<0.001**	**<0.001**
Incisal–Id Final	**<0.001**	**<0.001**	>0.999
Incisal–I' Final	**<0.001**	**<0.001**	0.245
Intercondylar Final	>0.999	>0.999	>0.999

* *p*-values in bold indicate statistical significance at 5% level.

Table 5. Pairwise timing per group measurement comparisons' *p*-values derived from linear regression models, adjusted for multiple comparison (Bonferroni).

	p-Values *					
Measurement	**A1 vs. A2**	**A1 vs. A3**	**A2 vs. A3**	**B1 vs. B2**	**B1 vs. B3**	**B2 vs. B3**
Weight Initial	>0.999	>0.999	>0.999	>0.999	0.295	0.075
Go'–Menton Initial	**<0.001**	>0.999	**<0.001**	0.211	0.194	>0.999
Go–Menton Initial	0.116	>0.999	**0.031**	>0.999	>0.999	>0.999
Coronoid–Menton Initial	0.192	>0.999	**0.020**	>0.999	0.627	>0.999
Condylion/Go'–Menton Initial	>0.999	>0.999	0.189	0.153	0.089	>0.999
Condylion–Go' Initial	**0.016**	>0.999	**0.007**	>0.999	0.959	0.959
Condylion–Menton Initial	**0.027**	>0.999	**0.014**	0.711	>0.999	>0.999
Condylion–Id Initial	0.205	>0.999	0.831	>0.999	>0.999	0.595
Condylion–I' Initial	0.338	>0.999	0.128	>0.999	>0.999	>0.999
Incisal–Id Initial	>0.999	>0.999	>0.999	>0.999	0.726	>0.999
Incisal–I' Initial	>0.999	0.192	0.618	>0.999	>0.999	>0.999
Intercondylar Initial	0.099	0.415	>0.999	>0.999	>0.999	>0.999
Weight Final	**<0.001**	**<0.001**	0.999	**<0.001**	**<0.001**	**<0.001**
Go'–Menton Final	>0.999	**0.003**	0.212	0.266	**0.001**	0.152
Go–Menton Final	**<0.001**	**<0.001**	**0.009**	**<0.001**	**<0.001**	**<0.001**
Coronoid–Menton Final	**<0.001**	**<0.001**	**0.023**	**<0.001**	**<0.001**	**0.001**
Condylion/Go'–Menton Final	**<0.001**	**<0.001**	>0.999	**<0.001**	**<0.001**	**0.009**
Condylion–Go' Final	**<0.001**	**<0.001**	>0.999	**<0.001**	**<0.001**	**0.028**
Condylion–Menton Final	**<0.001**	**<0.001**	**<0.001**	**<0.001**	**<0.001**	**0.005**
Condylion–Id Final	**<0.001**	**<0.001**	**0.012**	**<0.001**	**<0.001**	0.117
Condylion–I' Final	**<0.001**	**<0.001**	**<0.001**	**<0.001**	**<0.001**	**0.011**
Incisal–Id Final	**<0.001**	**<0.001**	**<0.001**	**0.007**	**<0.001**	0.594
Incisal–I' Final	**0.003**	**<0.001**	**<0.001**	**0.032**	**0.002**	>0.999
Intercondylar Final	0.198	0.990	>0.999	>0.999	>0.999	>0.999

* *p*-values in bold indicate statistical significance at 5% level.

At baseline, only a few differences were evidenced among the subgroups, as expected, due to the sample's prior randomization. Contrarily, the final measurements revealed noteworthy changes (Tables 4–6).

Table 6. Estimated mean differences (final–initial), 95% Confidence Intervals and p-values (compared to 0 i.e., no change) per group and timing.

Final–Initial	Mean Difference (mm)	95% Conf. Interval		p-Value
Go′–Menton				
A1	2.29	1.74	2.84	**<0.001**
B1	4.19	3.63	4.74	**<0.001**
A2	1.82	1.27	2.38	**<0.001**
B2	5.23	4.66	5.79	**<0.001**
A3	3.15	2.60	3.70	**<0.001**
B3	6.26	5.70	6.82	**<0.001**
Go–Menton				
A1	2.73	2.22	3.24	**<0.001**
B1	3.96	3.45	4.47	**<0.001**
A2	3.67	3.16	4.18	**<0.001**
B2	5.34	4.83	5.85	**<0.001**
A3	4.75	4.23	5.26	**<0.001**
B3	6.23	5.72	6.74	**<0.001**
Coronoid–Menton				
A1	2.61	2.18	3.03	**<0.001**
B1	3.38	2.96	3.81	**<0.001**
A2	3.69	3.27	4.11	**<0.001**
B2	4.78	4.35	5.21	**<0.001**
A3	4.52	4.10	4.94	**<0.001**
B3	5.73	5.30	6.15	**<0.001**
Condylion/Go′–Menton				
A1	1.45	1.14	1.76	**<0.001**
B1	1.51	1.20	1.82	**<0.001**
A2	2.39	2.08	2.70	**<0.001**
B2	2.26	1.94	2.57	**<0.001**
A3	2.43	2.12	2.74	**<0.001**
B3	2.71	2.39	3.02	**<0.001**
CondylionGo				
A1	1.32	0.98	1.66	**<0.001**
B1	1.17	0.83	1.51	**<0.001**
A2	2.51	2.17	2.85	**<0.001**
B2	1.96	1.61	2.31	**<0.001**
A3	2.51	2.17	2.85	**<0.001**
B3	2.37	2.03	2.72	**<0.001**
Condylion–Menton				
A1	2.42	2.02	2.82	**<0.001**
B1	3.42	3.02	3.82	**<0.001**
A2	3.37	2.97	3.77	**<0.001**
B2	4.77	4.36	5.18	**<0.001**
A3	4.40	4.00	4.80	**<0.001**
B3	5.50	5.10	5.91	**<0.001**
Condylion–Id				
A1	2.51	2.04	2.99	**<0.001**
B1	3.47	2.99	3.95	**<0.001**
A2	3.57	3.10	4.05	**<0.001**
B2	5.29	4.80	5.78	**<0.001**
A3	4.40	3.93	4.88	**<0.001**
B3	5.94	5.45	6.42	**<0.001**
Condylion–I′				
A1	2.25	1.82	2.67	**<0.001**
B1	3.45	3.01	3.88	**<0.001**
A2	3.28	2.85	3.71	**<0.001**
B2	5.12	4.68	5.56	**<0.001**
A3	4.40	3.97	4.82	**<0.001**
B3	5.82	5.39	6.26	**<0.001**
Incisal–Id				
A1	−0.60	−1.37	0.17	0.235
B1	2.40	1.63	3.17	**<0.001**
A2	0.95	0.18	1.72	**0.007**
B2	3.25	2.48	4.02	**<0.001**
A3	4.40	3.63	5.17	**<0.001**
B3	3.85	3.08	4.62	**<0.001**

Table 6. *Cont.*

Final–Initial	Mean Difference (mm)	95% Conf. Interval		p-Value
Incisal–I′				
A1	−0.90	−1.57	−0.23	**0.002**
B1	1.85	1.18	2.52	**<0.001**
A2	0.70	0.03	1.37	**0.033**
B2	2.60	1.93	3.27	**<0.001**
A3	3.60	2.93	4.27	**<0.001**
B3	2.70	2.03	3.37	**<0.001**
Intercondylar				
A1	0.26	−0.09	0.62	0.261
B1	0.37	0.02	0.72	**0.033**
A2	0.79	0.44	1.14	**<0.001**
B2	0.55	0.19	0.91	**0.001**
A3	0.59	0.24	0.94	**<0.001**
B3	0.59	0.23	0.94	**<0.001**

Comparisons between the experimental and the respective control subgroups (A1 vs. B1, A2 vs. B2, A3 vs. B3) revealed a statistically significant restriction of mandibular body length a (Go′–Menton), mandibular body length b (Go–Menton), and mandibular length (Condylion–I′), across all subgroups ($p < 0.001$). The Coronoid–Menton, Condylion–Menton, Condylion–Id, measurements were also found significantly different. Incisal–Id and Incisal–I′ measurements measured significantly different between A1 vs. B1 and A2 vs. B2 subgroups, respectively, albeit differences did not persist after 90 days of experiment (A3 vs. B3 subgroups). The condylion height (Condylion/Go′–Menton), the ramus height (Condylion–Go′) dimensions and the Intercondylar distance did not show significant difference.

Differences within the subgroups in the experimental group (A1, A2, A3) and within the control subgroups (B1, B2, B3) are presented in Table 5. Table 6 presents the mean difference (Final minus Initial) for each subgroup regarding each measurement. It appears that the rate of mandibular growth was smaller in the experimental group in comparison to the control group. The comprehensive regression model regarding major measurements is depicted graphically in Figures 4–7.

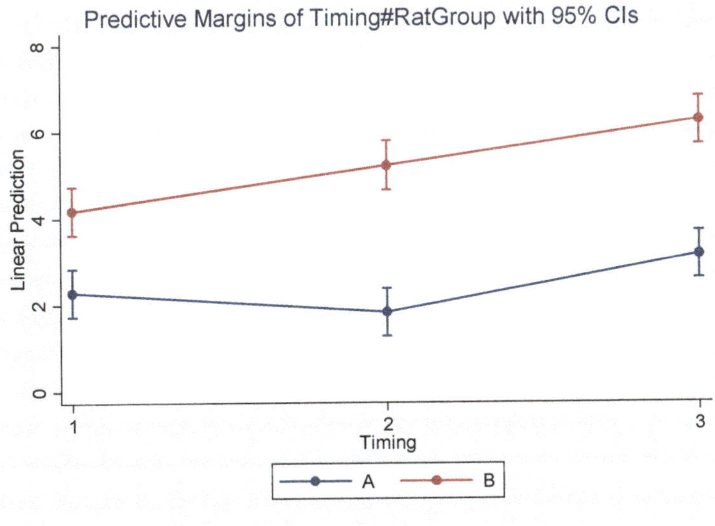

Figure 4. Estimated mean difference (Final–Initial) and 95% Confidence Interval per group and timing in Go′–Menton (mandibular body length a).

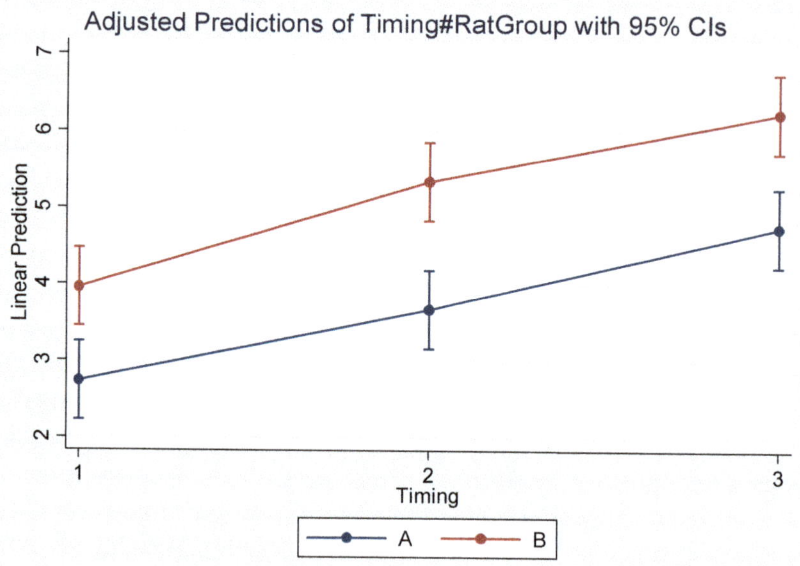

Figure 5. Estimated mean difference (Final–Initial) and 95% Confidence Interval per group and timing in Go–Menton (mandibular body length b).

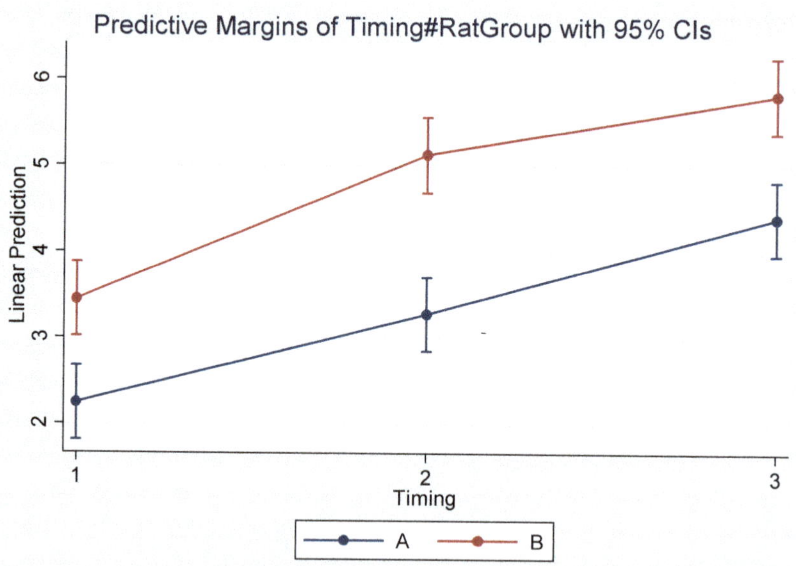

Figure 6. Estimated mean difference (Final–Initial) and 95% Confidence Interval per group and timing in Condylion–I′ (mandibular length).

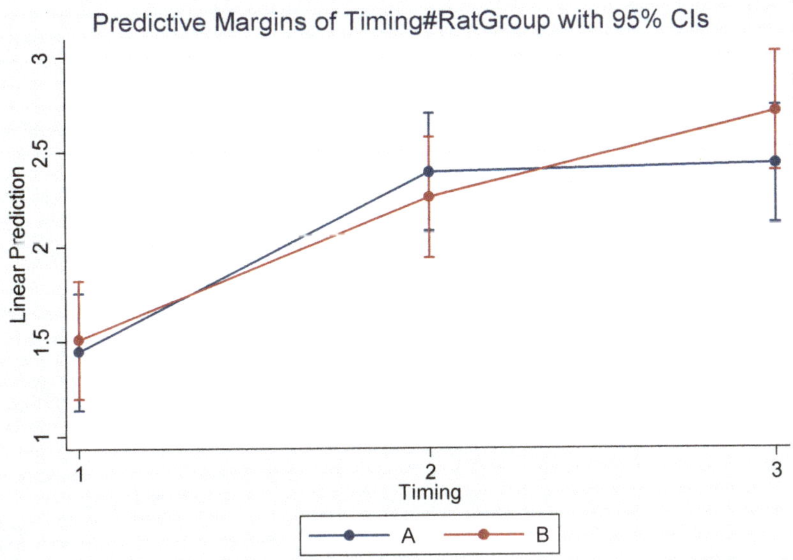

Figure 7. Estimated mean difference (Final–Initial) and 95% Confidence Interval per group and timing in Condylion/Go'–Menton (Condylion height).

4. Discussion

The present study investigates in rats an important feature of human face, the mandibular shape, which is a potential determinant of self-esteem [24,65], it affects the individual's social interaction [66,67] and professional success [68], and is important in the function of the orofacial complex [69,70]. Mandibular development is multifactorial and is regulated by genetic and environmental variables [21,71].

Currently, the rat is the most popular animal used in experimental studies involving anatomy and physiology [72,73], despite the existing differences with humans [55,56,74]. Thus, Wistar rats were used in this study with provision of eliminating potential confounding factors related to their characteristics. Therefore, the animals were all male and had no significant differences in measurements that might be connected to the variables of interest. Although the majority of similar studies also selected the rat, some past research has also reported on rabbits [15,62] and even monkeys [75,76].

Despite the randomization, some comparisons among subgroups regarding initial measurements were found to be statistically different. This might be attributed to the small size of each subgroup. Twelve rats were sufficient to detect the difference of interest, as was determined after power analysis, but they might not have been adequate to eliminate differences of initial characteristics. However, regression models concerning final measurements were adjusted for initial ones, when appropriate.

Every possible effort was made to breed the animals in a healthy, safe environment, to provide necessary nutrition, and to treat them with dignity. Assessing the rat final weight, there was no difference between A1-B1, contrary to the observations between A2-B2 and A3-B3 (although the device in the A3 subgroup was removed on the 60th experimental day). It could be hypothesized that, initially, the appliance did not seem to have caused any important difference, but subsequently the animals might have faced trouble with feeding. Nevertheless, the rats continued growing, as there were differences among A1, A2, and A3 (Table 5).

Rats live for 3 years, on average. They develop rapidly and their adolescence ends by the end of the second month of ontogenesis. Thus, a rat at 2 months of age (60 days) is considered a young adult. The period of rapid growth allegedly ends by 5 weeks, whereas at the period from 8 to 16 weeks, growth slows down [73,77–80]. The experimental period of the present study lasted for 90 days, and the rat age in the last subgroup was 120 days. The differences between subgroups A3 and B3 remained statistically significant 30 days following the removal of the device (from A3 subgroup), meaning that the mandible did not exhibit any post-treatment catch-up growth.

Calculations were not performed separately for the left and right mandibular sides. By contrast, the mean values of the contralateral sides were used and are reported. The orthodontic device that was used in the present study was full-cast, intraoral, and was attached to the maxillary incisors. It was introduced by Desai et al. [62], although their research did not report many details on osseous mandibular change, but focused mainly on the temporomandibular joint. It was adopted by Cholasueksa et al. [8] and Farias-Neto et al. [81], whereas Hua et al. [82] and Wang et al. [59] used modified upper/lower devices, unlike Asano [57], Tsolakis [15] and Teramoto et al. [10], who selected extra-oral appliances. Teramoto et al. mention that the magnitude of the traction was excessive, and thus ended up traumatic [10].

The mandibular distal displacement effected by the intraoral device that was cemented to the rat maxillary incisors caused a restriction in mandibular length and in mandibular body length. Both Go and Go′ landmarks were identified because the distal outline of the ramus in rodents appears to be particularly more concave related to human anatomy (Figure 3a,b). Asano appears to have faced the same challenge in highlighting similar landmarks [57]. Such developmental restriction is in agreement with the observations of Desai et al. [62] and Cholasueksa et al. [8], and the conclusions of Asano [57], Farias-Neto et al. [81], and Hua et al. [82].

In addition, the oblique osseous and dento-osseous measurements, as depicted by Coronoid–Menton, Condylion–Menton, Condylion–Id, were also found to be significantly different. Interestingly, Hua et al. found their respective oblique measurement, namely the angle between the axis of the condylar process to the mandibular plane, also changing spatial orientation [82].

The present study also identified dental alterations, as manifested by the statistically significant differences between experimental and control animals regarding measurements Incisal–Id and Incisal–I′ between A1 vs. B1 and A2 vs. B2 subgroups, respectively, although differences did not persist after 90 days of experiment (A3 vs. B3 subgroups). Dental attrition of the lower incisors in subgroups A1, A2 during the first 60 days of the experiment due to their contact with the device discontinued after debonding the apparatus. Subsequently, lower incisors resumed eruption and so no difference appeared in the dental crown length between experimental and control subgroups.

Interestingly, the vertical component of the mandibular structure as expressed by the condylion height (Condylion/Go′–Menton) and ramus height (Condylion–Go′) did not show significant difference across the experimental and control subgroups inthe final records. This observation agrees with that of Farias-Neto et al., who did not observe any significant difference considering the ramus height [81]. Asano's conclusions are similar regarding the condylar height, although he also noticed a thickening of the retromolar region and the condylar neck, potentially due to spatial remodeling [57]. This leads to the conclusion that mandibular retrusion might not be expected to cause unwanted side-outcomes affecting facial appearance. Lastly, the Intercondylar distance remained statistically unaffected, in agreement with the conclusions of Farias-Neto et al. [81].

To increase accuracy in identifying anatomical landmarks [46,47] and for increased consistency in measuring dimensions [52], CBCT, a 3D reconstruction method, was used in place of ordinary 2D lateral cephalometric radiography [29,32–34,38]. To the best of our knowledge, this is the first study to use CBCT for comprehensive cephalometric evaluation in rats, although various digital radiographic techniques have been used [59,81].

The present study was conducted in rodents with respect, and complying with established legislature and regulations. Such experimentation would be off-limits in humans because the interventions might inflict irreversible changes. On the other hand, the existing differences between rodents and humans call for caution when interpreting the results. The present study may be a contribution to evidence-based decision making in orthodontics when treating skeletal Class III malocclusion, and could intensify the call for further research on the long-lasting effects of such interventions aiming to alleviate facial deviations. A randomized controlled trial should be conducted to validate the present research.

5. Conclusions

Posterior mandibular displacement in growing rats alters the mandibular morphology and results in the development of a smaller mandible at a grown age. In the rat, it can be concluded that the effects of distal mandibular displacement follow a consistent temporal pattern and are statistically significant. The present study emphasized the long-term stability of the outcomes, revealing that the mandible does not show catch-up growth following treatment.

Supplementary Materials: The following supporting information can be downloaded at: https://www.mdpi.com/article/10.3390/vetsci9030144/s1, Table S1: Lin's concordance correlation coefficient (rc), Mean difference and Bland–Altman's 95% LOA (Limits of Agreement) between measurements of the first observer; Table S2: Lin's concordance correlation coefficient (rc), Mean difference and Bland–Altman's 95% LOA (Limits of Agreement) between measurements of the two observers.

Author Contributions: Conceptualization, I.L. and A.I.T.; methodology, I.L., E.F., D.H. and A.I.T.; CBCT acquisition, I.L., E.F., K.-E.A. and Z.Y.; animal sacrifice, I.L., M.G. and E.V.; software, D.H.; validation, I.L., D.H., E.V. and A.I.T.; formal analysis, I.L.; investigation, I.L. (1st observer) and E.F. (2nd observer); data curation, I.L., T.L., A.A., K.-E.A. and Z.Y.; writing—original draft preparation, I.L., E.F., T.L., K.-E.A., M.G. and Z.Y.; writing—review and editing, I.L., D.H., T.L. and A.I.T.; visualization, I.L., M.G., E.V. and A.I.T.; supervision, A.I.T.; project administration, I.L. and A.I.T.; All authors have read and agreed to the published version of the manuscript.

Funding: This research was funded by the Department of Orthodontics, School of Dentistry, National and Kapodistrian University of Athens, Greece.

Institutional Review Board Statement: The animal study protocol was approved by the Veterinary Directorate (protocol code 598742/04-10-2019).

Data Availability Statement: The data presented in this study are available on request from the corresponding author.

Conflicts of Interest: The authors declare no conflict of interest.

References

1. Bombonatti, R.; Aliaga-Del Castillo, A.; Bombonatti, J.F.S.; Garib, D.; Tompson, B.; Janson, G. Cephalometric and occlusal changes of Class III malocclusion treated with or without extractions. *Dent. Press J. Orthod.* **2020**, *25*, 24–32, Erratum in *Dent. Press J. Orthod.* **2020**, *25*, e11. [CrossRef]
2. Campbell, P.M. The dilemma of Class III treatment. Early or late? *Angle Orthod.* **1983**, *53*, 175–191. [CrossRef] [PubMed]
3. Rabie, A.B.; Wong, R.W.; Min, G.U. Treatment in Borderline Class III Malocclusion: Orthodontic Camouflage (Extraction) Versus Orthognathic Surgery. *Open Dent. J.* **2008**, *2*, 38–48. [CrossRef]
4. Zere, E.; Chaudhari, P.K.; Sharan, J.; Dhingra, K.; Tiwari, N. Developing Class III malocclusions: Challenges and solutions. *Clin. Cosmet. Investig. Dent.* **2018**, *10*, 99–116. [CrossRef] [PubMed]
5. Almeida, R.R.; Alessio, L.E.; Almeida-Pedrin, R.R.; Almeida, M.R.; Pinzan, A.; Vieira, L.S. Management of the Class III malocclusion treated with maxillary expansion, facemask therapy and corrective orthodontic. A 15-year follow-up. *J. Appl. Oral Sci.* **2015**, *23*, 101–109. [CrossRef] [PubMed]
6. Singh, S.P.; Kumar, V.; Verma, R.; Singh, S. Management of Developing Skeletal Class III Malocclusion in a Prepubertal Girl with Prognathic Mandible in Late Mixed Dentition. *Contemp. Clin. Dent.* **2017**, *8*, 139–144. [CrossRef] [PubMed]
7. Nardoni, D.N.; Siqueira, D.F.; Cardoso Mde, A.; CapelozzaFilho, L. Cephalometric variables used to predict the success of interceptive treatment with rapid maxillary expansion and face mask. A longitudinal study. *Dent. Press J. Orthod.* **2015**, *20*, 85–96. [CrossRef] [PubMed]

8. Cholasueksa, P.; Warita, H.; Soma, K. Alterations of the Rat Temporomandibular Joint in Functional Posterior Displacement of the Mandible. *Angle Orthod.* **2004**, *74*, 677–683. [CrossRef] [PubMed]
9. Meikle, M.C. Remodeling the Dentofacial Skeleton: The Biological Basis of Orthodontics and Dentofacial Orthopedics. *J. Dent. Res.* **2007**, *86*, 12–24. [CrossRef]
10. Teramoto, M.; Kaneko, S.; Shibata, S.; Yanagishita, M.; Soma, K. Effect of compressive forces on extracellular matrix in rat mandibular condylar cartilage. *J. Bone Miner. Metab.* **2003**, *21*, 276–286. [CrossRef] [PubMed]
11. Von den Hoff, J.W.; Delatte, M. Interplay of mechanical loading and growth factors in the mandibular condyle. *Arch. Oral Biol.* **2008**, *53*, 709–715. [CrossRef]
12. Wendell, P.D.; Nanda, R.; Sakamoto, T.; Nakamura, S. The effects of chin cup therapy on the mandible: A longitudinal study. *Am. J. Orthod.* **1985**, *87*, 265–274. [CrossRef]
13. Lyros, I.; Makrygiannakis, M.A.; Lykogeorgos, T.; Ferdianakis, E.; Tsolakis, A.I. Posterior Mandibular Displacement—A Systematic Review Based on Animal Studies. *Animals* **2021**, *11*, 823. [CrossRef] [PubMed]
14. Du, W.; Bhojwani, A.; Hu, J.K. FACEts of mechanical regulation in the morphogenesis of craniofacial structures. *Int. J. Oral Sci.* **2021**, *13*, 4. [CrossRef] [PubMed]
15. Tsolakis, A.I. Effects of Posterior Mandibular Traction in the Rabbit: A Cephalometric, Histologic, Electromyographic Study. Master's Thesis, Case Western Reserve University, Cleveland, OH, USA, 1981.
16. Carlson, D.S. Evolving concepts of heredity and genetics in orthodontics. *Am. J. Orthod. Dent. Orthop.* **2015**, *148*, 922–938. [CrossRef] [PubMed]
17. Hersberger-Zurfluh, M.A.; Papageorgiou, S.N.; Motro, M.; Kantarci, A.; Will, L.A.; Eliades, T. Genetic and environmental components of vertical growth in mono- and dizygotic twins up to 15–18 years of age. *Angle Orthod.* **2021**, *91*, 384–390. [CrossRef] [PubMed]
18. Anusuya, V.; Nagar, A.; Tandon, P.; Singh, G.K.; Singh, G.P.; Mahdi, A.A. Serum DHEA-S levels could be used as a comparable diagnostic test to assess the pubertal growth spurt in dentofacial orthopedics. *Prog. Orthod.* **2020**, *21*, 15. [CrossRef]
19. Kelly, M.P.; Vorperian, H.K.; Wang, Y.; Tillman, K.K.; Werner, H.M.; Chung, M.K.; Gentry, L.R. Characterizing mandibular growth using three-dimensional imaging techniques and anatomic landmarks. *Arch. Oral Biol.* **2017**, *77*, 27–38. [CrossRef]
20. Yu, S.; Tang, Q.; Xie, M.; Zhou, X.; Long, Y.; Xie, Y.; Guo, F.; Chen, L. Circadian BMAL1 regulates mandibular condyle development by hedgehog pathway. *Cell Prolif.* **2020**, *53*, e12727. [CrossRef]
21. Kahn, S.; Ehrlich, P.; Feldman, M.; Sapolsky, R.; Wong, S. The Jaw Epidemic: Recognition, Origins, Cures, and Prevention. *Bioscience* **2020**, *70*, 759–771. [CrossRef] [PubMed]
22. Liu, C.; Kaneko, S.; Soma, K. Effects of a mandibular lateral shift on the condyle and mandibular bone in growing rats. *Angle Orthod.* **2007**, *77*, 787–793. [CrossRef] [PubMed]
23. Maspero, C.; Cavagnetto, D.; Abate, A.; Cressoni, P.; Farronato, M. Effects on the Facial Growth of Rapid Palatal Expansion in Growing Patients Affected by Juvenile Idiopathic Arthritis with Monolateral Involvement of the Temporomandibular Joints: A Case-Control Study on Posteroanterior and Lateral Cephalograms. *J. Clin. Med.* **2020**, *9*, 1159. [CrossRef] [PubMed]
24. Alabdulrazaq, R.S.; Al-Haj Ali, S.N. Parental Reported Bullying among Saudi Schoolchildren: Its Forms, Effect on Academic Abilities, and Associated Sociodemographic, Physical, and Dentofacial Features. *Int. J. Pediatr.* **2020**, *2020*, 8899320. [CrossRef] [PubMed]
25. Langlois, J.H.; Kalakanis, L.; Rubenstein, A.J.; Larson, A.; Hallam, M.; Smoot, M. Maxims or myths of beauty? A meta-analytic and theoretical review. *Psychol. Bull.* **2000**, *126*, 390–423. [CrossRef] [PubMed]
26. Proffit, W.R.; Jackson, T.H.; Turvey, T.A. Changes in the pattern of patients receiving surgical-orthodontic treatment. *Am. J. Orthod. Dentofac. Orthop.* **2013**, *143*, 793–798. [CrossRef] [PubMed]
27. Al-Taai, N.; LevringJäghagen, E.; Persson, M.; Ransjö, M.; Westerlund, A. A Superimposition-Based Cephalometric Method to Quantitate Craniofacial Changes. *Int. J. Environ. Res. Public Health* **2021**, *18*, 5260. [CrossRef] [PubMed]
28. Devereux, L.; Moles, D.; Cunningham, S.J.; McKnight, M. How important are lateral cephalometric radiographs in orthodontic treatment planning? *Am. J. Orthod. Dentofac. Orthop.* **2011**, *139*, e175–e181. [CrossRef] [PubMed]
29. Chadwick, J.W.; Prentice, R.N.; Major, P.W.; Lam, E.W. Image distortion and magnification of 3 digital CCD cephalometric systems. *Oral Surg. Oral Med. Oral Pathol. Oral Radiol. Endod.* **2009**, *107*, 105–112. [CrossRef] [PubMed]
30. Houston, W.J. The analysis of errors in orthodontic measurements. *Am. J. Orthod.* **1983**, *83*, 382–390. [CrossRef]
31. Paula, L.K.; Solon-de-Mello Pde, A.; Mattos, C.T.; Ruellas, A.C.; Sant'Anna, E.F. Influence of magnification and superimposition of structures on cephalometric diagnosis. *Dent. Press J. Orthod.* **2015**, *20*, 29–34. [CrossRef] [PubMed]
32. Durão, A.P.; Morosolli, A.; Pittayapat, P.; Bolstad, N.; Ferreira, A.P.; Jacobs, R. Cephalometric landmark variability among orthodontists and dentomaxillofacial radiologists: A comparative study. *Imaging Sci. Dent.* **2015**, *45*, 213–220. [CrossRef] [PubMed]
33. Ghafari, J.; Baumrind, S.; Efstratiadis, S.S. Misinterpreting growth and treatment outcome from serial cephalographs. *Clin. Orthod. Res.* **1998**, *1*, 102–106. [CrossRef] [PubMed]
34. Botticelli, S.; Verna, C.; Cattaneo, P.M.; Heidmann, J.; Melsen, B. Two- versus three-dimensional imaging in subjects with unerupted maxillary canines. *Eur. J. Orthod.* **2011**, *33*, 344–349. [CrossRef]
35. Kau, C.H.; Richmond, S.; Palomo, J.M.; Hans, M.G. Three-dimensional cone beam computerized tomography in orthodontics. *J. Orthod.* **2005**, *32*, 282–293. [CrossRef] [PubMed]

36. Tulunoglu, O.; Esenlik, E.; Gulsen, A.; Tulunoglu, I. A comparison of three-dimensional and two-dimensional cephalometric evaluations of children with cleft lip and palate. *Eur. J. Dent.* **2011**, *5*, 451–458. [CrossRef]
37. Wei, D.; Zhang, L.; Li, W.; Jia, Y. Quantitative Comparison of Cephalogram and Cone-Beam Computed Tomography in the Evaluation of Alveolar Bone Thickness of Maxillary Incisors. *Turk. J. Orthod.* **2020**, *33*, 85–91. [CrossRef] [PubMed]
38. American Academy of Oral and Maxillofacial Radiology. Clinical recommendations regarding use of cone beam computed tomography in orthodontics. [corrected]. Position statement by the American Academy of Oral and Maxillofacial Radiology. *Oral Surg. Oral Med. Oral Pathol. Oral Radiol.* **2013**, *116*, 238–257, Erratum in *Oral Surg. Oral Med. Oral Pathol. Oral Radiol.* **2013**, *116*, 661. [CrossRef]
39. Coşkun, İ.; Kaya, B. Cone Beam Computed Tomography in Orthodontics. *Turk. J. Orthod.* **2018**, *31*, 55–61. [CrossRef] [PubMed]
40. Kapila, S.D.; Nervina, J.M. CBCT in orthodontics: Assessment of treatment outcomes and indications for its use. *Dentomaxillofac. Radiol.* **2015**, *44*, 20140282. [CrossRef] [PubMed]
41. Abdelkarim, A. Cone-Beam Computed Tomography in Orthodontics. *Dent. J.* **2019**, *7*, 89. [CrossRef] [PubMed]
42. Theodorakou, C.; Walker, A.; Horner, K.; Pauwels, R.; Bogaerts, R.; Jacobs, R. SEDENTEXCT Project Consortium. Estimation of paediatric organ and effective doses from dental cone beam CT using anthropomorphic phantoms. *Br. J. Radiol.* **2012**, *85*, 153–160. [CrossRef]
43. Grünheid, T.; KolbeckSchieck, J.R.; Pliska, B.T.; Ahmad, M.; Larson, B.E. Dosimetry of a cone-beam computed tomography machine compared with a digital X-ray machine in orthodontic imaging. *Am. J. Orthod. Dentofac. Orthop.* **2012**, *141*, 436–443. [CrossRef] [PubMed]
44. Pauwels, R.; Beinsberger, J.; Collaert, B.; Theodorakou, C.; Rogers, J.; Walker, A.; Cockmartin, L.; Bosmans, H.; Jacobs, R.; Bogaerts, R.; et al. SEDENTEXCT Project Consortium. Effective dose range for dental cone beam computed tomography scanners. *Eur. J. Radiol.* **2012**, *81*, 267–271. [CrossRef] [PubMed]
45. Kapetanović, A.; Oosterkamp, B.C.M.; Lamberts, A.A.; Schols, J.G.J.H. Orthodontic radiology: Development of a clinical practice guideline. *Radiol. Med.* **2021**, *126*, 72–82. [CrossRef] [PubMed]
46. Navarro Rde, L.; Oltramari-Navarro, P.V.; Fernandes, T.M.; Oliveira, G.F.; Conti, A.C.; Almeida, M.R.; Almeida, R.R. Comparison of manual, digital and lateral CBCT cephalometric analyses. *J. Appl. Oral Sci.* **2013**, *21*, 167–176. [CrossRef] [PubMed]
47. Nalçaci, R.; Oztürk, F.; Sökücü, O. A comparison of two-dimensional radiography and three-dimensional computed tomography in angular cephalometric measurements. *Dentomaxillofac Radiol.* **2010**, *39*, 100–106. [CrossRef]
48. Kumar, V.; Ludlow, J.; SoaresCevidanes, L.H.; Mol, A. In vivo comparison of conventional and cone beam CT synthesized cephalograms. *Angle Orthod.* **2008**, *78*, 873–879. [CrossRef] [PubMed]
49. Maspero, C.; Farronato, M.; Bellincioni, F.; Cavagnetto, D.; Abate, A. Assessing mandibular body changes in growing subjects: A comparison of CBCT and reconstructed lateral cephalogram measurements. *Sci. Rep.* **2020**, *10*, 11722. [CrossRef]
50. Sfogliano, L.; Abood, A.; Viana, G.; Kusnoto, B. Cephalometric evaluation of posteroanterior projection of reconstructed three-dimensional Cone beam computed tomography, two-dimensional conventional radiography, and direct measurements. *J. World Fed. Orthod.* **2016**, *5*, 22–27. [CrossRef]
51. Li, N.; Hu, B.; Mi, F.; Song, J. Preliminary evaluation of cone beam computed tomography in three-dimensional cephalometry for clinical application. *Exp. Ther. Med.* **2017**, *13*, 2451–2455. [CrossRef]
52. Lin, H.S.; Li, J.D.; Chen, Y.J.; Lin, C.C.; Lu, T.W.; Chen, M.H. Comparison of measurements of mandible growth using cone beam computed tomography and its synthesized cephalograms. *Biomed. Eng. Online* **2014**, *13*, 133. [CrossRef] [PubMed]
53. Ludlow, J.B.; Gubler, M.; Cevidanes, L.; Mol, A. Precision of cephalometric landmark identification: Cone-beam computed tomography vs. conventional cephalometric views. *Am. J. Orthod. Dentofac. Orthop.* **2009**, *136*, 312.e1–312.e10, discussion 312–313. [CrossRef]
54. Park, C.S.; Park, J.K.; Kim, H.; Han, S.S.; Jeong, H.G.; Park, H. Comparison of conventional lateral cephalograms with corresponding CBCT radiographs. *Imaging Sci. Dent.* **2012**, *42*, 201–205. [CrossRef]
55. Herring, S.W. TMJ anatomy and animal models. *J. Musculoskelet. Neuronal. Interact.* **2003**, *3*, 391–394. [PubMed]
56. Suzuki, A.; Iwata, J. Mouse genetic models for temporomandibular joint development and disorders. *Oral Dis.* **2016**, *22*, 33–38. [CrossRef] [PubMed]
57. Asano, T. The effects of mandibular retractive force on the growing rat mandible. *Am. J. Orthod. Dentofac. Orthop.* **1986**, *90*, 464–474. [CrossRef]
58. Ingervall, B.; Fredén, H.; Heyden, G. Histochemical study of mandibular joint adaptation in experimental posterior mandibular displacement in the rat. *Arch. Oral Biol.* **1972**, *17*, 661–671. [CrossRef]
59. Wang, S.; Sun, Y.; Xia, L.; Li, H.; Xu, Y.; Hua, X. Effects of Twin Inclined Plane Device on Adaptation and Ultrastructure Variations in Condyle of Growing Rats. *BioMed Res. Int.* **2019**, *2019*, 3069347. [CrossRef]
60. Padzys, G.S.; Tankosic, C.; Trabalon, M.; Martrette, J.M. Craniofacial development and physiological state after early oral breathing in rats. *Eur. J. Oral Sci.* **2012**, *120*, 21–28. [CrossRef]
61. Tsolakis, A.I.; Spyropoulos, M.N.; Katsavrias, E.; Alexandridis, K. Effects of altered mandibular function on mandibular growth after condylectomy. *Eur. J. Orthod.* **1997**, *19*, 9–19. [CrossRef]
62. Desai, S.; Johnson, D.L.; Howes, R.I.; Rohrer, M.D. Changes in the rabbit temporomandibular joint associated with posterior displacement of the mandible. *Int. J. Prosthodont.* **1996**, *9*, 46–57.

63. Bland, J.M.; Altman, D.G. Statistical methods for assessing agreement between two methods of clinical measurement. *Lancet* **1986**, *1*, 307–310. [CrossRef]
64. Lin, L.I. A concordance correlation coefficient to evaluate reproducibility. *Biometrics* **1989**, *45*, 255–268. [CrossRef] [PubMed]
65. Lin, F.; Ren, M.; Yao, L.; He, Y.; Guo, J.; Ye, Q. Psychosocial impact of dental esthetics regulates motivation to seek orthodontic treatment. *Am. J. Orthod. Dentofac. Orthop.* **2016**, *150*, 476–482. [CrossRef]
66. Shen, H.; Chau, D.K.; Su, J.; Zeng, L.L.; Jiang, W.; He, J.; Fan, J.; Hu, D. Brain responses to facial attractiveness induced by facial proportions: Evidence from an fMRI study. *Sci. Rep.* **2016**, *6*, 35905. [CrossRef] [PubMed]
67. Sofer, C.; Dotsch, R.; Wigboldus, D.H.; Todorov, A. What is typical is good: The influence of face typicality on perceived trustworthiness. *Psychol. Sci.* **2015**, *26*, 39–47. [CrossRef]
68. Frieze, I.H.; Olson, J.E.; Russell, J. Attractiveness and Income for Men and Women in Management. *J. Appl. Soc. Psychol.* **1991**, *21*, 1039–1057. [CrossRef]
69. Lathrop-Marshall, H.; Keyser, M.M.B.; Jhingree, S.; Giduz, N.; Bocklage, C.; Couldwell, S.; Edwards, H.; Glesener, T.; Moss, K.; Frazier-Bowers, S.; et al. Orthognathic speech pathology: Impacts of Class III malocclusion on speech. *Eur. J. Orthod.* **2021**, *25*, cjab067. [CrossRef] [PubMed]
70. Rezaei, F.; Masalehi, H.; Golshah, A.; Imani, M.M. Oral health related quality of life of patients with class III skeletal malocclusion before and after orthognathic surgery. *BMC Oral Health* **2019**, *19*, 289. [CrossRef] [PubMed]
71. Vieira, A.R. Orthodontics and Genetics. *Dent. Press J. Orthod.* **2019**, *24*, 92–97. [CrossRef] [PubMed]
72. Logan, C.A. Commercial Rodents in America: Standard Animals, Model Animals, and Biological Diversity. *Brain Behav. Evol.* **2019**, *93*, 70–81. [CrossRef]
73. Sengupta, P. The Laboratory Rat: Relating Its Age with Human's. *Int. J. Prev. Med.* **2013**, *4*, 624–630. [PubMed]
74. Bolker, J.A. Selection of Models: Evolution and the Choice of Species for Translational Research. *Brain Behav. Evol.* **2019**, *93*, 82–91. [CrossRef] [PubMed]
75. Baume, L.J.; Derichsweiler, H. Is the condylar growth center responsive to orthodontic therapy? An Experimental study in macacamulatta. *Oral Surg. Oral Med. Oral Path.* **1961**, *14*, 347–362. [CrossRef]
76. Janzen, E.K.; Bluher, J.A. The cephalometric, anatomic, and histologic changes in Macacamulatta after application of a continuous-acting retraction force on the mandible. *Am. J. Orthod.* **1965**, *51*, 803–878. [CrossRef]
77. Roach, H.I.; Mehta, G.; Oreffo, R.O.; Clarke, N.M.; Cooper, C. Temporal analysis of rat growth plates: Cessation of growth with age despite presence of a physis. *J. Histochem. Cytochem.* **2003**, *51*, 373–383. [CrossRef]
78. Andreollo, N.A.; Santos, E.F.; Araújo, M.R.; Lopes, L.R. Rat's age versus human's age: What is the relationship? *Arq. Bras. Cir. Dig.* **2012**, *25*, 49–51, (In English, Portuguese). [CrossRef] [PubMed]
79. Quinn, R. Comparing rat's to human's age: How old is my rat in people years? *Nutrition* **2005**, *21*, 775–777. [CrossRef] [PubMed]
80. Klein, Z.A.; Romeo, R.D. Changes in hypothalamic-pituitary-adrenal stress responsiveness before and after puberty in rats. *Horm. Behav.* **2013**, *64*, 357–363. [CrossRef] [PubMed]
81. Farias-Neto, A.; Varela Brown Martins, A.P.; Figueroba, S.R.; Groppo, F.C.; de Almeidad, S.M.; Rizzatti-Barbosa, C.M. Altered mandibular growth under functional posterior displacement in rats. *Angle Orthod.* **2012**, *82*, 3–7. [CrossRef] [PubMed]
82. Hua, X.; Xiong, H.; Han, G.; Cheng, X. The effects of gradually induced backward movement of the mandible by a Twin Inclined Plane Device in rats. *Angle Orthod.* **2012**, *82*, 839–845. [CrossRef]

veterinary sciences

MDPI

Review

Molecular Markers in Urinary Bladder Cancer: Applications for Diagnosis, Prognosis and Therapy

Ana Mafalda Rasteiro [1,2], Eva Sá e Lemos [1], Paula A. Oliveira [3] and Rui M. Gil da Costa [3,4,5,6,*]

1 CEDIVET, Laboratório Clínico Veterinário, 4200-071 Porto, Portugal; ana.rasteiro@gardenvets.co.uk (A.M.R.); evasalemos@gmail.com (E.S.e.L.)
2 Garden Veterinary Group, Chippenham SN15 1NQ, UK
3 Centre for the Research and Technology of Agro-Environmental and Biological Sciences (CITAB), Inov4Agro, University of Trás-os-Montes e Alto Douro (UTAD), Quinta de Prados, 5000-801 Vila Real, Portugal; pamo@utad.pt
4 Molecular Oncology and Viral Pathology Group, Research Center of IPO Porto (CI-IPOP)/RISE@CI-IPOP (Health Research Network), Portuguese Oncology Institute of Porto (IPO Porto)/Porto Comprehensive Cancer Center (Porto. CCC), 4200-072 Porto, Portugal
5 Postgraduate Programme in Adult Health (PPGSAD), Department of Morphology, University Hospital (HUUFMA), Federal University of Maranhão (UFMA), São Luís 65080-805, Brazil
6 LEPABE—Laboratory for Process Engineering, Environment, Biotechnology and Energy, Faculty of Engineering, University of Porto, Rua Dr. Roberto Frias, 4200-465 Porto, Portugal
* Correspondence: rmcosta@fe.up.pt

Abstract: Cancer of the urinary bladder is a neoplasm with considerable importance in veterinary medicine, given its high incidence in several domestic animal species and its life-threatening character. Bladder cancer in companion animals shows a complex and still poorly understood biopathology, and this lack of knowledge has limited therapeutic progress over the years. Even so, important advances concerning the identification of tumour markers with clinical applications at the diagnosis, prognosis and therapeutic levels have recently been made, for example, the identification of pathological *BRAF* mutations. Those advances are now facilitating the introduction of targeted therapies. The present review will address such advances, focusing on small animal oncology and providing the reader with an update on this field. When appropriate, comparisons will be drawn with bladder cancer in human patients, as well as with experimental models of the disease.

Keywords: transitional cell carcinoma; urothelial carcinoma; histology; therapeutic marker; prognosis

Citation: Rasteiro, A.M.; Sá e Lemos, E.; Oliveira, P.A.; Gil da Costa, R.M. Molecular Markers in Urinary Bladder Cancer: Applications for Diagnosis, Prognosis and Therapy. *Vet. Sci.* **2022**, *9*, 107. https://doi.org/10.3390/vetsci9030107

Academic Editor: David J. Argyle

Received: 15 December 2021
Accepted: 22 February 2022
Published: 28 February 2022

1. Prevalence and Aetiology

Urinary bladder cancer represents about 2 percent of all reported canine malign neoplasms [1,2], and it is rare in cats [3,4]. The majority of canine bladder tumours are malignant and of epithelial origin [5]. Transitional cell carcinoma (TCC), also referred to as urothelial carcinoma (UC) is the most frequent canine urinary bladder tumour [6]. The aetiology of the canine disease is thought to be multifactorial. Several risk factors have been proposed to play a role, such as exposure to older topical insecticides for flea and tick control, obesity, female sex, herbicides and breed predisposition (e.g., Scottish Terrier, West Highland White Terrier, Shetland Sheepdog, Beagle and others) [7–9].

Urinary bladder tumours are also frequently observed in cattle grazing on pastures infested by toxic ferns (mainly *Pteridium* spp.) (reviewed by Gil da Costa et al., 2012 [10]) and have been reported in other ruminant species [11]. The aetiology of bladder cancer in ruminants is much clearer than in companion animals. Grazing on poisonous ferns has been identified as a decisive risk factor since the mid-1900s [12,13]. The occurrence of bladder tumours in cattle is closely related to the geographical distribution of toxic ferns and originates in a syndrome known as bovine enzootic haematuria. Bladder lesions have also been reproduced experimentally in multiple laboratory animal models, by administering the fern or its toxin ptaquiloside [14–18]. Ptaquiloside is a DNA-alkylating agent, which causes point mutations, as well as structural and numeric chromosomal aberrations [19–21]. This toxin also has immunotoxic properties, contributing to reducing immune surveillance against newly arising neoplasms [22–24]. Other ferns containing ptaquiloside (e.g., *Pteris* spp. and *Dryopteris* spp.) or structurally related illudane toxins (for a review of illudane toxins, see Gil da Costa et al., 2013 [25]) do occur and have been reported to cause bladder cancer in cattle in various locations worldwide [26,27]. Bracken consumption has been proposed to facilitate a persistent abortive infection of the bovine bladder by bovine papillomavirus (BPV) types 1, 2, 13 and 14 [28–31]. These Delta BPV types are hypothesised to contribute to bladder carcinogenesis, by activating the platelet-derived growth factor receptor beta (PDGFR-β) through their oncoprotein E5 [32,33].

2. Histology and Grading

The urothelium is a hierarchically organised tissue, comprising basal, intermediate and umbrella cells, and the development of urothelial cancers progressively subverts this normal hierarchical structure [34]. Data obtained from human patients and from laboratory animals have helped trace different types of urothelial carcinoma to specific cell populations of origin and different differentiation pathways [35–38]. Although the pathogenesis of canine bladder cancer is less clear, it seems that numerous genetic changes involving key genes are shared between human and dogs, reflecting a conserved mechanism of pathogenesis [39]. Current laboratory models of bladder cancer, based on rats and mice, are out of the scope of the present work, but several comprehensive reviews have been recently published [40,41].

About 90 percent of all urinary bladder tumours in dogs are epithelial and malignant, and 50 to 90 percent of these will metastasise. Urinary bladder carcinomas include transitional cell carcinoma, squamous cell carcinoma, adenocarcinoma and undifferentiated carcinoma (Table 1) [2]. Among primary epithelial neoplasms of the urinary bladder, TCC represents 75 to 90 percent in dogs [2]. More than 50 percent of overall malignant tumours show involvement of both the bladder and the urethra [5]. Benign epithelial tumours are rare, and only 10 percent of canine urinary bladder tumours are of mesenchymal origin, with smooth muscle neoplasms being the most common [2].

Table 1. Primary bladder tumours in dogs, types and percentages (the most common tumours reported are shown; adapted from Meuten and Meuten, 2016 [2]).

Primary Canine Urinary Bladder Tumours			
Epithelial			
Malignant	(%)	**Benign**	(%)
Transitional cell carcinoma	75–90	Papilloma	2
Undifferentiated carcinoma	6		
Adenocarcinoma	4	Adenoma	0.2
Squamous cell carcinoma	3		
Mesenchymal			
Malignant		**Benign**	
Leiomyosarcoma	2	Leiomyoma	2
Sarcoma	1.5		
Rhabdomyosarcoma	1.3		
Haemangiosarcoma	1	Haemangioma	0.2
Fibrosarcoma	1	Fibroma	1

Canine transitional cell carcinomas are classified based on their growth patterns (Table 2). They can be divided into papillary (papillary or cauliflower exophytic growths projected into the lumen) and non-papillary (plaques, flat nodules or masses), and into infiltrating or non-infiltrating tumours [2]. The consistent observation is that the majority

(90 percent) of canine TCC shows an infiltrating growth pattern [2]. Papillary infiltrating TCC is one of the most common variants and likely to metastasise. The non-papillary and infiltrating type is the second or the most common variant in dogs, depending on the study, and shows a high tendency to metastasise [2].

In cattle, the histological features of bladder tumours are quite different, with 51.2% of purely epithelial tumours, 17.4% of purely mesenchymal tumours and 31.4% of coexisting epithelial and mesenchymal tumours, and numerous benign tumours (papillomas, haemangiomas, etc.) [42].

Table 2. Classification of canine TCC subtypes based on growth pattern (adapted from Meuten and Meuten, 2016 [2]).

Canine Transitional Cell Carcinoma Classification
Papillary infiltrating
Often multiple and may cover large regions of the mucosa. Form papillary or exophytic growths that project into the lumen of the bladder. Invade the stalk and wall of the bladder, lamina propria, and muscle layers and may be transmural. Mild to marked cellular atypia. Likely to metastasise.
Papillary non-infiltrating
Do not invade the stroma of their own stalk, do not go beyond the lamina propria, so unlikely to metastasise. Differentiation from papilloma is subjective and based on criteria such as overall size, cellular atypia, small branches off the main lesion, among others. Non-invasive tumours may be adjacent to invasive TCC, and additional sections should be searched for invasion.
Non-papillary infiltrating
Form plaques and flat nodules, which can cover large regions of the mucosa. Surfaces are often ulcerated, tumour infiltrates into muscle layers, so high tendency to metastasise. Marked histological and cytological variability.
Non-papillary non-infiltrating
Rare. Additionally, defined as carcinoma in situ; confined to the epithelium and do not form papillae. Neoplastic epithelium more intensely eosinophilic than non-neoplastic cells; cells may be dysplastic to mildly anaplastic. Loss of intercellular cohesion. Usually located adjacent to invasive carcinoma; if seen, additional section analysis recommended to look for invasion.

Over the years, several different grading systems for urothelial carcinomas in humans have been proposed and applied to veterinary tumours, looking for a better approach on the evaluation of the tumours' biologic behaviour [2,43]. Meuten and Meuten (2016) proposed a more simplified classification of TCC into low or high grade (Figure S1, Supplementary Material). The majority of canine TCC are invasive, high grade and at an advanced stage when diagnosed [2]. In affected dogs, high-grade tumours seem to be more common in terriers than in non-terrier breeds [44].

3. Diagnosis

3.1. Clinical Signs and Differential Diagnosis

Clinical signs in dogs with TCC are usually nonspecific, many of which, such as dysuria, haematuria and pollakiuria, are commonly observed with urinary tract disease [2,5]. Concurrent urinary tract infections (UTI) are often present [2]. Tumour growth can lead to obstruction of the ureters or urethra and invasion and disruption of the normal functioning of the urethral sphincter [45]. On physical examination, a thickening of the urethra and of the trigone region of the bladder and enlargement of iliac lymph nodes may be found and, occasionally, a mass in the bladder or a distended bladder [1]. Urinary tract obstruction can occur prior to the development of lethal metastasis and is a common cause of death in dogs with TCC [6]. However, a normal physical examination does not exclude the presence of a TCC [46]. Differential diagnoses of canine TCC comprise other neoplasia, chronic cystitis, polypoid cystitis, fibroepithelial polyps, granulomatous cystitis/urethritis, calculi, among others [1].

In cattle, haematuria and weight loss are the main symptoms of bladder cancer and often present as part of the previously mentioned syndrome, known as bovine enzootic haematuria [26,47].

3.2. Diagnostic Procedures and Staging

Diagnostic procedures for TCC should include a complete blood cell count, serum biochemistry profile, urinalysis, urine culture (to rule out lower urinary tract infection) and cancer staging [1,46].

Definitive diagnosis of TCC can be established via histopathologic examination of tumour tissue and/or cytology of a representative sample [2]. Biopsies can be collected by means of cystotomy, cystoscopy and traumatic catheterisation, invasive procedures that usually involve general anaesthesia [1,39]. Cytological samples may be obtained by direct or ultrasound-guided percutaneous mass fine-needle aspirate or traumatic catheterisation [48].

The risk of tumour implantation or seeding/dissemination throughout other tissues following diagnostic or therapeutic procedures has been reported, especially after surgical manipulation of the tumour [49–52]. Even though reports are scarce, these should be carefully interpreted. Where possible, less invasive techniques should be favoured.

A less invasive technique consists of performing a cytology from urine sediment. If tumour cells are present, a diagnosis can be achieved [2]. However, negative results do not rule out TCC. In one study, malignant cells were seen in only 30% of dogs with lower urinary tract tumours [5]. Thus, cytological results must be interpreted with caution, especially upon the presence of inflammation of the urinary tract, and correlation with clinical data is essential for reaching a diagnosis.

Clinical staging of canine TCC includes thoracic and abdominal radiography, abdominal ultrasonography and specific urinary tract imaging [1,46]. Computer tomography (CT) has increasingly been used to aid in diagnostics and staging, particularly for more accurately evaluating the urethra and to detect metastases [53]. Figure S2 (Supplementary Material) shows the TNM (tumour, node, metastasis) classification for clinical staging of canine bladder cancer [54]. The TNM stage at diagnosis for TCC has shown to be strongly related to prognosis. More advanced TNM stage at the time of diagnosis was significantly associated with shorter survival [7,53]. Tumours located in the urethra were also associated with shorter survival time than ones in the bladder [53]. TCC has rarely been curable; however, with current therapies, many dogs will achieve stable disease for several months after diagnosis [1].

3.3. Recent Advances in Diagnostic Techniques for UC

As mentioned above, clinical presentation of canine TCC is comparable to several other (and far more common) urinary tract disorders. Consequently, the diagnosis of TCC is often delayed, allowing the tumour to grow, infiltrate and metastasise [2]. In fact, most TCCs are currently not diagnosed until they reach an advanced stage and thus present poor prognosis [55]. As such, effective (and preferably less invasive) methods for the early identification of UC are needed, which could improve responses to treatment and survival rates among affected dogs [56]. In particular, because TCC tumour cells and metabolites may be shed into urine, this body fluid is likely to present tumour-specific molecules that could be used as biomarkers for tumour detection using easily accessible samples collected through non-invasive techniques [46,57]. Over recent years, several potential biomarkers for canine TCC have been investigated for diagnostic/screening or prognostic purposes, and a few of them are currently available for commercial use.

These and other potential biomarkers are summarised in Table 3 and detailed below.

Table 3. Current and potential markers for clinical applications in canine transitional cell carcinoma.

Biomarker	Sample	Method	Diagnostic Utility, Commercial Availability	Utility as a Prognostic and/or Therapeutic Target	Power of the Test	
					Sensitivity	Specificity
***BRAF* mutation**	**Tissue, urine, blood** [58–67]	Determination of *cBRAF*^{V595E} mutation status in DNA retrieved from cells, using ddPCR analysis or other molecular methods.	Highly sensitive test for detecting TCC cells bearing the *BRAF* mutation. Could be used as a first, non-invasive screening test. Commercially available for dogs, for use in free-catch urine samples—CADET® *BRAF* mutation detection assay. Provides qualitative results (positive vs. negative for V595E) and quantitative data of tumour-derived mutation load in urine DNA. Reported to detect TCC in free-catch urine samples up to several months before development of clinical signs. The test is not affected by the presence of blood or bacteria in the urine. ~20% of tumours of canine TCC and PC patients do not possess the mutation, which limits the sensitivity of the ddPCR assay to ~80%. A more recent test that detects chromosomal copy number variation can be added in *BRAF* mutation-negative patients, increasing combined sensitivity to ~95% (CADET® *BRAF-PLUS)*.	BRAF mutation was not a predictor for histological grade, nor for survival. Measuring levels of BRAF mutation in urine or blood samples may be useful for monitoring treatment response and relapse. Potential target for treatment.	67–88% (TCC, tissue) 83–100% (TCC, urine)	100% (TCC, tissue and urine)

Table 3. *Cont.*

Biomarker	Sample	Method	Diagnostic Utility, Commercial Availability	Utility as a Prognostic and/or Therapeutic Target	Power of the Test	
					Sensitivity	Specificity
BTA	**Urine** [68–72]	Rapid latex agglutination dipstick colorimetric test for qualitative detection of tumour analytes in urine. The test uses antibodies to detect a urinary bladder tumour-associated glycoprotein complex.	Useful as a screening test to rule out TCC, especially in dogs at high risk of developing TCC. False positive test results reported in dogs with non-neoplastic urinary tract disease, e.g., in the presence of significant glycosuria, proteinuria, and pyuria or haematuria. Presence of lower urinary tract malignant tumours other than TCC may yield positive results. Discrepancies with results may be observed over time, while reading the test. Commercially available—V-BTA Test. Results are either positive or negative. Not recommended as a confirmatory/definitive diagnostic test for urinary tract TCC in dogs, and should not be indiscriminately used in every patient presenting clinical signs of urinary tract disease.	N.A.	88–90%	35–41% in dogs with non-malignant urinary tract disease; 84–94% in healthy dogs or unhealthy dogs due to non-urinary tract diseases
bFGF	**Urine** [73–78]	ELISA urine test for human and canine bFGF. A quantitative sandwich enzyme immunoassay technique has also been developed using an antibody for canine bFGF.	Urine bFGF could be useful as a diagnostic tumour marker, helping to distinguish dogs with UTI from those with TCC. Commercially available (for research use, only): Quantikine® HS ELISA, Human FGF basic Immunoassay, Canine BFGF ELISA Kit, Nori® Canine FGF Basic ELISA Kit.	Quantification of urine bFGF could be useful as a non-invasive indicator of treatment response.	N.S.	N.S.

Table 3. *Cont.*

Biomarker	Sample	Method	Diagnostic Utility, Commercial Availability	Utility as a Prognostic and/or Therapeutic Target	Power of the Test	
					Sensitivity	Specificity
Chromosomal CNAs	**Tissue, urine** [39,79]	Assessment of urothelial cell ploidy/DNA copy number status in biopsy sections and in urine sediment by FISH.	Non-invasive method for canine TCC diagnosis. Potentially high-sensitivity and high-specificity FISH-based method/assay for the detection of canine TCC diagnosis utilising low-volume, free-catch urine specimens. Expensive and high effort method/labour intensive, expertise, time-consuming, increased cost, which may limit its application as routine diagnostics in a clinical environment. Not commercially available for canine TCC. Available for in vitro diagnostic use in human samples. A multicolour FISH-based assay for detection of aneuploidy for chromosomes 3, 7, 17, and loss of the 9p21 locus through FISH in urine specimens—UroVysion Bladder Cancer Kit.	N.A.	N.S.	N.S.
	Tissue, urine [55,64]	Multiplexed ddPCR assay for the detection and quantification of DNA copy number imbalances/changes characteristic to canine TCC.	Accurate, high-throughput method for evaluation of copy number changes in dogs with TCC. In this study, changes in copy number were not detected in 33% of urine DNA samples from dogs with TCC, which was probably due to the presence of inflammatory cells. Thus, additional techniques to improve sensitivity in those samples may be required. In such cases, FISH will still provide a more accurate evaluation. Commercially available for dogs, for use in free-catch urine samples: CADET® *BRAF-PLUS*. Can be used in *BRAF* mutation-negative patients. Could be added to CADET® *BRAF*, increasing combined sensitivity to ~95%.	N.A.	N.S.	N.S.

Table 3. *Cont.*

Biomarker	Sample	Method	Diagnostic Utility, Commercial Availability	Utility as a Prognostic and/or Therapeutic Target	Power of the Test	
					Sensitivity	Specificity
Microsatellite instability	**Urine** [72]	PCR study of a panel of 22 microsatellite DNA sequences from exfoliated urothelial cells and blood cells; comparison of microsatellites genotypes.	The technique added little value as a diagnostic test for TCC in dogs. High rate of false positives (32%, 12 of 38).	N.A.	55% (48% *)	68% (76%, vs. V-BTA)
					* When compared with results of V-BTA from the same study.	
MicroRNAs	**Tissue, cell lines** [56]	QPCR of specific miRNAs involved in the pathophysiology of TCC in humans.	MiR-34a, miR-16, miR-103b and miR-106b could be useful diagnostic biomarkers for the identification of dogs with TCC. More studies are required, with a larger sample.	N.A.	N.S.	N.S.
	Blood, urine [80]		MiR-103b and miR-16 are potential non-invasive diagnostic biomarkers for TCC; particularly for distinguishing LUTD and TCC in canine urine samples. Urine tests seem to be superior in distinguishing TCC from LUTD.	N.A.	N.A.	N.S.
Telomerase	**Canine TCC cell line, urine** [81,82]	PCR-based telomeric repeat amplification protocol for detection/measurement of telomerase activity.	Telomerase activity may be useful in diagnosing canine TCC in urine samples in a clinical context. Results of the assay are either telomerase-positive or telomerase-negative. Urine samples containing other telomerase-positive cells may yield false-positive results (e.g., presence of activated lymphocytes in dogs with bacterial cystitis). False-negative results may occur with unappropriated urine samples storage.	N.A.	91%	89%
					Diagnostic sensitivity/specificity of the TRAP assay applied to clinical canine urine samples.	

Table 3. *Cont.*

Biomarker	Sample	Method	Diagnostic Utility, Commercial Availability	Utility as a Prognostic and/or Therapeutic Target	Power of the Test	
					Sensitivity	Specificity
Calgranulins	**Urine** [44,83]	Species-specific radioimmunoassays to measure urine concentrations of canine calgranulins S100A8/A9 and S100A12.	Results presented as normalised to urine specific gravity levels (S100A8/A9$_{USG}$) and as S100A8/A9-to-S100A12 ratio (UcalR). Provides quantitative results. S100A8/A9$_{USG}$ could be a good a screening test for TCC/PC in dogs, especially in those where a UTI has been ruled out as a cause of clinical signs of lower urinary tract disease (due to a moderate rate of false positives observed for dogs ≥6 years of age with UTI). UcalR can help differentiate patients with a UTI from those with TCC/PC, even though a moderate false negative rate was seen in dogs ≥ 6 y.o. with a UTI. A combination of S100A8/A9$_{USG}$ and uCalR improved diagnostic accuracy for the detection of canine TCC/PC. Test levels are not affected by haematuria.	N.A.	96% S100A8/A9$_{USG}$ * 91% UcalR **	66% S100A8/A9$_{USG}$ * 60% UcalR **
					* For detection of TCC/PC in dogs ≥ 6 y.o.; ** to distinguish dogs with TCC/PC from dogs with UTI in dogs ≥ 6 y.o.	
Proteomics	**Urine** [84]	Characterisation of the canine urinary proteome by using liquid chromatography tandem mass spectrometry and immunoblot.	A protein signature was identified, that could distinguish between healthy patients and those with TCC or UTIs. A statistical model using a biomarker multiplex for categorising samples as TCC or non-TCC was developed, predicting the presence of disease with 90% confidence. Potential relevance of the identified proteins as biomarkers for the diagnosis of TCC in dogs. Preliminary study, high-throughput technique. A more direct assay will be useful for clinical diagnosis.	N.A.	N.S.	N.S.

Table 3. *Cont.*

Biomarker	Sample	Method	Diagnostic Utility, Commercial Availability	Utility as a Prognostic and/or Therapeutic Target	Power of the Test	
					Sensitivity	Specificity
Metabolomics	**Urine** [55]	Nuclear magnetic resonance spectroscopy-based metabolite profiling analysis.	Six metabolites showed significantly higher levels in dogs with TCC compared to controls: urea, choline, methylguanidine, citrate, acetone and β-hydroxybutyrate. Good sensitivity to predict the healthy control and disease samples. Potential for early detection of bladder cancer. Preliminary study, high-throughput technique.	N.A.	86%	78%
Lipidomics	**Tissue** [85]	Imaging analysis to examine lipidome/lipid profiles, using desorption electrospray ionisation mass spectrometry.	Differentiation of canine cancerous bladder tissue and cutaneous metastasis from noncancerous canine bladder tissue samples. Different lipid distributions between healthy and diseased tissues. DESI-MS imaging could be useful in diagnosing TCC by using a multimarker approach based on the lipid profiles and intensities of tissue samples. Further studies are required with larger populations and additional control groups, i.e., with other lower urinary diseases. Still requires invasive techniques for tissue collection.	N.A.	N.S.	N.S.
	Urine [86]	Analysis of lipid profiles using liquid chromatography-mass spectrometry.	Unique lipid profiles were found among dogs with TCC, dogs with UTI, and healthy dogs. Specific statistical analyses allowed their differentiation. Concentrations of the specific lipids could not be determined, and thus the study did not conclude which lipid families were up or downregulated. Foundation for further research on urinary lipids as potential biomarkers for TCC. Non-invasive method.	N.A.	N.S.	N.S.

Table 3. *Cont.*

Biomarker	Sample	Method	Diagnostic Utility, Commercial Availability	Utility as a Prognostic and/or Therapeutic Target	Power of the Test	
					Sensitivity	Specificity
Survivin	**Tissue** [87]	Immunohistochemistry for detection of survivin, an apoptosis-inhibiting protein; RT-PCR analysis for the survivin gene.	Initial phases of investigational development with limited samples. Additional research needed to investigate potential role of nuclear survivin as an early marker for bladder tumours, as well as in the development, progression and as a therapeutic target.	N.A.	N.S.	N.S.
EGFR	**Tissue** [88]	IHC and qPCR analysis for EGFR.	EGFR expression could potentially be used as a marker to aid canine TCC diagnosis. It may improve the sensitivity of urine cytological diagnosis when provisional diagnosis is needed.	Not useful for predicting prognosis of TCC.	72%	100%
HER-2	**Tissue** [61,89]	IHC for HER-2.	N.A.	Potential maker of malignancy and therapeutic target in canine TCC.	N.S.	N.S.
VEGFR2, PDGFR-β, c-KIT	**Tissue, cell lines** [90–94]	IHC for expression of VEGFR2, PDGFR-β, c-KIT.	PDGFR-β could play a role in canine TCC tumourigenesis.	PDGFR-β and VEGFR2 might be involved in mediating clinical response of TCC to toceranib.	N.S.	N.S.
Granzyme B, CD3	**Tissue** [95]	IHC and PCR assay for CD3 and granzyme B.	N.A.	Granzyme B^+ tumour-infiltrating cells could be involved in inhibition of tumour progression, and a favourable prognosis. Presence of granzyme B^+ tumour-infiltrating cells might be an independent prognostic factor.	N.S.	N.S.

Table 3. *Cont.*

Biomarker	Sample	Method	Diagnostic Utility, Commercial Availability	Utility as a Prognostic and/or Therapeutic Target	Power of the Test	
					Sensitivity	Specificity
P63, Ki67, β-catenin	**Tissue** [96–102]	IHC for p63.	P63 could potentially be used as a clinical marker for diagnosing canine TCC.	P63 could potentially be used as a clinical marker for predicting prognosis in canine TCC.	N.S.	N.S.
UP III **CK 7** **CK 20** **COX-2**	**Tissue** [2,44,103–105]	IHC for UP III, CK 7 and CK20.	UP III is the most common marker of urothelial differentiation used in dogs. It was considered the marker of choice in canine urothelial neoplasms. Although UP III is not a specific marker for TCC itself (it does not differentiate neoplastic from non-neoplastic lesions), it can be useful e.g., to rule in TCC in a biopsy from a tumour of unknown origin and to identify metastatic carcinomas in the skin. CK 7 was more sensitive than UP III for canine TCC, but CK 7 is expressed in several non-urothelial tumours and also in normal tissues, as is CK 20. CK 7 should be used for tumours negative for UP III but suspected of being TCC. CK 20 alone did not prove to be useful for diagnosis of urothelial tumours. Some urothelial carcinomas might not be positively labelled when using UP III and CK 7 as diagnostic markers. COX-2 has been found to be expressed in canine TCCs but not by normal urothelium of the urinary bladder.	UP III, CK 7, COX-2: Significant associations between specific patterns of expression and tumour classification, depth of neoplastic cell infiltration. COX-2: Intensity of COX-2 expression did not correlate with grading. Nonselective COX and COX-2 specific inhibitors have been used for treating TCC. Still unclear whether it could be useful as a predictive factor for treatment response.	N.S.	N.S.

TCC: transitional cell carcinoma; PC: prostatic carcinoma; N.A.: Not available; N.S.: Not specified; y.o.: years old; vs.: versus.

3.3.1. BRAF Mutation

The somatic *BRAF*^{V595E} mutation in canine *Canis familiaris* chromosome 16 (CFA16) results in a valine to glutamic acid substitution at codon 595 of canine *BRAF* [58,59].

In one study, the *BRAF*^{V595E} mutation was detected across 87.9% of canine invasive TCC bladder tumour tissue samples (58 out of 66 samples). The mutation was also detected in the urine sediments of dogs tested with mutation-positive tumours (9/9), and none of the healthy control dogs were positive for the mutation [58]. Another study screened for the presence of the *BRAF* mutation in 667 canine tumours, including a series of haematopoietic tumours (n = 245), sarcomas (n = 160), carcinomas (n = 115), melanocytic tumours (n = 72), as well as other, less common cancers (n = 75). The *BRAF* mutation was identified in 64 primary tumours (9.6%) of different origins with various frequencies, with particularly high frequency in prostatic carcinoma (PC) (20/25, 80%) and urothelial carcinoma (30/45, 67%) tissue samples. No mutation was found in normal control tissues [59].

Furthermore, a droplet digital PCR (ddPCR) assay for detection of the canine *BRAF*^{V595E} mutation in canine urogenital tumours was developed. Tested samples included tissue obtained from TCC, PC and non-neoplastic bladder epithelium, and free-catch urine samples derived from dogs with UC, PC, with cystitis and healthy controls. In tissue samples, 75% of TCC (36/48), 85% of PC (23/27) and none of control samples were mutation positive. The mutation was also identified in urine samples from 83% (19/23) of canine TCC and 100% of PC patients (3/3), and none of control samples were positive for the mutation. The V595E mutation was consistently detected in matched tissue and urine specimens from six of these dogs [60].

Grassinger and their team (2019) studied the presence of *BRAF* mutation in TCC tissue samples from 65 dogs from different breeds (both terriers and non-terriers). Overall, the mutation was detected in over half of the specimens, with considerably higher prevalence in terriers compared to the other breeds (73% versus 36%) [44]. Parker et al. (2020) confirm this hypothesis of hereditary predisposition to TCC [61]. Histological grade, however, was not significantly correlated with the *BRAF* mutation [44].

A recent retrospective study looked at the prognostic significance of *BRAF* mutation in 79 canine TCCs. A total of 51 tumours (65%) were *BRAF*^{V595E}-positive, and the mutation was not correlated with survival [62].

Cells, both normal and tumoural, can release DNA into the bloodstream (circulating cell-free DNAs) via processes such as direct secretion, necrosis and apoptosis. Circulating tumour DNAs (ctDNA) contain the same genetic information as the original tumour, and they have been investigated in human and veterinary medicine for various applications, including studying tumour features and monitoring disease progression and response to treatment [106–110].

In a study including 15 dogs with TCC, 11 (73%) were positive for *BRAF* mutation, which was assessed in either tissue or urine samples. Blood samples were collected, and cfDNA was obtained from plasma. Results showed that concentrations of *BRAF*-mutated ctDNA were higher in *BRAF*^{V595E}-positive dogs than in wild-type dogs. However, they did not correlate with clinical stage or the presence of metastases. Furthermore, six of the dogs were monitored throughout treatment and course of the disease. Levels of mutated *BRAF* ctDNA increased with disease progression and decreased with response to treatment. It is noteworthy that *BRAF* mutation could be present in other neoplasms and therefore contribute to the overall levels of ctDNA detected. Additionally, treatment therapies varied between analysed patients [63]. Further validation studies are encouraged.

Overall, these findings suggest that detection and quantification of the *BRAF* mutation in canine TCC patients, especially through non-invasive techniques, could be very useful in the clinical setting, not only as a biomarker for diagnosis, but also for monitoring disease progression and treatment response. A test, which based on ddPCR, is currently commercially available for use in urine samples and could be particularly useful as a screening test (Table 3) [60,64].

3.3.2. Bladder Tumour-Associated Antigen Test (BTA)

The BTA was originally developed for use in humans (Bard BTA test). It was then assessed in canine patients, and a veterinary version (V-BTA) of the test was also developed. It consists of a rapid latex agglutination dipstick colorimetric test that allows the qualitative detection of tumour analytes in urine. The test uses antibodies to detect a urinary bladder tumour-associated glycoprotein complex. This complex contains basement membrane proteins that are degraded upon urinary tract invasion or tissue damage and shed into the urine and may also contain immunoglobulins [68,69]. Tests results are either positive or negative.

Studies performed with canine cohorts evaluated: 65 animals, including patients with TCC (associated with the bladder, urethra, prostate or vagina), as well as healthy controls and urologic controls (e.g., dogs with other urologic conditions or with systemic disease leading to abnormal urinalysis examination) [70]; 54 animals, including dogs with lower urinary tract neoplasia (TCC or other neoplasia affecting the bladder and/or other urogenital structures), dogs without urinary tract abnormalities and dogs with non-malignant urinary tract disease [71]; 229 animals, including dogs with TCC of the lower urinary tract (involving the urinary bladder and/or other structures), healthy control dogs, unhealthy control dogs with (non-TCC) urinary tract disease and unhealthy control dogs without urinary tract disease [68]; 73 animals, including dogs with TCC (of the bladder, prostate and/or urethra, with or without concomitant UTI), healthy dogs, proteinuric dogs and dogs with non-TCC associated lower urinary tract disease (LUTD) [72]. Overall, the test exhibited high sensitivity but lower specificity for the detection of TCC in dogs. Therefore, the V-BTA has not been recommended as a confirmatory/definitive diagnostic test for urinary tract TCC in dogs and should not be indiscriminately used in every patient presenting clinical signs of urinary tract disease. However, it can be useful as a screening test to rule out TCC, especially in dogs at high risk of developing the disease. As an example, considering the prevalence of TCC in a population of geriatric dogs, less than 3% of dogs with positive V-BTA test results would be expected to have TCC. On the other hand, 99.9% of dogs with a negative test would not have the disease [68].

3.3.3. Basic Fibroblast Growth Factor (bFGF)

The basic fibroblast growth factor (bFGF) is a proangiogenic peptide that has been associated with tumour progression in humans and has been detected at high levels in the urine of humans with urologic and non-urologic malignancies [73]. An ELISA test kit for human bFGF that recognises natural and recombinant bFGF has been developed and has also been used for the quantification of canine bFGF in urine [74–76].

In one study, bFGF levels were quantified in the urine of dogs with locally active TCC of the urinary bladder ($n = 7$) and compared with those in dogs with urinary tract infection ($n = 10$) and normal dogs ($n = 17$). bFGF concentrations were significantly higher in dogs with bladder cancer than in normal dogs and in dogs with UTI, while these two groups showed comparable results. The study suggested that urine bFGF could be useful as a diagnostic tumour marker, allowing for distinction between dogs with UTI and dogs with TCC, or as a non-invasive indicator of treatment response [73]. In another study, urinary bFGF concentrations were measured in 14 dogs with invasive TCC of the urinary bladder (with or without involvement of the urethra or the prostate) before and after treatment with piroxicam. Before treatment, urine bFGF levels were significantly higher than those of eight normal dogs. In 77% of dogs (11 out of 14), bFGF concentrations decreased with treatment, and tumour volume decreased by 9–75% in 10 of those 11 dogs. Overall, a positive correlation was seen between the change in urine bFGF concentration and the change in tumour size [77]. In a similar study, bFGF concentrations were assessed in the urine of dogs with bladder TCC before and after treatment with piroxicam/cisplatin. Before treatment, concentrations of bFGF were significantly higher than those in normal dogs (eight animals per group). Concentrations decreased with treatment in four out of eight dogs, and tumour volume decreased by 59–95% in these four dogs. On the other hand, concentrations increased in four dogs, two of them also showing an increase in tumour size. No significant association was observed between change in urine bFGF concentration and change in tumour size [78]. Additional studies are required for determining the sensitivity and specificity of this test in a larger canine population and its potential use to determine disease progression, clinical stage and response to treatment.

3.3.4. Chromosomal Copy Number Aberrations (CNAs)

Fluorescence in situ hybridisation (FISH) is considered the gold standard of copy number detection and enumeration. In biopsies from canine TCC, chromosomal copy number aberrations (CNAs) have been identified throughout several regions of the canine genome. In fact, three of those CNAs are highly recurrent features: *Canis familiaris* (CFA) chromosome 13 gain, CFA 36 gain and CFA 19 loss.

FISH was shown to be effective in identifying aberrant cells in urine sediment samples from dogs with TCC confirmed via urine cytology or histopathology. In a study involving 24 dogs diagnosed with UC, all animals showed at least one of these three aberrations in cells recovered from urine sediment specimens [39]. Thus, the detection of these aberrations may potentially be used as a molecular diagnostic test for TCC. In the future, validation of the technique with a larger canine population is needed, including dogs presenting with non-neoplastic urinary diseases, in order to establish true sensitivity and specificity values for a clinically valuable diagnostic assay [39]. An in vitro test is available for use in human samples [79].

A multiplexed droplet digital polymerase chain reaction (ddPCR) assay has also been developed for detection and quantification of copy number imbalances/changes characteristic of canine TCC. Aimed to detect CNAs of specific regions of canine chromosomes 13, 19 and 36, the assay proved effective at differentiating 31 neoplastic and 25 non-neoplastic bladder tissues (including normal bladder and bladder with non-neoplastic lesions), based on copy number, with 100% sensitivity and specificity. When evaluated in DNA isolated from free-catch urine samples from TCC and non-TCC dogs, copy number imbalance (for CFA 13 and CFA 36) was detected in 67% (12 out of 18) TCC specimens and was absent in non-TCC urine samples (n = 7, including clinically healthy dogs and those with urinary tract infections). This assay was shown to be a fast method with a potential clinical value for determination of copy imbalance not only in tissue samples but also in free-catch urine from dogs diagnosed with UC. Further evaluation of the ddPCR assay as a molecular diagnostic test for canine TCC in a clinical setting is required, with a larger sample including both TCC and non-neoplastic urinary diseases, in order to assess its sensitivity and specificity [55]. A test for use in urine clinical samples has recently been developed (Table 3) [64].

3.3.5. Microsatellite Instability

Microsatellites (MS) are short tandem repeats of DNA that occur mainly in noncoding regions [111]. Microsatellite instability (MSI) refers to the accumulation of mutations within MS, and it can promote tumourigenesis. A panel of 22 MS sequences was evaluated by PCR amplification of DNA samples extracted from exfoliated urothelial cells and blood cells. Samples included urine from dogs with TCC (of the bladder, prostate and urethra, diagnosed by fine-needle aspiration cytology or histologic evaluation) and from control dogs (proteinuric dogs, dogs with lower urinary tract disease not associated with TCC and healthy controls). The study showed that detection of these MS abnormalities could be achieved in canine urine samples. MSI was more frequently detected in urine samples from dogs with TCC (55%, n = 11 out of 20) compared with control dogs (32%, n= 12 out of 38). However, the difference was not statistically significant. Measurement of MSI in urine was poorly specific and not sensitive for identification of TCC in dogs [72].

Further research is needed to determine whether targeting other specific loci might improve the diagnostic utility of the test and to determine whether MSI represents an early preneoplastic change and can be an indicator of urothelial carcinogenesis in canine TCC [72].

3.3.6. MicroRNAs (miRNAs)

MiRNAs are small endogenous noncoding RNAs involved in post-transcriptional regulation of gene expression that take part in most biological processes in mammals, such as cell proliferation, differentiation and apoptosis. Dysregulation of miRNAs plays a significant role in cancer development [112,113].

A study measured the expression of specific miRNAs that have the ability to target components of the p53, Rb and Bcl-2 pathways, which are involved in the development of bladder tumours in humans. Five miRNAs (miR-34a, let-7c, miR-16, miR-103b and miR-106b) were quantified in bladder tissue samples from dogs with grossly normal urinary bladders, non-neoplastic inflammatory bladder disease and with TCC, as well as in five established TCC cell lines. Expression of miR-16, miR-34a, miR-103b and miR-106b was up-regulated in TCC samples, with significantly higher expression compared with bladder samples obtained from dogs with inflammatory lower urinary tract disease. The levels of miR-34a and miR-106b in TCC samples were also significantly increased compared to those in grossly normal bladder. The study concluded that miR-34a, miR-16, miR-103b and miR-106b could potentially be oncogenic in TCC and could be useful diagnostic biomarkers for the identification of dogs with the disease. One limitation of the study was the fact that only samples from muscle-invasive TCC were analysed, with many of the patients being at an advanced or metastatic stage [56].

In another study, expression levels of the five miRNAs were also evaluated by qPCR, in blood and urine (sediment) samples from 70 dogs with clinically normal bladders (n = 28), inflammatory or infectious lower urinary tract diseases (LUTD, n = 25) and TCC (n = 17). In blood samples, statistically significant differences were found in the expression levels of miR-103b in normal patients compared to LUTD and with TCC, but no significant differences were identified between LUTD and TCC patients in any of the five miRNAs levels. Conversely, in urine samples, significant differences were found in miR-103b and miR-16 levels between LUTD and TCC dogs. Additionally, differences were detected in urine levels of let-7c in normal versus LUTD and TCC patients, and in miR-103b and miR-106b in normal versus TCC patients. Data also revealed alterations in the coordinated expression (expression trends) of miRNAs in urine from patients with LUTD and TCC. Both miR-103b and miR-16 were suggested as potential non-invasive diagnostic biomarkers for TCC, particularly for distinguishing LUTD and TCC in canine urine samples [80]. However, additional research is required to evaluate diagnostic sensitivity of the method, to determine the potential role and changes in expression of miRNAs during development and progression of TCC, and to investigate their potential as therapeutic targets.

3.3.7. Telomerase

Telomeres are repeated DNA sequences located at the end of chromosomes, which, with each cellular division, are partly lost and become shorter, a process that is related to ageing and cell death. Telomerase reverse transcriptase is frequently up-regulated in cancer cells and is responsible for the preservation of telomere ends, an important mechanism by which tumour cells escape senescence [114,115].

The activity of this enzyme was measured in canine samples by using an in vitro PCR-based telomeric repeat amplification protocol (TRAP) developed by Kim et al., (1994) [81,82]. Telomerase activity was detected in a canine TCC cell line, as well as in urine samples collected from 10 out of 11 dogs with TCC and from 2 out of 10 dogs with benign lower urinary tract disease. Activity of this enzyme was not detected in any of the healthy dogs. Samples were obtained from dogs with TCC (located in the bladder or the prostate; confirmed by histology/cytology of the mass; some dogs were receiving treatment for TCC at the time of sample collection), dogs with benign lower urinary tract disease and healthy dogs (n = 30). The technique was shown to be highly sensitive in detecting telomerase activity. However, urine samples containing other telomerase-positive cells could yield false-positive results (e.g., presence of activated lymphocytes in dogs with bacterial cystitis) [82].

The study concluded that telomerase activity may be a useful diagnostic marker for canine urothelial carcinomas, and specifically, the TRAP assay may be useful in diagnosing canine TCC in clinical urine samples/in a clinical context. Additional studies are required to more accurately determine the performance of the test in larger populations, including more dogs with varying degrees of benign LUTDs, to further establish the specificity of the test. Further research is also needed on optimisation of urine storage and processing protocols in order to develop an accurate non-invasive diagnostic test for evaluation of clinical samples [82].

3.3.8. Calgranulins

Calgranulins are innate antimicrobial proteins belonging to the S100 family of calcium-binding proteins that include the S100A8/A9 complex (also called calgranulin A/B or calprotectin) and S100A12 (calgranulin C). They are expressed by cells of the innate immune system and have been associated with inflammatory disorders. Calprotectin is also expressed by epithelial cells upon malignant transformation and plays a role in the regulation of cell proliferation and metastasis. This protein complex is overexpressed in human bladder and prostate cancers, whereas results for S100A12 have not been clear [2,83].

A method based on species-specific radioimmunoassays that was established and validated by Heilmann and co-authors [46] was conducted to measure urine concentrations of canine calgranulins S100A8/A9 and S100A12 [83]. Urine samples ($n = 239$) were collected from dogs with TCC/PC (treated and treatment-naïve), non-neoplastic urinary tract disease, other neoplasms, UTI and healthy controls. Tumour locations included urinary bladder, urethra and the prostate, whose diagnoses were confirmed by either cytology or histopathology. Results were presented as normalised to urine specific gravity levels (S100A8/A9USG) and as S100A8/A9-to-S100A12 ratio (UCalR). The study concluded that S100A8/A9USG and UCalR could be useful for diagnosing canine TCC/PC. S100A8/A9$_{USG}$ could be a good a screening test for TCC/PC in dogs, especially in those where a UTI has been ruled out as a cause of clinical signs of lower urinary tract disease (due to a moderate rate of false positives observed for dogs $\geq$ 6 years of age with UTI). The uCalR can, in turn, help differentiate patients with a UTI from those with TCC/PC, even though a moderate false negative rate was seen in dogs $\geq$ 6 y.o. with a UTI. A combination of S100A8/A9$_{USG}$ and uCalR improved diagnostic accuracy for the detection of canine TCC/PC.

In this study, the possibility of occult urinary tract diseases, including TCC/PC, could not be completely excluded in the healthy control group. So, validation of these results in a larger cohort is encouraged. Further investigation is also required to explore a potential correlation between urinary expression/concentrations of the S100/calgranulins and tumour grade, staging, treatment response, progression and survival time [83].

3.3.9. Proteomics

A preliminary study characterised the proteome of canine urine samples by using liquid chromatography tandem mass spectrometry (LC-MS/MS). Three cohorts were included, comprising four animals each: healthy dogs, dogs with UTI and dogs with TCC (located in the bladder, some of them with urethral and ureter involvement; confirmed via cytology or histology; treated or untreated for TCC). Of the 379 proteins identified in urine samples, 96 were present exclusively in the TCC group. The study identified a protein signature that could distinguish between healthy patients and those with TCC or UTIs. A statistical model using a biomarker multiplex for categorising samples as TCC or non-TCC was developed, predicting the presence of disease with 90% confidence. Further analyses of larger cohorts are required to confirm the relevance of the identified proteins as biomarkers for the diagnosis of TCC in dogs. Development of a more direct assay for the detection of the specific proteins will be useful for clinical diagnosis [84].

3.3.10. Metabolomics

Metabolomics (also known as metabolic profiling) is an approach that involves the identification and global analysis of metabolite concentrations in cells, tissues or organisms [116]. Metabolite profiling analysis was performed on urine samples obtained from dogs with naturally occurring invasive TCC of the urinary bladder and/or urethra (n = 40, diagnosed by histopathology) and from healthy control dogs (n = 42). Proton nuclear magnetic resonance (NMR) spectroscopy-based metabolite profiling analysis and statistical analysis methods were used. Among the identified metabolites, six of them were found at significantly higher levels in dogs with TCC compared to control dogs, i.e., urea, choline, methylguanidine, citrate, acetone and β-hydroxybutyrate. These markers revealed good sensitivity to predict the healthy control and disease samples, suggesting a potential of urine metabolic profiling for early detection of bladder cancer [55]. Further studies involving larger cohorts will be needed to determine the accuracy and clinical usefulness of this approach as a screening, diagnostic or disease monitoring tool for canine TCC.

3.3.11. Lipidomics

In one study, cancerous and noncancerous tissues were analysed for potential differences in their lipid profiles. Imaging analysis of tissue sections was performed, by using desorption electrospray ionisation mass spectrometry (DESI-MS). Differentiation between tissues was made using multiple marker lipids and free fatty acids. Samples included tissue sections from canine spontaneous invasive TCC of the urinary bladder, a cutaneous TCC metastasis sample and matched adjacent noncancerous tissues (n = 4). The obtained imaging mass spectrometry data were subjected to statistical analysis to determine whether the results correlated with those obtained from H&E staining, and with DESI-MS images of specific lipids [85].

Results indicated different lipid distributions between healthy and diseased tissues, concerning glycerophospholipids, sphingolipids and free fatty acids. The imaging technique enabled the distinction of canine cancerous bladder tissue and cutaneous metastasis from noncancerous canine bladder tissue samples. The obtained images agreed with those generated by DESI-MS and with H&E-stained tissue. The study concluded that DESI-MS imaging could be useful in diagnosing TCC by using a multimarker approach based on the lipid profiles and intensities of tissue samples. Further studies are required with larger populations and additional control groups, i.e., with other lower urinary diseases [85].

Lipid profiles were also assessed in urine from dogs with TCC, by using liquid chromatography-mass spectrometry (LC-MS). Urine samples from dogs with TCC, dogs with UTI and healthy dogs were analysed for potential differences in their lipid profiles (n = 15). In this study, 208 lipids were identified, belonging to several lipid families. Unique lipid profiles were found among the three cohorts, and specific statistical analyses allowed their differentiation. However, concentrations of the specific lipids could not be determined, and thus the study did not conclude which lipid families were up- or down-regulated in the urine samples of TCC patients compared to the other groups. Nevertheless, the study presented a foundation for further research on urinary lipids as potential biomarkers for TCC [86].

3.3.12. Survivin

Survivin is an apoptosis-inhibiting protein expressed in both cancerous and noncancerous tissues. Expression of this protein was characterised by immunohistochemistry (IHC) in urinary bladder tissues from dogs with TCC, cystitis and dogs with normal urinary bladders. Cytoplasmic survivin was detected in all groups: 8% of cystitis tissues (2/24), 17% of TCC tissues (7/41) and 37% of normal tissues (17/46). However, significant differences were only seen between cystitis and normal groups. On the other hand, nuclear survivin was not detected in any of the normal bladder tissues (0/46), whereas 50% of cystitis tissues (12/24) and 68% of TCC tissues (28/41) were immunoreactive. The proportions of both TCC and cystitis positive specimens were statistically different compared to the normal group. However, TCC and cystitis tissues revealed comparable results. Furthermore, 57% of cystitic tissues (4/7) were positive for survivin mRNA, as well as 100% of TCC tissues (6/6) and 57% of normal tissues (11/22), even though these results were statistically comparable. It was hypothesised that this differential distribution of surviving within the cell in both cystitic and TCC tissues compared to normal tissues could be related to different functions of this protein depending on its location and that survivin could be implicated in tumour development in hyperplastic or inflamed tissues. Although survivin may not serve as a specific diagnostic marker for TCC of the urinary bladder in dogs, nuclear survivin could be detected in dogs with TCC or cystitis and not in normal dogs. Additional research is necessary to investigate whether nuclear survivin, particularly, can be involved in the development or progression of TCC and the possibility of nuclear survivin to be an early marker of bladder tumours and potential therapeutic target [87].

3.3.13. EGFR

Epidermal growth factor receptor (EGFR) is a receptor tyrosine kinase of the ErbB family (which also includes HER2, HER3 and HER4), and its overexpression has been reported in several human and canine tumours. EFGR protein expression was evaluated by IHC in tissues from dogs with TCC (n = 25), polypoid cystitis (n = 5) and normal healthy bladders (n = 5), and specimens were divided into two groups depending on their staining scores: low-expression group and high-expression group. MRNA expression levels were also determined by quantitative real-time PCR in TCC (n = 4) and normal bladder (n = 3) tissues. Protein levels of EGFR were significantly increased in TCC compared to normal bladder and to polypoid cystitis tissues, while no significant difference was found between normal urinary bladder and polypoid cystitis tissues. Moreover, high EGFR protein expression was significantly associated with TCC, with a sensitivity of 72% and specificity of 100%. TCC exhibited significantly higher mRNA levels than normal urinary bladder, positively correlating with protein levels. No significant association was observed between EGFR protein expression and malignant tumour behaviour (presence of vessel invasion or lymph node metastasis) or with survival time in canine TCC [88].

EGFR expression could be used as a marker to aid canine TCC diagnosis. It may improve the sensitivity of urine cytological diagnosis when provisional diagnosis is needed. Development of a qPCR-based method could also be useful for diagnosing this malignancy, as it represents a more sensitive analytical method than IHC, requiring small amounts of samples [88]. Further research is essential for clarifying the underlying mechanisms of EGFR overexpression and tumourigenesis in canine TCC.

3.3.14. HER-2

HER-2 is another member of the ErbB family that plays a role in the control of epithelial cells growth and differentiation and has been found to be overexpressed in human and canine cancers. Protein expression if HER-2 was assessed by IHC in bladder tissues from canine TCC (n = 23, of which 20 invasive TCC and 3 in situ papillary TCC) and compared with non-neoplastic canine urothelium (n = 5). Positivity to HER-2 was observed in 56% (13/23) of TCC specimens, and all control cases were considered negative. The receptor was found to be significantly overexpressed in TCC compared to non-neoplastic specimens, and expression did not statistically differ between invasive and in situ TCC [89]. HER-2 is the gene product of ERBB2, which was also found to be overexpressed in canine invasive TCC [61]. These results support the need for additional studies to further investigate the role of HER-2 as a potential marker of malignancy and therapeutic target in canine TCC [89].

3.3.15. VEGFR2, PDGFR-β, c-KIT

Vascular endothelial growth factor receptor 2 (VEGFR2), platelet-derived growth factor receptor beta (PDGFR-β) and v-kit cellular homologue (c-KIT) are receptor tyrosine kinases (RTK), whose expression has been reported in multiple canine tumours. Toceranib phosphate is a receptor tyrosine kinase inhibitor that exhibits activity against these RTK and has been used anecdotally to treat canine TCC [90].

In one study, expression of these RTK was evaluated by IHC in bladder tissue samples from dogs diagnosed with TCC (n = 30), cystitis (n = 10) and in normal urinary bladder (n = 10). All TCC samples stained positive for PDGFR-β and for VEGFR2. The number of TCC samples that expressed PDGFR-β was significantly different from that of cystitis and normal bladder samples, with a more intense and diffuse staining in tumour cells, while no differences were found between normal bladder and cystitis groups. Regarding intensity scores or staining distribution for VEGFR2, no significant differences were observed between the three groups. However, most TCC samples exhibited intense cytoplasmatic staining for this RTK in >50% of tumour cells. Although 11/36 (36.7%) tumour samples showed positive staining for c-KIT, in most of them only minimal staining was noted (<1% of cells). No positive staining was observed in non-neoplastic tissues, and neither were there significant differences between the three groups [90]. Based on those findings, it was suggested that PDGFR-β could play a role in canine TCC tumourigenesis and that PDGFR-β and VEGFR2 might be involved in mediating clinical response of TCC to toceranib but that this would be unlikely for c-KIT [90].

Korec et al. (2021) confirmed the presence of these markers in canine TCC cell lines and tumour samples, and further supported the variable patterns of expression. However, activation of these RTKs was only present in a small subset of cells, which suggests that signalling through these pathways would less likely contribute to the aggressiveness of urothelial cancer cells [91].

Further studies are needed to investigate the presence of mutations associated with these receptors and to clarify the role of these RTKs in tumourigenesis, response to therapy and clinical outcome in dogs with TCC.

3.3.16. Granzyme B, CD3

Infiltrates of immune cells can often be found in tumour tissues, and different types of infiltrating immune cells (e.g., tumour infiltrating lymphocytes) may affect tumour progression in different ways. IHC was used to investigate the localisation and number of CD3 (a T-cell maker) or granzyme B-positive cells in tissues from canine TCC and normal urinary bladder (n = 32 and n = 10, respectively). Granzyme B is a serine protease found in cytotoxic granules of cytotoxic T lymphocytes and NK cells that plays an important role in antitumour immunity. Both CD3 and granzyme B-positive cells were significantly increased in cancerous tissues compared with normal controls. The number of granzyme B^+ cells was associated with favourable prognosis, while this was not noted for $CD3^+$ cells. Primary tumour stage (WHO classification) was significantly associated with the overall survival. Additionally, no correlation was found between the number of CD3 and granzyme B-positive cells in TCC lesions and the primary tumour stage. It was therefore suggested that the presence of granzyme B^+ tumour-infiltrating cells might be an independent prognostic factor in dogs affected with this malignancy. Granzyme B^+ tumour-infiltrating cells could be involved in the inhibition of tumour progression in canine TCC. Additional studies are needed to identify and clarify the role of specific subsets of tumour-infiltrating immune cells in canine TCC and their potential association with prognosis [95].

3.3.17. P63, Ki67, β-Catenin

The potential usefulness of proteins p63, Ki67 and β-catenin as clinical markers for predicting biological behaviour and prognosis was investigated in canine TCC [96]. It is thought that p63 plays an important role in cellular development and differentiation of stratified epithelia in skin, prostate gland, mammary gland and urinary bladder [96–99]. In humans, loss of p63 expression has been associated with tumourigenesis and malignancy [100,101]. Ki67 is a nuclear protein expressed by proliferating cells, while membranous β-catenin is involved in cellular adherence, and a reduced expression of this protein has been linked to progression and poor prognosis in human urothelial carcinoma [96].

Expression of these proteins was evaluated by IHC in tissue samples from dogs with TCC (n = 25), polypoid cystitis (n = 5) and in normal urinary bladder (n = 5). In TCC samples, staining scores for p63 and β-catenin were significantly lower than in polypoid cystitis and in normal urinary bladder, whereas significantly higher staining scores for Ki67 were noted in TCC samples compared to both other tissue groups. Comparable results were observed for all proteins between cystitis and normal urinary bladder tissues [96].

Furthermore, low p63 expression in TCC was significantly related with the presence of vessel invasion and metastasis, and with short survival time [96]. Similar findings and correlations were observed with expression of ΔNp63, an isoform of p53 [102]. Based on this evidence, the authors suggested that p63 could be used as a clinical marker for diagnosing and predicting prognosis in canine TCC. Additional studies are needed, namely on investigating molecular features, such as p63 gene expression [96].

3.3.18. UIII, CK 7, CK 20, COX-2, Activated Caspase 3, GATA-3

Uroplakin III (UP III) belongs to a group of membrane-associated proteins expressed by urothelial cells [103]. It is the most common marker of urothelial differentiation used in dogs [104]. Cytokeratin 7 (CK 7) and cytokeratin 20 (CK 20) are cytokeratins expressed by simple epithelium, as well as urothelial cells, and have been used for characterising urothelial tumours.

The expressions of UP III, CK 7 and CK 20 were (separately) characterised by IHC in tissues from canine urinary bladder tumours (n = 72) and from normal urinary bladders. Tumours from the urinary bladder comprised TCC (n = 60 total, including 5 metastatic cases and 2 TCC from the ureter and renal pelvis), transitional cell papillomas, rhabdomyosarcomas, squamous cell carcinoma, lymphosarcoma and spindle cell sarcoma. Additionally, 285 canine tumours/lesions not associated with the urinary bladder were evaluated for UP III expression, as were multiple other normal tissues. UP III was identified in superficial (umbrella) cells and some intermediate cells of the normal urinary bladder, as well as in the majority of TCCs (91% of primary (50/55) and 80% of metastatic TCCs (4/5)) and in all transitional cell papillomas (7/7). The remaining tumours of the urinary bladder, as well as non-urothelial normal and neoplastic tissues, were negative for UP III. Staining for CK 7 was detected in 53 out of 54 TCC (98%), in all 5 metastatic TCC and all 7 transitional cell papillomas. Concerning CK 20, staining was observed in 37 out of 54 TCC, 1 metastatic TCC and in 1 out of 7 transitional cell papillomas. Concurrent expression (detection of several antigens in the same tumour) of UP III, CK 7 and CK 20 was identified in 67% (36) of primary TCC and in 1 metastatic TCC. The only anaplastic TCC evaluated was negative for both CK 7, CK 20 and UP III, although the adjacent normal urothelium was positive for UP III [104].

The authors suggested that the pattern of staining for UP III could help distinguish between carcinoma in situ and invasive TCCs. UP III was shown as a specific and sensitive marker of canine transitional epithelial neoplasms. Even though CK 7 was more sensitive than UP III for canine TCC, CK 7 is expressed in several non-urothelial tumours and also in normal tissues, as is CK 20. The authors considered UP III as the marker of choice in canine urothelial neoplasms and that CK 7 should be used for tumours negative for UP III but suspected of being TCC. Negative results could be obtained with anaplastic tumours. CK 20 alone did not prove to be useful for diagnosis of urothelial tumours [104].

In a subsequent study, Sledge and co-authors (2015) [105] evaluated the expression of UP III, CK 7 and also COX-2 and activated caspase 3 in canine urothelial lesions. Cycloocygenase-2 (COX-2) and prostaglandin E2 have been implicated in carcinogenesis at various levels. COX-2 has been found to be expressed in canine TCCs but not by normal urothelium of the urinary bladder. Detection of activated caspase 3 expression by IHC has been used to evaluate apoptotic rate, and this protein has been suggested to have prognostic significance in human urinary bladder cancers [105]. Expression of aforementioned markers was evaluated by IHC in canine urothelial lesions and compared with each lesion's classification and grade. Proliferative urothelial lesions of the urinary bladder from dogs (n = 99) were examined, including non-neoplastic lesions (urothelial polyps and polypoid cystitis), low-grade neoplasms (urothelial papillomas, papillary urothelial neoplasms of low malignant potential and grade 1 urothelial carcinomas) and grade 2 and 3 urothelial carcinomas [105].

Results showed overall strong differences in the patterns of UP III, CK 7 and COX-2 expression in canine urothelial proliferative lesions of the urinary bladder. Furthermore, significant associations were identified between tumour classification and overall UP III pattern, loss of UP III expression, overall CK 7 pattern and COX-2 pattern, as well as between depth of neoplastic cell infiltration into the urinary bladder wall and overall UP III pattern, loss of UP III, overall CK 7 pattern, loss of CK 7 expression and COX-2 pattern. Regarding the expression pattern of activated caspase 3, no significant association was observed with tumour classification, grade or infiltration. Based on these findings, it was hypothesised that loss of UP III and CK 7 in urothelial carcinomas could suggest a lack of differentiation or epithelial-mesenchymal transition favouring infiltration. Moreover, some urothelial carcinomas might not be positively labelled when using UP III and CK 7 as diagnostic markers [105].

It should be noted that UP III is not a specific marker of TCC itself, since it does not differentiate neoplastic from non-neoplastic lesions. However, it can be useful, for instance, to rule in TCC in a biopsy from a tumour of unknown origin and to identify metastatic carcinomas on the skin [2].

More recently, the expression of COX-2 was studied by IHC in TCC biopsies from 65 dogs from different breeds (both terriers and non-terriers) [44]. Samples were graded histologically into low- and high-grade. Both neoplastic and inflammatory cells expressed this marker, whilst it was not detected in normal transitional cell epithelium. The intensity of COX-2 expression was highly variable within neoplastic specimens, but it did not correlate with grading, which is different to what has been observed in humans. Samples were also tested for the *BRAF* mutation. A positive correlation was found between presence of the mutation and the intensity of expression of COX-2 in TCC from non-terrier breeds. This was not as clear for the terriers, although many fewer cases of this breed were analysed, so a significant result would not be completely ruled out. More studies would be encouraged to further determine the clinical and prognostic relevance of COX-2 expression, both on its own and in association with *BRAF* mutation and breed [44].

GATA-Binding Protein 3 (GATA3) is a zinc finger transcription factor that has been used in human medicine as a diagnostic and prognostic marker of urothelial carcinoma [117–119]. In a review article, Knapp et al., 2014 [114] showed the expression of GATA3 in a canine TCC bladder tissue sample. To the best of our knowledge, no other studies have yet looked at its role in canine TCC.

Further studies on the expression of these markers, especially concerning their association with grading and prognostic evaluation, are encouraged.

4. Therapies for UC

Nowadays, a wide range of treatments have been studied that could lead to remission of TCC or provide stable disease for several months. Therapeutic options for TCC include surgery, radiation therapy, systemic and localised medical therapy, and combinations of these [1].

4.1. Surgical Approaches

Surgery may be performed for collecting tissue samples for diagnosis and for tumour removal (if lesions are located outside of the trigone region). However, complete surgical excision is often not feasible due to the trigonal location of the tumour and/or multifocal distribution in the bladder, urethral involvement and presence of metastases [1,120]. In cases where excision is possible (e.g., tumours located in the apical region of the bladder), surgery may prolong survival, and chemotherapy following excision should be considered [121–123]. Caution should be taken during surgery to avoid cancer seeding along the surgery site and the abdominal wall [1,120].

It is noteworthy that incontinence is an expected outcome of total cystectomy, which is another reason why this procedure is infrequently carried out in dogs [124].

Surgical procedures may also be used for the management/palliation of urethral obstruction secondary to TCC, to maintain or restore urine flow. For this purpose, cystotomy tubes or catheters can be placed to allow bypassing urethral obstruction. Moreover, ureteral or urethral stents can be placed surgically or by using less invasive techniques [1,125–127].

Ureterocolonic anastomosis, a technique that consists of transecting the ureters and anastomosing them to the colon, has been reported in dogs with TCC, along with the complete excision of the bladder and other affected structures. The technique allowed diversion of urine into the colon with maintenance of continence by the anus. However, several neurological and gastrointestinal complications developed over time, with limited survival, and therefore the procedure has not been recommended [120,128]. More recently, a procedure involving total cystectomy and urinary diversion to the prepuce or vagina (ureteropreputial or ureterovaginal anastomosis) in dogs with TCC of the trigonal area was associated with fewer gastrointestinal and neurological complications, although high morbidity and complication rates were still seen compared with techniques used in humans [124]. Cutaneous ureterostomy with radical cystectomy has also been reported in dogs with invasive trigonal TCC, with both ureters being transected and anastomosed to the ventral abdominal skin. The procedure was viable, and results suggested survival benefits compared to other methods, even though further studies are required to evaluate the potential survival benefits in a larger population [129]. Urinary incontinence is invariably associated with these two latter procedures, requiring owner compliance with life-long diaper changes and hygiene protocols. Therefore, these options may be contraindicated in dogs that are intolerant of direct owner handling [124,129].

Recently, a laparoscopic technique was used for the removal of a TCC in the distal part of the urethra in a female dog. A pre-pubic urethrostomy and laparoscopic removal was performed. Patient remained clinically stable with no urinary incontinence for 2.5 months. The use of these minimally invasive procedures is therefore promising in some cases for the complete removal of urethral tumours where possible and, in inoperable cases, as a palliative alternative [130].

Liptak et al. (2004) [131] reported the use of electrosurgical transurethral resection in dogs with neoplastic obstruction of the lower urinary tract. Despite showing promising results for male dogs with prostatic carcinoma, the procedure was not recommended in female dogs with urethral TCC due to high complication rate, including urethral perforation. Laser-based techniques have also been described for the ablation of TCC and/or to relieve urethral obstruction caused by the tumour, combined with other therapeutic options [132,133]. However, further studies are necessary to clarify their potential benefits.

4.2. Radiation Therapy

The use of radiation therapy (intraoperative and/or external beam) has been reported in a few studies as a first-line or rescue therapy, occasionally combined with surgery, systemic chemotherapy or NSAIDs. The approach has been associated with several complications and is not routinely used in the management of TCC. One of the challenges is related to variations in size, position and shape of the bladder and surrounding structures throughout treatment, with surrounding normal tissues often being irradiated within the pelvic region. However, more recent protocols have shown improved tolerability, encouraging further studies [134–138]. The contribution of radiotherapy to antitumour activity in multimodal approaches requires better clarification, as well as its benefits compared to other therapies.

4.3. Chemotherapy/Systemic Medical Therapy

The current mainstay of treatment for canine TCC includes systemic medical therapy, usually with chemotherapy, non-steroidal anti-inflammatory drugs (NSAIDs) or combinations of these [1,123]. Medical therapy is not typically curative. However, several different agents can produce remission or stable disease, and most of them are well tolerated [1].

Numerous chemotherapeutic drugs and protocols have been reported for canine TCC, either as single agents or in combination therapies, including the genotoxic agents carboplatin [139,140], cisplatin [141,142], doxorubicin [122], gemcitabine [143], mitoxantrone [132,134,140,144], vinblastine [145,146], vinorelbine [147] and metronomic chlorambucil [148].

Concerning NSAIDs, nonselective cyclooxygenase (COX) inhibitors and COX-2 specific inhibitors have been used for their antitumour activity [142,149]. Piroxicam [77,122,132,134,140,143,144,150] is generally preferred as the first-line NSAID, but deracoxib [151] and firocoxib [142] have also been studied for the treatment of canine TCC [123,142,150,151]. The most frequently used chemotherapy protocol consists of a combination of mitoxantrone and piroxicam [1,140,144].

Adverse effects of NSAIDs include both nephro- and gastro-intestinal (GI) toxicity. COX-2 selective inhibitors are known to cause fewer GI effects; however, the effectiveness of these ones compared to nonselective COX inhibitors is still unclear. Assessment of renal and liver function prior to chemotherapy is recommended, as well as routine monitoring of these parameters and of GI signs during the course of treatment. Drugs such as omeprazole or famotidine could be used to reduce the risk of GI effects [1,123].

As mentioned in Section 3.3.18, COX-2 has been found to be expressed in canine TCC. However, it is still unclear whether expression of this molecule could be useful as a predictive factor for the response to COX-2 inhibitors.

Concerning cytotoxic chemotherapy, the risks and intensity of side effects vary with each drug, including GI signs and bone marrow suppression, likewise requiring frequent monitoring of overall health parameters. Overall, the majority of dogs (*circa* 80–85%) do tolerate these treatments with a good quality of life [1,123].

4.4. Localised Therapies

Regarding localised treatments, intravesical therapy is frequently used in humans with superficial TCC and has been investigated in dogs too [120].

Intravesical administration of mitomycin C has been carried out in dogs with naturally occurring TCC as a chemotherapeutic agent, showing promising antitumour activity. Severe adverse events were occasionally observed, possibly due to systemic absorption of the drug, thus requiring further investigation [152].

Bacillus Calmette-Guérin (BCG) has successfully been used in humans for the treatment of non-muscle invasive bladder cancer [153]. In one study with healthy dogs, published in 1975, severe local inflammatory reactions were caused after BCG was instilled in their bladders [154]. No recent data on clinical trials in dogs with TCC were found, and there are concerns about the risk of systemic absorption, among others.

IL-2 is a cytokine that is involved in several immunological processes. This protein is secreted by different types of immune cells and binds to IL-2 receptors, which are also expressed by multiple immune cell populations. Within its numerous functions, IL-2 is capable of recruiting cytotoxic T-lymphocytes selectively to tumours [155,156]. This cytokine has been used in cancer immunotherapy in both human and veterinary medicine, mostly via its local application into tumour lesions [156]. When locally applied, IL-2 is usually much more effective and causes fewer side effects compared with the systemic route [157]. Local administration of IL-2 induces vascular leakage and tumour necrosis, with subsequent stimulation of an immune response. Time required until tumour regression varies with the degree of tumour vascularisation, ranging generally from a week to several months [156].

The influence of intralesional treatment with interleukin-2 (IL-2) on the clinical course and tumour progression of canine TCC was evaluated in a retrospective clinical study. Twenty-five dogs with advanced TCC of the urinary bladder and/or urethra, in which curative surgery was unfeasible, were treated. This was achieved via either transabdominal ultrasound-guided intralesional injection of IL-2, endoscopically/transrectally assisted or by IL-2 injection into the tumour bed after cytoreductive (palliative) surgical tumour resection. Additionally, all dogs received long-term NSAIDs, and a chemotherapeutic agent was used in two dogs. Adverse effects associated with intralesional IL-2 treatment were not observed, and at the time of re-examination (which varied individually), 17 dogs showed marked clinical improvement and regression of tumour size, and 4 dogs were in complete remission. IL-2 intralesional application was considered a safe and minimally invasive palliative treatment modality for canine advanced TCC in the impossibility of a surgical cure. Further prospective studies are warranted for clarifying the efficacy of intralesional IL-2 treatment in dogs with TCC [156].

Photodynamic therapy has also been studied in healthy dogs and dogs with spontaneous TCC by orally administering 5-aminolevulinic acid (ALA), which is converted to a photosensitiser metabolite, protoporphyrin IX (PpIX). In healthy dogs, fluorescence of PpIX was confined to the mucosa and submucosa layers of the urinary bladder and substantially increased in the former. In diseased dogs, the observed long-term responses suggested the usefulness of this approach for treatment of canine TCC, requiring further research [158,159].

Nanoparticles carrying chemotherapeutic agents have been used experimentally for drug delivery to canine bladder TCC cells [160–165]. Gelatin nanoparticles containing paclitaxel have been delivered intravesically to dogs with TCC [166]. Furthermore, targeting nanomicelles coated with the ligand PLZ4, a peptide that specifically binds to both human and dog bladder TCC cells, were able to efficiently deliver chemotherapeutic drugs (daunorubicin or paclitaxel) and imaging agents into the tumour site in vivo, in a mouse model carrying canine bladder cancer xenografts [160–162]. More recently, Lin and coworkers (2016) developed a multifunctional platform of nanoporphyrins that was also coated with PLZ4 (PNPs), which are nanoparticles with the ability to emit fluorescent signals/heat/reactive oxygen species when illuminated with specific light. Doxorubicin-loaded PNPs were tested in vitro in both human bladder cancer cells and normal canine urothelial cells, and in vivo in a mouse model. The platform selectively targeted tumour cells for photodynamic diagnosis and was also effective against bladder cancer, integrating three therapeutic modalities in a single procedure (photodynamic, photothermal and chemotherapeutic), with either intravesical or systemic administration [164]. In the future, investigation is warranted on the efficacy and toxicity of these nanotechnology-based approaches to translate them into both veterinary and human clinical application.

Another targeted therapy that is being investigated in dogs is a treatment targeting folate (vitamin B9) receptors, which is based on the high uptake of folate and folate drug conjugates into certain cancers compared with normal tissues [114]. In one study evaluating canine and human tissues from invasive urothelial carcinomas, folate receptors were detected in most primary UC tissues and nodal and lung metastases from dogs, and folate uptake was detected in primary and metastatic lesions. Dogs with folate-receptors-positive invasive urothelial carcinoma were treated with a folate-targeted vinblastine compound in a dose escalation study, showing promising initial results with tumour responses, including partial remission and stable disease, and further study is ongoing [167].

4.5. Epigenetic-Based Therapies

Therapies targeted at epigenetic changes are also being developed.

Aberrant methylation in the promoter region of tumour suppressor genes, causing gene silencing, is an example of the epigenetic events that can lead to cancer development and progression in the absence of DNA mutations [168]. Several DNA methylation markers have been studied, with growing evidence supporting their usefulness for the diagnosis, prognosis and treatment of human bladder cancer. Aberrant DNA methylation in cancer-related genes has been reported in human TCC, cell lines and urine sediments [169–172].

DNA methyltransferase 1 (DNMT1), an important enzyme in DNA methylation, has been found to be overexpressed in both human and canine TCC and was identified as a potential target in the treatment of TCC, but data are lacking concerning its potential applications in canine TCC [173,174]. DNMT1 inhibitors have shown antiproliferative effects both in vitro and in vivo [174,175]. In a preclinical phase I trial, dogs with naturally occurring invasive UC, used as a model for human UC, were treated with 5-azacitidine subcutaneously, and this compound exhibited a promising clinical/antitumour activity, encouraging further studies [175]. Zebulatine, an orally bioavailable agent that also has demethylating activity with a similar mechanism of action to that of 5-azacitidine, was studied in three laboratory dogs and three tumour-bearing dogs, two of them with invasive TCC. In three laboratory dogs, the treatment resulted in some severe but reversible adverse effects, while no appreciable toxicity was observed in the tumour-bearing dogs, with those with invasive TCC showing stable disease. At the time of writing, a subsequent dose escalating trial is ongoing in dogs with invasive TCC [176].

Another example of an epigenetic change is histone acetylation. Histone deacetylases (HDAC) have been suggested to be overexpressed in both human and canine tumours. Excessive activity of HDAC could facilitate the deacetylation of histones, causing down-regulation of the expression of tumour suppressor genes. Molecules that inhibit the activity of HDAC have been evaluated as a possible antitumour therapy in both humans and dogs (reviewed by Xavier et al., 2020 and Goutas et al., 2021 [177,178]). It is thought that HDAC inhibitors (HDACi) induce the acetylation of deacetylated histones, thereby promoting the expression of tumour suppressor genes, which could result in an antitumour effect [179].

Vorinostat, a HDACi, was found to have an antitumour effect on canine UC cell lines, both in vitro, and in vivo in a xenograft mouse model. It inhibited cell growth and induced G0/G1 cell cycle arrest. This HDACi induced histone acetylation and the expression of cell-cycle related molecules [179]. Moreover, histone deacetylation was aberrantly observed in canine UC tissues compared to normal samples, suggesting that epigenetic dysregulation may play a role in canine UC progression, such that lower levels of deacetylation were related to a poor prognosis [179]. More studies will be required to further investigate these findings and the role of HDACi as a potential therapy for UC.

4.6. Other Emerging Targeted Therapies

Toceranib phosphate is a RTK inhibitor that has been used anecdotally to treat TCC in dogs. This small-molecule inhibitor targets c-KIT, VEGFR2 and PDGFR-β, among others. It is approved for the treatment of cutaneous mast cell tumours in dogs, and it is also active against several other cancer types. As mentioned in Section 3.3.15., it was recently suggested that PDGFR-β could play a role in canine TCC tumourigenesis and that PDGFR-β and VEGFR2 could be involved in mediating clinical response of TCC to toceranib. However, the mechanisms by which this could take place are not entirely understood [90,91].

In a phase I clinical trial with dogs with spontaneous malignancies, treatment with toceranib (SU11654) resulted in stable disease in three out of four dogs with TCC of the bladder [92]. A pilot study investigated the biological activity of the combination of toceranib and vinblastine for treatment of canine TCC of the bladder, but response to therapy was not improved when compared to treatment with single-agent vinblastine. However, lower doses of vinblastine were used in the combination protocol, sample size was limited, and a prospective control group was also lacking [93].

In a retrospective study, 37 dogs were treated for TCC with toceranib as a second- or third-line therapy. Of the 15 dogs in whom treatment response was evaluated, 13 were concomitantly treated with an NSAID. A partial response was achieved in 1 of the patients, while 12 (80%) dogs had stable disease for a median of around 128 days (34–310). The drug was overall well tolerated, although azotaemia progressed in over half of the patients [94].

In a more recent study with canine TCC cell lines that expressed these markers, treatment with toceranib had no significant effect on cell proliferation [91].

It is possible that stable disease, rather than tumour regression, is related to the mechanism of action of toceranib, by inhibiting growth factor signalling, and not directly damaging tumoural DNA [94]. Further prospective studies with larger samples and comparing different treatments are encouraged.

The identification of the $BRAF^{V595E}$ mutation in canine TCC and its high prevalence pointed to the possibility of targeting the BRAF/MAPK pathway via a therapeutic approach in those tumours carrying the mutation. In humans, several drugs, such as vemurafenib, a kinase inhibitor, have been developed to selectively target the $BRAF^{V600E}$ mutation in different malignancies, with high effectiveness. However, innate or acquired resistance to BRAF inhibitors can occur in some $BRAF^{V600E}$-positive cancers, and the presence of the mutation does not always correlate with clinical response to BRAF inhibitors [58,59,180].

Vemurafenib was studied in canine TCC cells and showed anti-proliferative effects [58]. In a subsequent phase I/II clinical trial, this drug was tested in dogs with naturally occurring invasive TCC, which were positive for the $BRAF^{V595E}$ mutation. In 9 of the 24 dogs (38%), partial remission was achieved, with a median progression-free interval of 181 days (53 to 608 days). Anorexia was the most common side effect. Similar to what has been observed in humans treated with this drug, new tumours developed in some of the patients (cutaneous SCC, papillomas). Resistance to therapy also occurred in some animals over time [65].

Sorafenib is another kinase inhibitor that targets multiple kinases, including RAF, VEGFRs and PDGFR-β, with anti-proliferative and anti-angiogenic activities. It is approved for the treatment of various cancers in humans, and it has been studied in veterinary medicine too, being well tolerated by dogs in preliminary studies [66,67,110,181].

In in vitro studies with established TCC cell lines derived from dogs harbouring the mutation, this drug inhibited the RAF/MAPK pathway and induced apoptosis. These effects were more pronounced with sorafenib than with vemurafenib. The reasons for these differences were not fully clear, so future studies would be needed [67].

Furthermore, sorafenib was used in a dog with metastatic TCC of the urethra, in whom complete surgical excision was not achievable. Tumour tissue was $BRAF^{V595E}$-positive and showed overexpression of VEFGR. After unsuccessful response to chemotherapy with mitoxantrone, sorafenib was trialled in association with piroxicam. To monitor treatment response, *BRAF* mutation levels were measured in ctDNA obtained from the patient's serial plasma samples. Genetic sequencing techniques were used. During treatment, levels of mutated *BRAF* varied; they tended to increase with the severity of the clinical signs (dysuria) and decrease when clinical signs improved after sorafenib was escalated to the maximum tolerated oral dose. Dysuria was well managed, and a decrease in thickness of urethral wall was seen. The patient remained stable for over 3 months of treatment, with few side effects [110].

The effect of sorafenib on VEGFR expression, as well as the effects of sorafenib alone versus the combined treatment, would require further study, but overall, these results were promising, and further clinical research is encouraged.

Additional in vitro and in vivo evaluations of the effects of therapies targeted at these and other canine TCC markers with potential applications in a clinical setting are underway.

5. Conclusions

The last few years have witnessed great progress concerning our understanding of bladder cancer. In dogs, bladder cancer consists mainly of TCC with poorly characterised aetiology, a complex molecular landscape, heterogeneous morphology and diverse biological behaviour, requiring effective prognostic and therapeutic markers. In contrast, cattle, who also show a significant incidence of bladder cancer, are affected by well-defined etiologic factors that deregulate specific signalling pathways.

Histological analysis of canine TCC is essential but insufficient to determine patient prognosis and reach a tailored therapeutic approach. In this context, the recent development and validation of TCC molecular markers is of great importance for scientists and clinicians alike. Somatic and hereditary BRAF mutations received much attention and can now be detected via multiple types of tests, sometimes in useful combinations with CNA tests. Urine-based tests for detecting BRAF may allow the early detection of post-treatment relapse. Other urine-based tests (e.g., for bFGF and calgranulins levels) have also found interesting clinical applications. A number of tissue-based markers have also been put forward (e.g., immunohistochemical detection of COX-2), but their use to predict prognosis or response to therapy is still under study. Additional work on blood-based liquid biopsies is expected in the coming years, aiming to match similar developments obtained for human patients. Minimally invasive techniques have proved valuable and more effective as biomarkers than tissue-based approaches. A more detailed knowledge of the molecular signalling pathways involved in canine TCC will help design more effective targeted therapies and new tests with enhanced predictive value for canine TCC patients.

Supplementary Materials: The following supporting information can be downloaded at: https://www.mdpi.com/article/10.3390/vetsci9030107/s1, Figure S1: Evolution of histological classification and grading schemes for urothelial lesions, Figure S2: TNM clinical staging of canine TCC.

Author Contributions: Conceptualisation, A.M.R. and R.M.G.d.C.; writing—original draft preparation, A.M.R. and R.M.G.d.C.; writing—review and editing, A.M.R., E.S.e.L., P.A.O. and R.M.G.d.C. All authors have read and agreed to the published version of the manuscript.

Funding: This study was supported by the Research Center of the Portuguese Oncology Institute of Porto (project no. PI127-CI-IPOP-118-2019) and by European Investment Funds by FEDER/COMPETE/POCI- Operational Competitiveness and Internationalization Program and national funds by FCT-Portuguese Foundation for Science and Technology under projects UID/AGR/04033/2020 and UIDB/CVT/00772/2020 and by Base Funding-UIDB/00511/2020 of the Laboratory for Process Engineering, Environment, Biotechnology, and Energy—LEPABE—funded by national funds through the FCT/MCTES (PIDDAC) and Project 2SMART—engineered Smart materials for Smart citizens, with reference NORTE-01-0145-FEDER-000054, supported by Norte Portugal Regional Operational Programme (NORTE 2020), under the PORTUGAL 2020 Partnership Agreement, through the European Regional Development Fund (ERDF).

Institutional Review Board Statement: Not applicable.

Informed Consent Statement: Not applicable.

Data Availability Statement: No data were generated in this study.

Conflicts of Interest: The authors declare no conflict of interest.

References

1. Knapp, D.W.; McMillan, S.K. Tumors of the urinary system. In *Withrow and MacEwen's Small Animal Clinical Oncology*, 5th ed.; Withrow, S.J., Vail, D.M., Eds.; Elsevier–Saunders: St. Louis, MO, USA, 2013; pp. 572–582.
2. Meuten, D.J.; Meuten, T.L. Tumors of the urinary system. In *Tumors in Domestic Animals*, 5th ed.; Meuten, D.J., Ed.; John Willey & Sons Inc.: Ames, IA, USA, 2016; pp. 632–688.
3. Schwarz, P.D.; Greene, R.W.; Patnaik, A.K. Urinary bladder tumors in the cat: A review of 27 cases. *J. Am. Anim. Hosp. Assoc.* **1985**, *21*, 237–245.
4. Wilson, H.M.; Chun, R.; Larson, V.S.; Kurzman, I.D.; Vail, D.M. Clinical signs, treatments, and outcome in cats with transitional cell carcinoma of the urinary bladder: 20 cases (1990–2004). *J. Am. Vet. Med. Assoc.* **2007**, *231*, 101–106. [CrossRef] [PubMed]
5. Norris, A.M.; Laing, E.J.; Valli, V.E.O.; Withrow, S.J.; Macy, D.W.; Ogilvie, G.K.; Tomlinson, J.; McCaw, D.; Pidgeon, G.; Jacobs, R.M. Canine Bladder and Urethral Tumors: A Retrospective Study of 115 Cases (1980–1985). *J. Vet. Intern. Med.* **1992**, *6*, 145–153. [CrossRef]
6. Mutsaers, A.J.; Widmer, W.R.; Knapp, D.W. Canine Transitional Cell Carcinoma. *J. Vet. Intern. Med.* **2003**, *17*, 136–144. [CrossRef] [PubMed]
7. Knapp, D.W.; Glickman, N.W.; DeNicola, D.B.; Bonney, P.L.; Lin, T.L.; Glickman, L.T. Naturally-occurring canine transitional cell carcinoma of the urinary bladder A relevant model of human invasive bladder cancer. *Urol. Oncol. Semin. Orig. Investig.* **2000**, *5*, 47–59. [CrossRef]

8. Glickman, L.T.; Schofer, F.S.; McKee, L.J.; Reif, J.S.; Goldschmidt, M. Epidemiologic study of insecticide exposures, obesity, and risk of bladder cancer in household dogs. *J. Toxicol. Environ. Health Part A* **1989**, *28*, 407–414. [CrossRef]
9. Glickman, L.T.; Raghavan, M.; Knapp, D.W.; Bonney, P.L.; Dawson, M.H. Herbicide exposure and the risk of transitional cell carcinoma of the urinary bladder in Scottish Terriers. *J. Am. Vet. Med. Assoc.* **2004**, *224*, 1290–1297. [CrossRef]
10. Gil da Costa, R.; Bastos, M.; Oliveira, P.; Lopes, C. Bracken-associated human and animal health hazards: Chemical, biological and pathological evidence. *J. Hazard. Mater.* **2012**, *203–204*, 1–12. [CrossRef]
11. Somvanshi, R. Papillomatosis in Buffaloes: A Less-Known Disease. *Transbound. Emerg. Dis.* **2011**, *58*, 327–332. [CrossRef]
12. Rosenberger, G.; Heeschen, W. Adler-farn (*Pteris aquiline*)—Die ursache des sog. *Dtsch. Tierarztl. Wochenschr.* **1960**, *67*, 201–208.
13. Rosenberger, G. Längere aufnahme von adlerfarn (*Pteris aquilina*)—Die ursache der chronischen vesikalen haematurie des rindes. *Wien. Tierarztl. Monatsschr.* **1965**, *52*, 415–421.
14. Evans, I.A.; Mason, J. Carcinogenic Activity of Bracken. *Nature* **1965**, *208*, 913–914. [CrossRef]
15. Pamukcu, A.M.; Price, J.M. Induction of Intestinal and Urinary Bladder Cancer in Rats by Feeding Bracken Fern (*Pteris aquilina*). *JNCI J. Natl. Cancer Inst.* **1969**, *43*, 275–281. [CrossRef] [PubMed]
16. Hirono, I.; Ogino, H.; Fujimoto, M.; Yamada, K.; Yoshida, Y.; Ikagawa, M.; Okumura, M. Induction of Tumors in ACI Rats Given a Diet Containing Ptaquiloside, a Bracken Carcinogen2. *JNCI J. Natl. Cancer Inst.* **1987**, *79*, 1143–1149. [CrossRef] [PubMed]
17. da Gil Costa, R.M.; Oliveira, P.A.; Vilanova, M.; Bastos, M.M.; Lopes, C.C.; Lopes, C. Ptaquiloside-induced, B-cell lymphoproliferative and early-stage urothelial lesions in mice. *Toxicon* **2011**, *58*, 543–549. [CrossRef]
18. da Gil Costa, R.M.; Oliveira, P.A.; Bastos, M.M.S.M.; Lopes, C.C.; Lopes, C. Ptaquiloside-induced early-stage urothelial lesions show increased cell proliferation and intact β-catenin and E-cadherin expression. *Environ. Toxicol.* **2012**, *29*, 763–769. [CrossRef]
19. Ojika, M.; Wakamatsu, K.; Niwa, H.; Yamada, K. Ptaquiloside, a potent carcinogen isolated from bracken fern var: Structure elucidation based on chemical and spectral evidence, and reactions with amino acids, nucleosides, and nucleotides. *Tetrahedron* **1987**, *43*, 5261–5274. [CrossRef]
20. Pereira, L.O.; Bicalho, L.S.; Lopes, M.C.-D.; De Sousa, T.M.M.; Báo, S.N.; Santos, M.D.F.M.A.; Fonseca, M.J.P. DNA damage and apoptosis induced by *Pteridium aquilinum* aqueous extract in the oral cell lines HSG and OSCC-3. *J. Oral Pathol. Med.* **2008**, *38*, 441–447. [CrossRef] [PubMed]
21. da Gil Costa, R.M.; Coelho, P.; Sousa, R.; Bastos, M.M.; Porto, B.; Teixeira, J.P.; Malheiro, I.; Lopes, C. Multiple genotoxic activities of ptaquiloside in human lymphocytes: Aneugenesis, clastogenesis and induction of sister chromatid exchange. *Mutat. Res. Toxicol. Environ. Mutagen.* **2012**, *747*, 77–81. [CrossRef] [PubMed]
22. Latorre, A.O.; Furlan, M.S.; Sakai, M.; Fukumasu, H.; Hueza, I.M.; Haraguchi, M.; Górniak, S.L. Immunomodulatory effects of *Pteridium aquilinumon* natural killer cell activity and select aspects of the cellular immune response of mice. *J. Immunotoxicol.* **2009**, *6*, 104–114. [CrossRef]
23. Caniceiro, B.D.; Latorre, A.O.; Fukumasu, H.; Sanches, D.S.; Haraguchi, M.; Górniak, S. Immunosuppressive effects of *Pteridium aquilinumenhance* susceptibility to urethane-induced lung carcinogenesis. *J. Immunotoxicol.* **2014**, *12*, 74–80. [CrossRef] [PubMed]
24. Santos, C.; Ferreirinha, P.; Sousa, H.; Ribeiro, J.; Bastos, M.M.; Neto, T.; Oliveira, P.A.; Medeiros, R.; Vilanova, M.; da Gil Costa, R.M. Ptaquiloside from bracken (*Pteridium* spp.) inhibits tumour-infiltrating CD8+ T cells in HPV-16 transgenic mice. *Food Chem. Toxicol.* **2016**, *97*, 277–285. [CrossRef] [PubMed]
25. da Gil Costa, R.M.; Lopes, C.; Oliveira, P.A.; Bastos, M.M.S.M. Illudane-Type Sesquiterpenes: Challenginges and Opportunities for Toxicology and Chemotherapy. In *New Development in Terpene Research*; Hu, J., Ed.; Nova Publishers, Inc.: New York, NY, USA, 2014; pp. 135–184. ISBN 978-1-62948-760-1.
26. Micheloud, J.F.; Caro, L.A.C.; Martínez, O.G.; Gimeno, E.J.; Ribeiro, D.D.S.F.; Blanco, B.S. Bovine enzootic haematuria from consumption of *Pteris deflexa* and *Pteris plumula* in northwestern Argentina. *Toxicon* **2017**, *134*, 26–29. [CrossRef] [PubMed]
27. Rai, S.K.; Sharma, R.; Kumari, A.; Rasmussen, L.H.; Patil, R.D.; Bhar, R. Survey of ferns and clinico-pathological studies on the field cases of Enzootic bovine haematuria in Himachal Pradesh, a north-western Himalayan state of India. *Toxicon* **2017**, *138*, 31–36. [CrossRef]
28. Roperto, S.; Russo, V.; Ozkul, A.; Sepici-Dincel, A.; Maiolino, P.; Borzacchiello, G.; Marcus, I.; Esposito, I.; Riccardi, M.G.; Roperto, F. Bovine papillomavirus type 2 infects the urinary bladder of water buffalo (*Bubalus bubalis*) and plays a crucial role in bubaline urothelial carcinogenesis. *J. Gen. Virol.* **2013**, *94*, 403–408. [CrossRef]
29. Roperto, F.; Russo, V.; Leonardi, L.; Martano, M.; Corrado, F.; Riccardi, M.G.; Roperto, F. Bovine Papillomavirus Type 13 Expression in the Urothelial Bladder Tumours of Cattle. *Transbound. Emerg. Dis.* **2016**, *63*, 628–634. [CrossRef]
30. Roperto, S.; Munday, J.S.; Corrado, F.; Goria, M.; Roperto, F. Detection of bovine papillomavirus type 14 DNA sequences in urinary bladder tumors in cattle. *Vet. Microbiol.* **2016**, *190*, 1–4. [CrossRef]
31. Gil Da Costa, R.M.; Medeiros, R. Bovine papillomavirus: Opening new trends for comparative pathology. *Arch. Virol.* **2013**, *159*, 191–198. [CrossRef]
32. Corteggio, A.; Di Geronimo, O.; Roperto, S.; Roperto, F.; Borzacchiello, G. Activated platelet-derived growth factor β receptor and Ras–mitogen-activated protein kinase pathway in natural bovine urinary bladder carcinomas. *Vet. J.* **2012**, *191*, 393–395. [CrossRef]
33. Gil Da Costa, R.M.; Peleteiro, M.C.; Pires, M.A.; DiMaio, D. An Update on Canine, Feline and Bovine Papillomaviruses. *Transbound. Emerg. Dis.* **2016**, *64*, 1371–1379. [CrossRef]

34. Gil Da Costa, R.M.; Oliveira, P.A.; Vasconcelos-Nóbrega, C.; Arantes-Rodrigues, R.; Pinto-Leite, R.; Colaço, A.A.; De La Cruz, L.F.; Lopes, C. Altered expression of CKs 14/20 is an early event in a rat model of multistep bladder carcinogenesis. *Int. J. Exp. Pathol.* **2015**, *96*, 319–325. [CrossRef]
35. Volkmer, J.-P.; Sahoo, D.; Chin, R.K.; Ho, P.L.; Tang, C.; Kurtova, A.V.; Willingham, S.B.; Pazhanisamy, S.K.; Contreras-Trujillo, H.; Storm, T.A.; et al. Three differentiation states risk-stratify bladder cancer into distinct subtypes. *Proc. Natl. Acad. Sci. USA* **2012**, *109*, 2078–2083. [CrossRef] [PubMed]
36. Choi, W.; Czerniak, B.; Ochoa, A.; Su, X.; Siefker-Radtke, A.; Dinney, C.P.N.; McConkey, D.J. Intrinsic basal and luminal subtypes of muscle-invasive bladder cancer. *Nat. Rev. Urol.* **2014**, *11*, 400–410. [CrossRef] [PubMed]
37. Shin, K.; Lim, A.; Odegaard, J.I.; Honeycutt, J.D.; Kawano, S.; Hsieh, M.H.; Beachy, P.A. Cellular origin of bladder neoplasia and tissue dynamics of its progression to invasive carcinoma. *Nat. Cell Biol.* **2014**, *16*, 469–478. [CrossRef] [PubMed]
38. Robertson, A.G.; Kim, J.; Al-Ahmadie, H.; Bellmunt, J.; Guo, G.; Cherniack, A.D.; Hinoue, T.; Laird, P.W.; Hoadley, K.A.; Akbani, R.; et al. Comprehensive Molecular Characterization of Muscle-Invasive Bladder Cancer. *Cell* **2018**, *174*, 1033. [CrossRef]
39. Shapiro, S.G.; Raghunath, S.; Williams, C.; Motsinger-Reif, A.; Cullen, J.M.; Liu, T.; Albertson, D.; Ruvolo, M.; Lucas, A.B.; Jin, J.; et al. Canine urothelial carcinoma: Genomically aberrant and comparatively relevant. *Chromosom. Res.* **2015**, *23*, 311–331. [CrossRef]
40. John, B.A.; Said, N. Insights from animal models of bladder cancer: Recent advances, challenges, and opportunities. *Oncotarget* **2017**, *8*, 57766–57781. [CrossRef]
41. Oliveira, P.A.; Vasconcelos-Nóbrega, C.; da Gil Costa, R.M.; Arantes-Rodrigues, R. The N-butyl-N-4-hydroxybutyl nitrosamine mouse urinary bladder cancer model. In *Urothelial Carcinoma: Methods and Protocols*; Schulz, W.A., Hoffmann, M.J., Niegisch, G., Eds.; Humana Press: New York, NY, USA, 2018; Volume 1655, pp. 155–167. [CrossRef]
42. Carvalho, T.; Pinto, C.; Peleteiro, M. Urinary Bladder Lesions in Bovine Enzootic Haematuria. *J. Comp. Pathol.* **2006**, *134*, 336–346. [CrossRef]
43. Cheng, L.; MacLennan, G.T.; Lopez-Beltran, A. Histologic grading of urothelial carcinoma: A reappraisal. *Hum. Pathol.* **2012**, *43*, 2097–2108. [CrossRef]
44. Grassinger, J.M.; Merz, S.; Aupperle-Lellbach, H.; Erhard, H.; Klopfleisch, R. Correlation of BRAF Variant V595E, Breed, Histological Grade and Cyclooxygenase-2 Expression in Canine Transitional Cell Carcinomas. *Vet. Sci.* **2019**, *6*, 31. [CrossRef]
45. Chun, R.; Garrett, L.D. Urogenital and mammary gland tumors. In *Textbook of Veterinary Internal Medicine: Diseases of the Dog and the Cat*, 7th ed.; Ettinger, S.J., Feldman, E.C., Eds.; Saunders Elsevier: St. Louis, MI, USA, 2010; pp. 2208–2212.
46. Heilmann, R.M.; Wright, Z.M.; Lanerie, D.J.; Suchodolski, J.S.; Steiner, J.M. Measurement of urinary canine S100A8/A9 and S100A12 concentrations as candidate biomarkers of lower urinary tract neoplasia in dogs. *J. Vet. Diagn. Investig.* **2014**, *26*, 104–112. [CrossRef] [PubMed]
47. Cota, J.; Peleteiro, M.C.; Petti, L.; Tavares, L.; Duarte, A. Detection and quantification of bovine papillomavirus type 2 in urinary bladders and lymph nodes in cases of Bovine Enzootic Hematuria from the endemic region of Azores. *Vet. Microbiol.* **2015**, *178*, 138–143. [CrossRef] [PubMed]
48. Borjesson, D.L.; DeJong, K. Urinary tract. In *Canine and Feline Cytology: A Color Atlas and Interpretation Guide*, 3rd ed.; Raskin, R.E., Meyer, D.J., Eds.; Saunders Elsevier: St. Louis, MI, USA, 2016; pp. 288–294.
49. Anderson, W.I.; Dunham, B.M.; King, J.M.; Scott, D.W. Presumptive subcutaneous surgical transplantation of a urinary bladder transitional cell carcinoma in a dog. *Cornell Vet.* **1989**, *79*, 263–266.
50. Nyland, T.G.; Wallack, S.T.; Wisner, E.R. Needle-tract implantation following us-guided fine-needle aspiration biopsy of transitional cell carcinoma of the bladder, urethra, and prostate. *Vet. Radiol. Ultrasound* **2002**, *43*, 50–53. [CrossRef] [PubMed]
51. Vignoli, M.; Rossi, F.; Chierici, C.; Terragni, R.; De Lorenzi, D.; Stanga, M.; Olivero, D. Needle tract implantation after fine needle aspiration biopsy (FNAB) of transitional cell carcinoma of the urinary bladder and adenocarcinoma of the lung. *Schweiz. Arch. Tierheilkd.* **2007**, *149*, 314–318. [CrossRef] [PubMed]
52. Higuchi, T.; Burcham, G.N.; Childress, M.O.; Rohleder, J.J.; Bonney, P.L.; Ramos-Vara, J.A.; Knapp, D.W. Characterization and treatment of transitional cell carcinoma of the abdominal wall in dogs: 24 cases (1985–2010). *J. Am. Vet. Med. Assoc.* **2013**, *242*, 499–506. [CrossRef] [PubMed]
53. Iwasaki, R.; Shimosato, Y.; Yoshikawa, R.; Goto, S.; Yoshida, K.; Murakami, M.; Kawabe, M.; Sakai, H.; Mori, T. Survival analysis in dogs with urinary transitional cell carcinoma that underwent whole-body computed tomography at diagnosis. *Vet. Comp. Oncol.* **2019**, *17*, 385–393. [CrossRef]
54. Owen, L.N.; World Health Organization. TNM Classification of tumours in domestic animals. In *Veterinary Public Health Unit & WHO Collaborating Center for Comparative Oncology*; World Health Organization: Geneva, Switzerland, 1980; 52p, Available online: http://www.who.int/iris/handle/10665/68618 (accessed on 18 November 2021).
55. Mochizuki, H.; Shapiro, S.G.; Breen, M. Detection of Copy Number Imbalance in Canine Urothelial Carcinoma with Droplet Digital Polymerase Chain Reaction. *Vet. Pathol.* **2015**, *53*, 764–772. [CrossRef] [PubMed]
56. Vinall, R.L.; Kent, M.S.; White, R.W.D. Expression of microRNAs in urinary bladder samples obtained from dogs with grossly normal bladders, inflammatory bladder disease, or transitional cell carcinoma. *Am. J. Vet. Res.* **2012**, *73*, 1626–1633. [CrossRef]
57. Zhang, J.; Wei, S.; Liu, L.; Gowda, G.N.; Bonney, P.; Stewart, J.; Knapp, D.W.; Raftery, D. NMR-based metabolomics study of canine bladder cancer. *Biochim. Biophys. Acta (BBA) Mol. Basis Dis.* **2012**, *1822*, 1807–1814. [CrossRef]

58. Decker, B.; Parker, H.G.; Dhawan, D.; Kwon, E.M.; Karlins, E.; Davis, B.W.; Ramos-Vara, J.A.; Bonney, P.L.; McNiel, E.A.; Knapp, D.W.; et al. Homologous Mutation to Human BRAF V600E Is Common in Naturally Occurring Canine Bladder Cancer—Evidence for a Relevant Model System and Urine-Based Diagnostic Test. *Mol. Cancer Res.* **2015**, *13*, 993–1002. [CrossRef] [PubMed]
59. Mochizuki, H.; Kennedy, K.; Shapiro, S.G.; Breen, M. BRAF Mutations in Canine Cancers. *PLoS ONE* **2015**, *10*, e0129534. [CrossRef] [PubMed]
60. Mochizuki, H.; Shapiro, S.G.; Breen, M. Detection of BRAF Mutation in Urine DNA as a Molecular Diagnostic for Canine Urothelial and Prostatic Carcinoma. *PLoS ONE* **2015**, *10*, e0144170. [CrossRef]
61. Parker, H.G.; Dhawan, D.; Harris, A.C.; Ramos-Vara, J.A.; Davis, B.W.; Knapp, D.W.; Ostrander, E.A. RNAseq expression patterns of canine invasive urothelial carcinoma reveal two distinct tumor clusters and shared regions of dysregulation with human bladder tumors. *BMC Cancer* **2020**, *20*, 251. [CrossRef]
62. Gedon, J.; Kehl, A.; Aupperle-Lellbach, H.; von Bomhard, W.; Schmidt, J.M. BRAF mutation status and its prognostic significance in 79 canine urothelial carcinomas: A retrospective study (2006–2019). *Vet. Comp. Oncol.* **2021**. [CrossRef] [PubMed]
63. Tagawa, M.; Tambo, N.; Maezawa, M.; Tomihari, M.; Watanabe, K.-I.; Inokuma, H.; Miyahara, K. Quantitative analysis of the BRAF V595E mutation in plasma cell-free DNA from dogs with urothelial carcinoma. *PLoS ONE* **2020**, *15*, e0232365. [CrossRef]
64. Antech Diagnostics. *CADET® BRAF and CADET® BRAF PLUS [Brochure on the Internet]*; Antech Diagnostics, Inc.: Fountain Valley, CA, USA, 2021; Available online: https://www.antechdiagnostics.com/laboratory-diagnostics/molecular-diagnostics/cadet-braf-plus (accessed on 14 November 2021).
65. Rossman, P.; Zabka, T.S.; Ruple, A.; Tuerck, D.; Ramos-Vara, J.A.; Liu, L.; Mohallem, R.; Merchant, M.; Franco, J.; Fulkerson, C.M.; et al. Phase I/II Trial of Vemurafenib in Dogs with Naturally Occurring, BRAF-mutated Urothelial Carcinoma. *Mol. Cancer Ther.* **2021**, *20*, 2177–2188. [CrossRef]
66. Marconato, L.; Sabattini, S.; Marisi, G.; Rossi, F.; Leone, V.F.; Casadei-Gardini, A. Sorafenib for the Treatment of Unresectable Hepatocellular Carcinoma: Preliminary Toxicity and Activity Data in Dogs. *Cancers* **2020**, *12*, 1272. [CrossRef]
67. Jung, H.; Bae, K.; Lee, J.Y.; Kim, J.-H.; Han, H.-J.; Yoon, H.-Y.; Yoon, K.-A. Establishment of Canine Transitional Cell Carcinoma Cell Lines Harboring BRAF V595E Mutation as a Therapeutic Target. *Int. J. Mol. Sci.* **2021**, *22*, 9151. [CrossRef]
68. Henry, C.J.; Tyler, J.W.; McEntee, M.C.; Stokol, T.; Rogers, K.S.; Chun, R.; Garrett, L.D.; McCaw, D.L.; Higginbotham, M.L.; Flessland, K.A.; et al. Evaluation of a bladder tumor antigen test as a screening test for transitional cell carcinoma of the lower urinary tract in dogs. *Am. J. Vet. Res.* **2003**, *64*, 1017–1020. [CrossRef]
69. Polymedco. *V-BTA Test*; Package Insert on the Internet; Polymedco, Inc.: Cortland Manor, NY, USA, 2010; Available online: https://www.vetbta.com/veterinarians (accessed on 14 November 2021).
70. Borjesson, D.L.; Christopher, M.M.; Ling, G.V. Detection of canine transitional cell carcinoma using a bladder tumor antigen urine dipstick test. *Vet. Clin. Pathol.* **1999**, *28*, 33–38. [CrossRef] [PubMed]
71. Billet, J.-P.H.; Moore, A.H.; Holt, P.E. Evaluation of a bladder tumor antigen test for the diagnosis of lower urinary tract malignancies in dogs. *Am. J. Vet. Res.* **2002**, *63*, 370–373. [CrossRef] [PubMed]
72. Sotirakopoulos, A.; Armstrong, P.; Heath, L.; Madrill, N.; McNiel, E. Evaluation of Microsatellite Instability in Urine for the Diagnosis of Transitional Cell Carcinoma of the Lower Urinary Tract in Dogs. *J. Vet. Intern. Med.* **2010**, *24*, 1445–1451. [CrossRef] [PubMed]
73. Allen, D.K.; Waters, D.J.; Knapp, D.W.; Kuczek, T. High Urine Concentrations of Basic Fibroblast Growth Factor in Dogs with Bladder Cancer. *J. Vet. Intern. Med.* **1996**, *10*, 231–234. [CrossRef] [PubMed]
74. R&D Systems. *Quantikine® HS ELISA—Human FGF Basic Immunoassay*; Package Insert on the Internet; R&D Systems, Inc.: Minneapolis, MN, USA, 2020. Available online: https://www.rndsystems.com/products/human-fgf-basic-quantikine-hs-elisa-kit_hsfb00d (accessed on 18 November 2021).
75. Creative Diagnostics. *Canine Basic Fibroblast Growth Factor ELISA Kit*; Package Insert on the Internet; Creative Diagnostics: Shirley, NY, USA, 2021; Available online: https://www.creative-diagnostics.com/FGF2-EIA-Kit-246601-463.htm (accessed on 18 November 2021).
76. Genorise Scientific, Inc. *Nori® Canine FGF Basic ELISA Kit—DataSheet*; Package Insert on the Internet; Genorise Scientific, Inc.: Glen Mills, PA, USA, 2021; Available online: http://www.genorise.com/Canine-FGF-ELISA-Kits.html (accessed on 18 November 2021).
77. Mohammed, S.I.; Bennett, P.F.; Craig, B.A.; Glickman, N.W.; Mutsaers, A.J.; Snyder, P.W.; Widmer, W.R.; de Gortari, A.E.; Bonney, P.L.; Knapp, D.W. Effects of the cyclooxygenase inhibitor, piroxicam, on tumor response, apoptosis, and angiogenesis in a canine model of human invasive urinary bladder cancer. *Cancer Res.* **2002**, *15*, 356–358. Available online: http://cancerres.aacrjournals.org/content/62/2/356.abstract (accessed on 23 September 2020).
78. Mohammed, S.I.; Craig, B.A.; Mutsaers, A.J.; Glickman, N.W.; Snyder, P.W.; De Gortari, A.E.; Schlittler, D.L.; Coffman, K.T.; Bonney, P.L.; Knapp, D.W. Effects of the cyclooxygenase inhibitor, piroxicam, in combination with chemotherapy on tumor response, apoptosis, and angiogenesis in a canine model of human invasive urinary bladder cancer. *Mol. Cancer Ther.* **2003**, *2*, 183–188. Available online: https://mct.aacrjournals.org/content/2/2/183 (accessed on 23 September 2020).
79. Abbott Laboratories. *UroVysion Bladder Cancer Kit*; Package Insert on the Internet; Abbott Molecular, Inc.: Des Plaines, IL, USA, 2021. Available online: https://www.molecular.abbott/us/en/products/oncology/urovysion-bladder-cancer-kit (accessed on 18 November 2021).

80. Kent, M.S.; Zwingenberger, A.; Westropp, J.L.; Barrett, L.E.; Durbin-Johnson, B.P.; Ghosh, P.; Vinall, R.L. MicroRNA profiling of dogs with transitional cell carcinoma of the bladder using blood and urine samples. *BMC Vet. Res.* **2017**, *13*, 339. [CrossRef]
81. Kim, N.W.; Piatyszek, M.A.; Prowse, K.R.; Harley, C.B.; West, M.D.; Ho, P.D.L.; Coviello, G.M.; Wright, W.E.; Weinrich, S.L.; Shay, J.W. Specific association of human telomerase activity with immortal cells and cancer. *Science* **1994**, *266*, 2011–2015. [CrossRef]
82. McCleary-Wheeler, A.L.; Williams, L.E.; Hess, P.R.; Suter, S.E. Evaluation of an in vitro telomeric repeat amplification protocol assay to detect telomerase activity in canine urine. *Am. J. Vet. Res.* **2010**, *71*, 1468–1474. [CrossRef]
83. Heilmann, R.M.; McNiel, E.A.; Grützner, N.; Lanerie, D.J.; Suchodolski, J.S.; Steiner, J.M. Diagnostic performance of the urinary canine calgranulins in dogs with lower urinary or urogenital tract carcinoma. *BMC Vet. Res.* **2017**, *13*, 112. [CrossRef]
84. Bracha, S.; McNamara, M.; Hilgart, I.; Milovancev, M.; Medlock, J.; Goodall, C.; Wickramasekara, S.; Maier, C.S. A multiplex biomarker approach for the diagnosis of transitional cell carcinoma from canine urine. *Anal. Biochem.* **2014**, *455*, 41–47. [CrossRef] [PubMed]
85. Dill, A.L.; Ifa, D.R.; Manicke, N.E.; Costa, A.B.; Ramos-Vara, J.A.; Knapp, D.W.; Cooks, R.G. Lipid Profiles of Canine Invasive Transitional Cell Carcinoma of the Urinary Bladder and Adjacent Normal Tissue by Desorption Electrospray Ionization Imaging Mass Spectrometry. *Anal. Chem.* **2009**, *81*, 8758–8764. [CrossRef] [PubMed]
86. Pope, G.B. Urinary Lipid Biomarkers for Detecting Canine Transitional Cell Carcinoma Pilot Study. Bachelor's Thesis, Oregon State University, Corvallis, OR, USA, 2016. Available online: http://ir.library.oregonstate.edu/concern/undergraduate_thesis_or_projects/bn999c19c (accessed on 20 September 2020).
87. Rankin, W.V.; Henry, C.J.; Turnquist, S.E.; Turk, J.R.; Beissenherz, M.E.; Tyler, J.W.; Green, J.A. Comparison of distributions of survivin among tissues from urinary bladders of dogs with cystitis, transitional cell carcinoma, or histologically normal urinary bladders. *Am. J. Vet. Res.* **2008**, *69*, 1073–1078. [CrossRef] [PubMed]
88. Hanazono, K.; Fukumoto, S.; Kawamura, Y.; Endo, Y.; Kadosawa, T.; Iwano, H.; Uchide, T. Epidermal growth factor receptor expression in canine transitional cell carcinoma. *J. Vet. Med. Sci.* **2015**, *77*, 1–6. [CrossRef]
89. Millanta, F.; Impellizeri, J.; McSherry, L.; Rocchigiani, G.; Aurisicchio, L.; Lubas, G. Overexpression of HER-2 via immunohistochemistry in canine urinary bladder transitional cell carcinoma—A marker of malignancy and possible therapeutic target. *Vet. Comp. Oncol.* **2017**, *16*, 297–300. [CrossRef]
90. Walters, L.; Martin, O.; Price, J.; Sula, M.M. Expression of receptor tyrosine kinase targets PDGFR-β, VEGFR2 and KIT in canine transitional cell carcinoma. *Vet. Comp. Oncol.* **2017**, *16*, E117–E122. [CrossRef]
91. Korec, D.I.; Louke, D.S.; Breitbach, J.T.; Geisler, J.A.; Husbands, B.D.; Fenger, J.M. Characterization of receptor tyrosine kinase activation and biological activity of toceranib phosphate in canine urothelial carcinoma cell lines. *BMC Vet. Res.* **2021**, *17*, 320. [CrossRef]
92. London, C.A.; Hannah, A.L.; Zadovoskaya, R.; Chien, M.B.; Kollias-Baker, C.; Rosenberg, M.; Downing, S.; Post, G.; Boucher, J.; Shenoy, N.; et al. Phase I dose-escalating study of SU11654, a small molecule receptor tyrosine kinase inhibitor, in dogs with spontaneous malignancies. *Clin. Cancer Res.* **2003**, *9*, 2755–2768.
93. Rippy, S.B.; Gardner, H.L.; Nguyen, S.M.; Warry, E.E.; Portela, R.A.; Drost, W.T.; Hostnik, E.T.; Green, E.M.; Chew, D.J.; Peng, J.; et al. A pilot study of toceranib/vinblastine therapy for canine transitional cell carcinoma. *BMC Vet. Res.* **2016**, *12*, 257. [CrossRef]
94. Gustafson, T.L.; Biller, B. Use of Toceranib Phosphate in the Treatment of Canine Bladder Tumors: 37 Cases. *J. Am. Anim. Hosp. Assoc.* **2019**, *55*, 243–248. [CrossRef]
95. Inoue, A.; Maeda, S.; Kinoshita, R.; Tsuboi, M.; Yonezawa, T.; Matsuki, N. Density of tumor-infiltrating granzyme B-positive cells predicts favorable prognosis in dogs with transitional cell carcinoma. *Vet. Immunol. Immunopathol.* **2017**, *190*, 53–56. [CrossRef] [PubMed]
96. Hanazono, K.; Nishimori, T.; Fukumoto, S.; Kawamura, Y.; Endo, Y.; Kadosawa, T.; Uchide, T. Immunohistochemical expression of p63, Ki67 andβ-catenin in canine transitional cell carcinoma and polypoid cystitis of the urinary bladder. *Vet. Comp. Oncol.* **2016**, *14*, 263–269. [CrossRef]
97. Barbareschi, M.; Pecciarini, L.; Cangi, M.G.; Macrì, E.; Rizzo, A.; Viale, G.; Doglioni, C. p63, a p53 Homologue, Is a Selective Nuclear Marker of Myoepithelial Cells of the Human Breast. *Am. J. Surg. Pathol.* **2001**, *25*, 1054–1060. [CrossRef] [PubMed]
98. Signoretti, S.; Waltregny, D.; Dilks, J.; Isaac, B.; Lin, D.; Garraway, L.; Yang, A.; Montironi, R.; McKeon, F.; Loda, M. p63 Is a Prostate Basal Cell Marker and Is Required for Prostate Development. *Am. J. Pathol.* **2000**, *157*, 1769–1775. [CrossRef]
99. Di Como, C.J.; Urist, M.J.; Babayan, I.; Drobnjak, M.; Hedvat, C.V.; Teruya-Feldstein, J.; Pohar, K.; Hoos, A.; Cordon-Cardo, C. p63 expression profiles in human normal and tumor tissues. *Clin. Cancer Res.* **2002**, *8*, 494–501. Available online: http://clincancerres.aacrjournals.org/content/8/2/494.long (accessed on 20 September 2018).
100. Urist, M.J.; DiComo, C.; Lu, M.-L.; Charytonowicz, E.; Verbel, D.; Crum, C.P.; Ince, T.A.; McKeon, F.D.; Cordon-Cardo, C. Loss of p63 Expression Is Associated with Tumor Progression in Bladder Cancer. *Am. J. Pathol.* **2002**, *161*, 1199–1206. [CrossRef]
101. Koga, F.; Kawakami, S.; Fujii, Y.; Saito, K.; Ohtsuka, Y.; Iwai, A.; Ando, N.; Takizawa, T.; Kageyama, Y.; Kihara, K. Impaired p63 expression associates with poor prognosis and uroplakin III expression in invasive urothelial carcinoma of the bladder. *Clin. Cancer Res.* **2003**, *9*, 5501–5507.
102. Nishimori, T.; Hanazono, K.; Matsuda, K.; Kawamura, Y.; Kadosawa, T.; Endo, Y.; Uchide, T. Prognostic role of and ΔNp63 expression in canine transitional cell carcinoma of the urinary bladder. *Open Vet. J.* **2021**, *11*, 700. [CrossRef]
103. Wu, X.; Lin, J.; Walz, T.; Häner, M.; Yu, J.; Aebi, U.; Sun, T. Mammalian uroplakins. A group of highly conserved urothelial differentiation-related membrane proteins. *J. Biol. Chem.* **1994**, *269*, 13716–13724. [CrossRef]

104. Ramos-Vara, J.A.; Miller, M.A.; Boucher, M.; Roudabush, A.; Johnson, G.C. Immunohistochemical Detection of Uroplakin III, Cytokeratin 7, and Cytokeratin 20 in Canine Urothelial Tumors. *Vet. Pathol.* **2003**, *40*, 55–62. [CrossRef]
105. Sledge, D.G.; Patrick, D.J.; Fitzgerald, S.D.; Xie, Y.; Kiupel, M. Differences in Expression of Uroplakin III, Cytokeratin 7, and Cyclooxygenase-2 in Canine Proliferative Urothelial Lesions of the Urinary Bladder. *Vet. Pathol.* **2014**, *52*, 74–82. [CrossRef]
106. Jahr, S.; Hentze, H.; Englisch, S.; Hardt, D.; Fackelmayer, F.O.; Hesch, R.D.; Knippers, R. DNA fragments in the blood plasma of cancer patients: Quantitations and evidence for their origin from apoptotic and necrotic cells. *Cancer Res.* **2001**, *61*, 1659–1665. [PubMed]
107. Grivas, P.; Lalani, A.-K.A.; Pond, G.R.; Nagy, R.J.; Faltas, B.; Agarwal, N.; Gupta, S.V.; Drakaki, A.; Vaishampayan, U.N.; Wang, J.; et al. Circulating Tumor DNA Alterations in Advanced Urothelial Carcinoma and Association with Clinical Outcomes: A Pilot Study. *Eur. Urol. Oncol.* **2020**, *3*, 695–699. [CrossRef]
108. Chibuk, J.; Flory, A.; Kruglyak, K.M.; Leibman, N.; Nahama, A.; Dharajiya, N.; Boom, D.V.D.; Jensen, T.J.; Friedman, J.S.; Shen, M.R.; et al. Horizons in Veterinary Precision Oncology: Fundamentals of Cancer Genomics and Applications of Liquid Biopsy for the Detection, Characterization, and Management of Cancer in Dogs. *Front. Vet. Sci.* **2021**, *8*, 235. [CrossRef]
109. Kruglyak, K.M.; Chibuk, J.; McLennan, L.; Nakashe, P.; Hernandez, G.E.; Motalli-Pepio, R.; Fath, D.M.; Tynan, J.A.; Holtvoigt, L.E.; Chorny, I.; et al. Blood-Based Liquid Biopsy for Comprehensive Cancer Genomic Profiling Using Next-Generation Sequencing: An Emerging Paradigm for Non-invasive Cancer Detection and Management in Dogs. *Front. Vet. Sci.* **2021**, *8*, 704835. [CrossRef] [PubMed]
110. Kim, J.; Bae, H.; Ahn, S.; Shin, S.; Cho, A.; Cho, K.-W.; Jung, D.-I.; Yu, D. Cell-Free DNA as a Diagnostic and Prognostic Biomarker in Dogs with Tumors. *Front. Vet. Sci.* **2021**, *8*, 735682. [CrossRef] [PubMed]
111. Lawes, D.; SenGupta, S.; Boulos, P. The clinical importance and prognostic implications of microsatellite instability in sporadic cancer. *Eur. J. Surg. Oncol. (EJSO)* **2003**, *29*, 201–212. [CrossRef]
112. Cowland, J.B.; Hother, C.; Grønbæk, K. MicroRNAs and cancer. *APMIS* **2007**, *115*, 1090–1106. [CrossRef]
113. Santos, J.M.; Gil da Costa, R.; Medeiros, R. Dysregulation of cellular microRNAs by human oncogenic viruses—Implications for tumorigenesis. *Biochim. Biophys. Acta* **2018**, *1861*, 95–105. [CrossRef]
114. Knapp, D.W.; Ramos-Vara, J.A.; Moore, G.E.; Dhawan, D.; Bonney, P.L.; Young, K.E. Urinary Bladder Cancer in Dogs, a Naturally Occurring Model for Cancer Biology and Drug Development. *ILAR J.* **2014**, *55*, 100–118. [CrossRef]
115. Lamarca, A.; Barriuso, J. Urine Telomerase for Diagnosis and Surveillance of Bladder Cancer. *Adv. Urol.* **2012**, *2012*, 693631. [CrossRef] [PubMed]
116. Van, Q.N.; Veenstra, T.D.; Issaq, H.J. Metabolic Profiling for the Detection of Bladder Cancer. *Curr. Urol. Rep.* **2010**, *12*, 34–40. [CrossRef] [PubMed]
117. Inoue, S.; Mizushima, T.; Fujita, K.; Meliti, A.; Ide, H.; Yamaguchi, S.; Fushimi, H.; Netto, G.J.; Nonomura, N.; Miyamoto, H. GATA3 immunohistochemistry in urothelial carcinoma of the upper urinary tract as a urothelial marker and a prognosticator. *Hum. Pathol.* **2017**, *64*, 83–90. [CrossRef] [PubMed]
118. Rana, C.; Agarwal, H.; Babu, S.; Kumar, M.; Singhai, A.; Shankhwar, S.N.; Singh, V.; Sinha, R.J. Diagnostic utility of GATA3 immunohistochemical expression in urothelial carcinoma. *Indian J. Pathol. Microbiol.* **2019**, *62*, 244–250. [CrossRef]
119. Vitti Gambim, V.; Laufer-Amorim, R.; Fonseca Alves, R.H.; Grieco, V.; Fonseca-Alves, C.E. A comparative meta-analysis and in silico analysis of differentially expressed genes and proteins in canine and human bladder cancer. *Front. Vet. Sci.* **2020**, *7*, 558978. Available online: https://www.frontiersin.org/articles/10.3389/fvets.2020.558978/full (accessed on 29 January 2022). [CrossRef]
120. Fulkerson, C.M.; Knapp, D.W. Management of transitional cell carcinoma of the urinary bladder in dogs: A review. *Vet. J.* **2015**, *205*, 217–225. [CrossRef]
121. Molnár, T.; Vajdovich, P. Clinical factors determining the efficacy of urinary bladder tumour treatments in dogs: Surgery, chemotherapy or both? *Acta Vet. Hung.* **2012**, *60*, 55–68. [CrossRef]
122. Robat, C.; Burton, J.; Thamm, D.; Vail, D. Retrospective evaluation of doxorubicin-piroxicam combination for the treatment of transitional cell carcinoma in dogs. *J. Small Anim. Pract.* **2013**, *54*, 67–74. [CrossRef]
123. Cannon, C.; Allstadt, S.D. Lower Urinary Tract Cancer. *Vet. Clin. N. Am. Small Anim. Pract.* **2015**, *45*, 807–824. [CrossRef]
124. Saeki, K.; Fujita, A.; Fujita, N.; Nakagawa, T.; Nishimura, R. Total cystectomy and subsequent urinary diversion to the prepuce or vagina in dogs with transitional cell carcinoma of the trigone area: A report of 10 cases (2005–2011). *Can. Vet. J. La Rev. Vet. Can.* **2015**, *56*, 73–80.
125. Weisse, C.; Berent, A.; Todd, K.; Clifford, C.; Solomon, J. Evaluation of palliative stenting for management of malignant urethral obstructions in dogs. *J. Am. Vet. Med. Assoc.* **2006**, *229*, 226–234. [CrossRef] [PubMed]
126. McMillan, S.K.; Knapp, D.W.; Ramos-Vara, J.A.; Bonney, P.L.; Adams, L.G. Outcome of urethral stent placement for management of urethral obstruction secondary to transitional cell carcinoma in dogs: 19 cases (2007–2010). *J. Am. Vet. Med. Assoc.* **2012**, *241*, 1627–1632. [CrossRef] [PubMed]
127. Radhakrishnan, A. Urethral Stenting for Obstructive Uropathy Utilizing Digital Radiography for Guidance: Feasibility and Clinical Outcome in 26 Dogs. *J. Vet. Intern. Med.* **2017**, *31*, 427–433. [CrossRef]
128. Stone, E.A.; Withrow, S.J.; Page, R.L.; Schwarz, P.D.; Wheeler, S.L.; Seim, H.B., III. Ureterocolonlc Anastomosis in Ten Dogs with Transitional Cell Carcinoma. *Vet. Surg.* **1988**, *17*, 147–153. [CrossRef] [PubMed]

129. Huppes, R.R.; De Nardi, A.B.; Lima, B.R.; Cintra, C.A.; Castro, J.L.C.; Adin, C.A.; Crivellenti, L.Z. Radical Cystectomy and Cutaneous Ureterostomy in 4 Dogs with Trigonal Transitional Cell Carcinoma: Description of Technique and Case Series. *Vet. Surg.* **2016**, *46*, 111–119. [CrossRef]
130. Prządka, P.; Liszka, B.; Lachowska, S.; Dzimira, S.; Ciaputa, R.; Tunikowska, J.; Juźwiak, Ł.; Kucharski, P.; Rudno-Rudzińska, J.; Kiełbowicz, Z. Case report laparoscopy-assisted pre-pubic urethrostomy as a palliative procedure for resection of distal urethral tumor in a female dog. *BMC Vet. Res.* **2021**, *17*, 309. [CrossRef]
131. Liptak, J.M.; Brutscher, S.P.; Monnet, E.; Dernell, W.S.; Twedt, D.C.; Kazmierski, K.J.; Walter, C.U.; Mullins, M.N.; LaRue, S.M.; Withrow, S.J. Transurethral Resection in the Management of Urethral and Prostatic Neoplasia in 6 Dogs. *Vet. Surg.* **2004**, *33*, 505–516. [CrossRef]
132. Upton, M.L.; Tangner, C.H.; Payton, M.E. Evaluation of carbon dioxide laser ablation combined with mitoxantrone and piroxicam treatment in dogs with transitional cell carcinoma. *J. Am. Vet. Med. Assoc.* **2006**, *228*, 549–552. [CrossRef]
133. Cerf, D.J.; Lindquist, E.C. Palliative ultrasound-guided endoscopic diode laser ablation of transitional cell carcinomas of the lower urinary tract in dogs. *J. Am. Vet. Med. Assoc.* **2012**, *240*, 51–60. [CrossRef]
134. Poirier, V.J.; Forrest, L.J.; Adams, W.M.; Vail, D.M. Piroxicam, Mitoxantrone, and Coarse Fraction Radiotherapy for the Treatment of Transitional Cell Carcinoma of the Bladder in 10 Dogs: A Pilot Study. *J. Am. Anim. Hosp. Assoc.* **2004**, *40*, 131–136. [CrossRef]
135. Nolan, M.W.; Kogan, L.; Griffin, L.R.; Custis, J.T.; Harmon, J.F.; Biller, B.J.; LaRue, S.M. Intensity-Modulated and Image-Guided Radiation Therapy for Treatment of Genitourinary Carcinomas in Dogs. *J. Vet. Intern. Med.* **2012**, *26*, 987–995. [CrossRef] [PubMed]
136. Nieset, J.R.; Harmon, J.F.; Johnson, T.E.; LaRue, S.M. Comparison of adaptive radiotherapy techniques for external radiation therapy of canine bladder cancer. *Vet. Radiol. Ultrasound* **2014**, *55*, 644–650. [CrossRef] [PubMed]
137. Choy, K.; Fidel, J. Tolerability and tumor response of a novel low-dose palliative radiation therapy protocol in dogs with transitional cell carcinoma of the bladder and urethra. *Vet. Radiol. Ultrasound* **2016**, *57*, 341–351. [CrossRef] [PubMed]
138. Malfassi, L.; Fidanzio, F.; Sala, M.; Marcarini, S.; Mazza, G.; Carrara, N.; Pavesi, S.; Gnudi, G.; Urso, G.; Dolera, M. A combined protocol with piroxicam, chemotherapy, and whole pelvic irradiation with simultaneous boost volumetric modulated arc radiotherapy for muscle-invasive canine urinary transitional cell carcinoma: First clinical experience. *J. Vet. Med. Sci.* **2021**, *83*, 695–704. [CrossRef]
139. Chun, R.; Knapp, D.W.; Widmer, W.R.; DelNicola, D.B.; Glickman, N.W.; Kuczek, T.; DeGortari, A.; Han, C.M. Phase II Clinical Trial of Carboplatin in Canine Transitional Cell Carcinoma of the Urinary Bladder. *J. Vet. Intern. Med.* **1997**, *11*, 279–283. [CrossRef]
140. Allstadt, S.; Rodriguez, C.; Boostrom, B.; Rebhun, R.; Skorupski, K. Randomized Phase III Trial of Piroxicam in Combination with Mitoxantrone or Carboplatin for First-Line Treatment of Urogenital Tract Transitional Cell Carcinoma in Dogs. *J. Vet. Intern. Med.* **2015**, *29*, 261–267. [CrossRef]
141. Chun, R.; Knapp, D.W.; Widmer, W.R.; Glickman, N.W.; DeNicola, D.B.; Bonney, P.L. Cisplatin treatment of transitional cell carcinoma of the urinary bladder in dogs: 18 cases (1983–1993). *J. Am. Vet. Med. Assoc.* **1996**, *209*.
142. Knapp, D.; Henry, C.; Widmer, W.; Tan, K.; Moore, G.; Ramos-Vara, J.; Lucroy, M.; Greenberg, C.; Greene, S.; Abbo, A.; et al. Randomized Trial of Cisplatin versus Firocoxib versus Cisplatin/Firocoxib in Dogs with Transitional Cell Carcinoma of the Urinary Bladder. *J. Vet. Intern. Med.* **2012**, *27*, 126–133. [CrossRef]
143. Marconato, L.; Zini, E.; Lindner, D.; Suslak-Brown, L.; Nelson, V.; Jeglum, A.K. Toxic effects and antitumor response of gemcitabine in combination with piroxicam treatment in dogs with transitional cell carcinoma of the urinary bladder. *J. Am. Vet. Med. Assoc.* **2011**, *238*, 1004–1010. [CrossRef]
144. Henry, C.J.; McCaw, D.L.; Turnquist, S.E.; Tyler, J.W.; Bravo, L.; Sheafor, S.; Straw, R.C.; Dernell, W.S.; Madewell, B.R.; Jorgensen, L.; et al. Clinical evaluation of mitoxantrone and piroxicam in a canine model of human invasive urinary bladder carcinoma. *Clin. Cancer Res.* **2003**, *9*, 906–911.
145. Arnold, E.; Childress, M.; Fourez, L.; Tan, K.; Stewart, J.; Bonney, P.; Knapp, D. Clinical Trial of Vinblastine in Dogs with Transitional Cell Carcinoma of the Urinary Bladder. *J. Vet. Intern. Med.* **2011**, *25*, 1385–1390. [CrossRef]
146. Knapp, D.W.; Ruple-Czerniak, A.; Ramos-Vara, J.A.; Naughton, J.F.; Fulkerson, C.M.; Honkisz, S.I. A Nonselective Cyclooxygenase Inhibitor Enhances the Activity of Vinblastine in a Naturally-Occurring Canine Model of Invasive Urothelial Carcinoma. *Bladder Cancer* **2016**, *2*, 241–250. [CrossRef] [PubMed]
147. Kaye, M.E.; Weißhaar, K.; Lawrence, J.A.; Thamm, D.H. Vinorelbine rescue therapy for dogs with primary urinary bladder carcinoma. *Vet. Comp. Oncol.* **2013**, *13*, 443–451. [CrossRef]
148. Schrempp, D.R.; Childress, M.O.; Stewart, J.C.; Leach, T.N.; Tan, K.M.; Abbo, A.H.; de Gortari, A.E.; Bonney, P.L.; Knapp, D.W. Metronomic administration of chlorambucil for treatment of dogs with urinary bladder transitional cell carcinoma. *J. Am. Vet. Med. Assoc.* **2013**, *242*, 1534–1538. [CrossRef] [PubMed]
149. Mutsaers, A.; Mohammed, S.I.; DeNicola, D.; Snyder, P.; Glickman, N.; Bennett, P.; De Gortari, A.; Bonney, P.; Knapp, D. Pretreatment tumor prostaglandin E2 concentration and cyclooxygenase-2 expression are not associated with the response of canine naturally occurring invasive urinary bladder cancer to cyclooxygenase inhibitor therapy. *Prostaglandins Leukot. Essent. Fat. Acids* **2005**, *72*, 181–186. [CrossRef]
150. Knapp, D.W.; Richardson, R.C.; Chan, T.C.; Bottoms, G.D.; Widmer, W.R.; DeNicola, D.B.; Teclaw, R.; Bonney, P.L.; Kuczek, T. Piroxicam Therapy in 34 Dogs with Transitional Cell Carcinoma of the Urinary Bladder. *J. Vet. Intern. Med.* **1994**, *8*, 273–278. [CrossRef] [PubMed]

151. McMillan, S.K.; Boria, P.; Moore, G.E.; Widmer, W.R.; Bonney, P.L.; Knapp, D.W. Antitumor effects of deracoxib treatment in 26 dogs with transitional cell carcinoma of the urinary bladder. *J. Am. Vet. Med. Assoc.* **2011**, *239*, 1084–1089. [CrossRef]
152. Abbo, A.; Jones, D.; Masters, A.; Stewart, J.; Fourez, L.; Knapp, D. Phase I Clinical Trial and Pharmacokinetics of Intravesical Mitomycin C in Dogs with Localized Transitional Cell Carcinoma of the Urinary Bladder. *J. Vet. Intern. Med.* **2010**, *24*, 1124–1130. [CrossRef] [PubMed]
153. Alhunaidi, O.; Zlotta, A.R. The use of intravesical BCG in urothelial carcinoma of the bladder. *Ecancermedicalscience* **2019**, *13*, 905. [CrossRef]
154. Bloomberg, S.D.; Brosman, S.A.; Hausman, M.S.; Cohen, A.; Battenberg, J.D. The effects of BCG on the dog bladder. *Investig. Urol.* **1975**, *12*, 423–427. Available online: https://pubmed.ncbi.nlm.nih.gov/1091593/ (accessed on 30 January 2022).
155. Sim, G.C.; Radvanyi, L. The IL-2 cytokine family in cancer immunotherapy. *Cytokine Growth Factor Rev.* **2014**, *25*, 377–390. [CrossRef] [PubMed]
156. Konietschke, U.; Teske, E.; Jurina, K.; Stockhaus, C. Palliative intralesional interleukin-2 treatment in dogs with urinary bladder and urethral carcinomas. *In Vivo* **2012**, *26*, 931–935. [PubMed]
157. Ziekman, P.G.P.M.; Otter, W.D.E.N.; Tan, J.F.V.; Teske, E.; Kirpensteijn, J.; Koten, J.-W.; Jacobs, J.J.L. Intratumoural interleukin-2 therapy can induce regression of non-resectable mastocytoma in dogs. *Anticancer Res.* **2013**, *33*, 161–165.
158. Ridgway, T.D.; Lucroy, M.D. Phototoxic effects of 635-nm light on canine transitional cell carcinoma cells incubated with 5-aminolevulinic acid. *Am. J. Vet. Res.* **2003**, *64*, 131–136. [CrossRef] [PubMed]
159. Lucroy, M.D.; Bowles, M.H.; Higbee, R.G.; Blaik, M.A.; Ritchey, J.W.; Ridgway, T.D. Photodynamic Therapy for Prostatic Carcinoma in a Dog. *J. Vet. Intern. Med.* **2003**, *17*, 235–237. [CrossRef]
160. Lin, T.-Y.; Zhang, H.; Wang, S.; Xie, L.; Li, B.; Rodriguez, C.O.; White, R.D.V.; Pan, C.-X. Targeting canine bladder transitional cell carcinoma with a human bladder cancer-specific ligand. *Mol. Cancer* **2011**, *10*, 9. [CrossRef]
161. Zhang, H.; Aina, O.H.; Lam, K.S.; White, R.D.V.; Evans, C.; Henderson, P.; Lara, P.N.; Wang, X.; Bassuk, J.A.; Pan, C.-X. Identification of a bladder cancer-specific ligand using a combinatorial chemistry approach. *Urol. Oncol. Semin. Orig. Investig.* **2010**, *30*, 635–645. [CrossRef]
162. Pan, C.-X.; Lin, T.-Y.; Zhang, H.; Luo, J.; Li, Y.; Gao, T.; Lara, P.N., Jr.; White, R.D.V.; Lam, K.S. Multifunctional targeting micelle nanocarriers with both imaging and therapeutic potential for bladder cancer. *Int. J. Nanomed.* **2012**, *7*, 2793–2804. [CrossRef]
163. Lin, T.-Y.; Li, Y.-P.; Zhang, H.; Luo, J.; Goodwin, N.; Gao, T.; White, R.D.V.; Lam, K.S.; Pan, C.-X. Tumor-targeting multifunctional micelles for imaging and chemotherapy of advanced bladder cancer. *Nanomedicine* **2013**, *8*, 1239–1251. [CrossRef]
164. Lin, T.-Y.; Li, Y.; Liu, Q.; Chen, J.-L.; Zhang, H.; Lac, D.; Zhang, H.; Ferrara, K.; Wachsmann-Hogiu, S.; Li, T.; et al. Novel theranostic nanoporphyrins for photodynamic diagnosis and trimodal therapy for bladder cancer. *Biomaterials* **2016**, *104*, 339–351. [CrossRef]
165. Pan, A.; Zhang, H.; Li, Y.; Lin, T.-Y.; Wang, F.; Lee, J.; Cheng, M.; Dall'Era, M.; Li, T.; White, R.D.; et al. Disulfide-crosslinked nanomicelles confer cancer-specific drug delivery and improve efficacy of paclitaxel in bladder cancer. *Nanotechnology* **2016**, *27*, 425103. [CrossRef] [PubMed]
166. Lu, Z.; Yeh, T.-K.; Wang, J.; Chen, L.; Lyness, G.; Xin, Y.; Wientjes, M.G.; Bergdall, V.; Couto, G.; Alvarez-Berger, F.; et al. Paclitaxel Gelatin Nanoparticles for Intravesical Bladder Cancer Therapy. *J. Urol.* **2011**, *185*, 1478–1483. [CrossRef] [PubMed]
167. Dhawan, D.; Ramos-Vara, J.A.; Naughton, J.F.; Cheng, L.; Low, P.; Rothenbuhler, R.; Leamon, C.P.; Parker, N.; Klein, P.J.; Vlahov, I.R.; et al. Targeting Folate Receptors to Treat Invasive Urinary Bladder Cancer. *Cancer Res.* **2012**, *73*, 875–884. [CrossRef]
168. Fulkerson, C.M.; Dhawan, D.; Ratliff, T.L.; Hahn, N.M.; Knapp, D.W. Naturally Occurring Canine Invasive Urinary Bladder Cancer: A Complementary Animal Model to Improve the Success Rate in Human Clinical Trials of New Cancer Drugs. *Int. J. Genom.* **2017**, *2017*, 6589529. [CrossRef]
169. Besaratinia, A.; Cockburn, M.; Tommasi, S. Alterations of DNA methylome in human bladder cancer. *Epigenetics* **2013**, *8*, 1013–1022. [CrossRef] [PubMed]
170. Kim, Y.-J.; Kim, W.-J. Can we use methylation markers as diagnostic and prognostic indicators for bladder cancer? *Investig. Clin. Urol.* **2016**, *57*, S77–S88. [CrossRef] [PubMed]
171. Nunes, S.P.; Henrique, R.; Jerónimo, C.; Paramio, J.M. DNA Methylation as a Therapeutic Target for Bladder Cancer. *Cells* **2020**, *9*, 1850. [CrossRef] [PubMed]
172. Tian, Z.; Meng, L.; Long, X.; Diao, T.; Hu, M.; Wang, M.; Liu, M.; Wang, J. DNA methylation-based classification and identification of bladder cancer prognosis-associated subgroups. *Cancer Cell Int.* **2020**, *20*, 1–11. [CrossRef]
173. Nakagawa, T.; Kanai, Y.; Ushijima, S.; Kitamura, T.; Kakizoe, T.; Hirohashi, S. DNA hypermethylation on multiple cpg islands associated with increased dna methyltransferase dnmt1 protein expression during multistage urothelial carcinogenesis. *J. Urol.* **2005**, *173*, 1767–1771. [CrossRef]
174. Dhawan, D.; Ramos-Vara, J.A.; Hahn, N.M.; Waddell, J.; Olbricht, G.R.; Zheng, R.; Stewart, J.C.; Knapp, D.W. DNMT1: An emerging target in the treatment of invasive urinary bladder cancer. *Urol. Oncol. Semin. Orig. Investig.* **2013**, *31*, 1761–1769. [CrossRef]
175. Hahn, N.M.; Bonney, P.L.; Dhawan, D.; Jones, D.R.; Balch, C.; Guo, Z.; Hartman-Frey, C.; Fang, F.; Parker, H.G.; Kwon, E.M.; et al. Subcutaneous 5-Azacitidine Treatment of Naturally Occurring Canine Urothelial Carcinoma: A Novel Epigenetic Approach to Human Urothelial Carcinoma Drug Development. *J. Urol.* **2012**, *187*, 302–309. [CrossRef] [PubMed]

176. Fulkerson, C.M.; Dhawan, D.; Jones, D.R.; Marquez, V.E.; Jones, P.A.; Wang, Z.; Wu, Q.; Klaunig, J.E.; Fourez, L.M.; Bonney, P.L.; et al. Pharmacokinetics and toxicity of the novel oral demethylating agent zebularine in laboratory and tumor bearing dogs. *Vet. Comp. Oncol.* **2015**, *15*, 226–236. [CrossRef] [PubMed]
177. Xavier, P.L.P.; Müller, S.; Fukumasu, H. Epigenetic Mechanisms in Canine Cancer. *Front. Oncol.* **2020**, *10*, 591843. [CrossRef]
178. Goutas, D.; Theocharis, S.; Tsourouflis, G. Unraveling the Epigenetic Role and Clinical Impact of Histone Deacetylases in Neoplasia. *Diagnostics* **2021**, *11*, 1346. [CrossRef] [PubMed]
179. Eto, S.; Saeki, K.; Yoshitake, R.; Yoshimoto, S.; Shinada, M.; Ikeda, N.; Kamoto, S.; Tanaka, Y.; Kato, D.; Maeda, S.; et al. Anti-tumor effects of the histone deacetylase inhibitor vorinostat on canine urothelial carcinoma cells. *PLoS ONE* **2019**, *14*, e0218382. [CrossRef]
180. Mochizuki, H.; Breen, M. Sequence analysis of RAS and RAF mutation hot spots in canine carcinoma. *Vet. Comp. Oncol.* **2016**, *15*, 1598–1605. [CrossRef]
181. Foskett, A.; Manley, C.; Naramore, R.; Gordon, I.K.; Stewart, B.M.; Khanna, C. Tolerability of oral sorafenib in pet dogs with a diagnosis of cancer. *Vet. Med. Res. Rep.* **2017**, *8*, 97–102. [CrossRef]

veterinary sciences

MDPI

Article

Predictive Value of Grooming Behavior for Development of Dermatitis in Selectively Bred P Rats as a Model of Trichotillomania Hair Pulling Disorder

Debra Hickman [1,2,*], Anjali Prakash [1] and Richard Bell [3]

1 Laboratory Animal Resource Center, School of Medicine, Indiana University, Indianapolis, IN 46202, USA; prakasha@indiana.edu
2 Department of Comparative Pathobiology, College of Veterinary Medicine, Purdue University, West Lafayette, IN 47907, USA
3 Department of Psychiatry, School of Medicine, Indiana University, Indianapolis, IN 46202, USA; ribell@iupui.edu
* Correspondence: hickmand@purdue.edu or hickmand@iupui.edu; Tel.: +1-764-494-9128

Abstract: Trichotillomania (TTM) is a body-focused repetitive disorder affecting as much as 0.5 to 2% of the population, with women four times more likely to be affected than men. This disorder causes impairment in daily function and significant distress. A potential animal model for this disorder is the inbred C57BL/6J mouse which displays clinical signs and behavioral characteristics similar to those described for people affected by this disorder. Because alcohol-preferring P rats also display similar clinical signs and behavioral characteristics, it was hypothesized that this selectively bred stock could be an additional animal model. In this study, 112 female P rats were recorded on digital media for 15 min after being sprayed with a mist of water and assessed for grooming patterns—oral, manual, and scratching. Significant elevations in scratching and oral grooming behavior were predictive of the future development of skin lesions. These findings suggest that P rats may be an additional model to study TTM, with the advantage of increased genetic variation (i.e., non-inbred) which mirrors the human population. The use of this model may help to identify preventative and therapeutic interventions for humans and other animals with similar body-focused repetitive disorders.

Keywords: animal model; trichotillomania; hair-pulling; body-focused repetitive disorder

Citation: Hickman, D.; Prakash, A.; Bell, R. Predictive Value of Grooming Behavior for Development of Dermatitis in Selectively Bred P Rats as a Model of Trichotillomania Hair Pulling Disorder. *Vet. Sci.* **2022**, *9*, 89. https://doi.org/10.3390/vetsci9020089

Academic Editors: Ana Faustino and Paula A. Oliveira

Received: 9 October 2021
Accepted: 11 February 2022
Published: 18 February 2022

1. Introduction

Trichotillomania (TTM, "hair-pulling disorder") is defined by the American Psychiatric Association as an obsessive compulsive disorder (OCD) [1]. People who are affected by TTM pull their hair out, creating bald patches, resulting in exposed skin and tissue, with the disorder being typically diagnosed in childhood or adolescence. The affected areas can include anywhere there is hair, with the scalp, eyebrows, eyelashes, pubic area, and beard representing the most common locations [1]. The true incidence rate is not known due to the social impairment induced by the condition, but it is estimated as high as 3% of the population [2]. Females are affected more frequently than males, with females representing 92.5% of the adult patients presenting to a TTM clinic, though it has been hypothesized that men may be better able to hide the behavior through interventions such as shaving the head. The diagnostic criteria for TTM include the recurrent pulling out of one's hair that results in noticeable hair loss, repeated attempts to decrease or stop the hair-pulling behavior, and clinically significant distress or impairment in social, occupational, or other areas of functioning [1]. Additionally, the condition must not be accounted for by another mental disorder or be due to a general medical condition (e.g., dermatologic condition). Often, these individuals exhibit more than one body-focused repetitive disorder, including nail biting, skin picking, thumb sucking, knuckle cracking, or nose picking [2–8].

Because the disorder often results in chastisement or teasing, individuals work to hide the disorder and frequently develop low self-esteem. Individuals also suffer from other psychiatric disorders, such as depression (57%), generalized anxiety (27%), simple phobia (19%), alcohol abuse (19%), obsessive compulsive disorder (13%), social phobia (11%), and eating disorders (11%) [4,8–15]. In addition to social impairment concerns, people suffering from this disorder may also ingest the hair, leading to secondary complications, such as trichobezoars and intestinal obstruction [12].

Treatment of TTM is challenging with behavioral, dietary, and pharmaceutical interventions having limited efficacy [8,9,16–21]. Human trials have the limitation of a high amount of genetic and environmental variation, in addition to a long lifespan and relatively low incidence of diagnosed candidates for clinical trials. The development of an animal model that mimics the clinical presentation of humans would be valuable by providing an experimental condition with fewer confounding factors (e.g., genetics and/or environment), leading to a better understanding of the underlying pathology and developing strategies to treat or prevent this disorder. There are a number of different animals that engage in similar body-focused repetitive disorders, such as feather picking in birds [22–31], psychogenic alopecia in cats [32–36] and nonhuman primates [37–39], acral lick dermatitis in dogs [35,40–42], tail and ear biting in swine [43–45], and flank biting in horses [46–48].

The C57BL/6J mouse, an inbred strain, has a common clinical presentation of "barbering" (hair loss) and idiopathic necrotizing dermatitis, especially prevalent in stressful situations [49–53]. Although hair loss has been noted as a precursor to idiopathic dermatitis, the phenomenon of barbering has historically been attributed to behavioral stressors in social dominance interactions [51,54]. However, recent work suggests that barbering is more appropriately diagnosed as a stereotypic behavior and can be self-directed [50,51,53]. Building on these assessments, the C57BL/6J mouse has been proposed as a potential model for TTM and a similar disorder, dermatotillomania [50]. These authors determined that they could predict the likelihood of developing hair loss and skin lesions through a simple behavioral test that was performed before the mouse exhibited clinical symptoms. They found that if they gently sprayed the mouse with water and observed its grooming pattern for 15 min, the proportion of time that the mouse spent scratching during grooming was predictive of an increased chance of developing hair loss and skin lesions in the future with an odds ratio of 1.20 [50]. This inbred mouse model has already been used to evaluate the effect of diet on the development of TTM and dermatotillomania [50].

In the Indiana University (IU) breeding colonies that support the production of the alcohol-preferring "P" rat, we have also noticed an exhibition of a clinical predisposition to hair loss and ulcerative dermatitis. The clinical presentation is variable, with hair loss occurring over the entire body, especially on the ventrum, neck, axillary, and inguinal regions (Figure 1A) as well as dermatitis lesions generally focused around the neck and ears (Figure 1B). Because this rat stock is not inbred [55,56], the use of this animal model has the potential to better approximate the human condition, with the genetic diversity of these populations. Although inbred mouse models and tightly controlled environmental conditions have been used extensively for the characterization and understanding of human diseases, the current debate has suggested that this could be contributing to the issues noted when translating findings from rodent models to humans [57–59]. For example, scientists evaluating an outbred model for vaccine development and testing noted that reliance on inbred mouse strains could lead to scientists being misled regarding the effectiveness of various vaccines in the diverse human population [60]. Identification of animal models with more genetic diversity could be invaluable models for human diseases.

The C57BL/6J mouse and the P rat are currently used as models for addiction studies [61–66]. They also consistently test as more anxious in assays to measure depression and anxiety when compared to other rodent models [56]. The comorbidity of these conditions with the expression of TTM is similar to the presentation seen in humans, further supporting the potential applicability of these rodents as animal models of TTM.

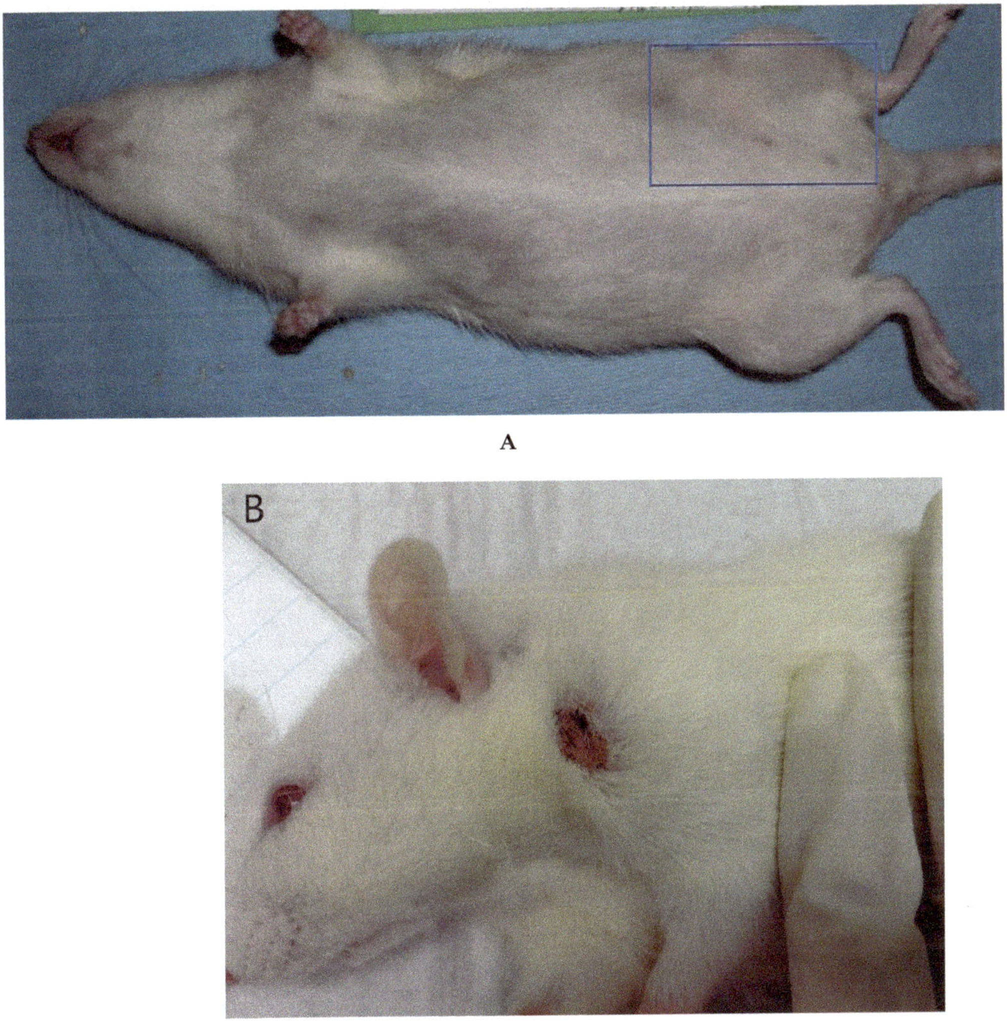

Figure 1. Examples of clinical presentations of P rats with (**A**) hair loss and (**B**) dermatitis. Note that the hair loss is most extensive in the inguinal region (box), extending down the hindlimbs, though there is thinning over the entire ventrum (**A**). The development of dermatitis (**B**) is typically around the head and neck region.

In this study, the spray test and subsequent grooming behavior assessment were used to determine if there was a behavioral pattern that was predictive of future development of hair loss and skin lesions in the P rat, to determine the applicability of this rat as a potential animal model of TTM.

2. Materials and Methods

2.1. Ethical Statement

All procedures were reviewed and approved by the IU School of Medicine IACUC prior to the initiation of the project. The program is accredited by AAALAC International and compliant with all applicable federal regulations.

2.2. Animals

The alcohol-preferring P rat was the line of rat used in this study. The P rat was developed through bidirectional (vs the alcohol-non-preferring NP rat) mass selection from a closed-colony of Wistar rats at the Walter Reed Army Hospital [66]. P rats were transferred to the Indiana University School of Medicine, and they have been maintained by the Indiana Alcohol Research Center, Indianapolis, IN, USA, since then. Two criteria determined the alcohol-preferring P phenotype. First, the animal preferred an unadulterated 10% ethanol solution over water by a ratio of at least 2:1; and second, the animals consumed more than 5 g of ethanol/kg body weight/day [55,65,66]. Five g/kg/day is equivalent to a 70 kg person consuming approximately a fifth of 90-proof whiskey per day. Alcohol-naïve P and NP rats display similar levels of alcohol metabolism [67]; after chronic free-choice alcohol-drinking (6–8 weeks) P rats displayed both metabolic and functional tolerance to the motor impairing and aversive effects of ethanol [68]. Moreover, similarly drinking P rats displayed withdrawal signs [69]. In addition, P rats display relapse-like drinking by exhibiting a robust alcohol deprivation effect (ADE) [70]. The ADE is a transient increase in alcohol intake after a period of ethanol withdrawal [70]. Regarding initial sensitivity, compared with NP rats, P rats are less sensitive to the ataxic [71] and hypothermic [72] effects of alcohol; and P rats develop tolerance quicker to the ataxic [71] and hypnotic effects [73]. During chronic alcohol drinking or operant self-administration, P rats achieve pharmacologically relevant blood alcohol concentrations (BACs: 80 to 250 mg%) [74,75]. These BACs parallel those observed in alcoholics. Thus, the P rat meets all criteria associated with an animal model of alcoholism [55,56,58,66].

One hundred and twelve rats were used for this study. The initial estimate of numbers of animals required for this study proposed the use of approximately 100 pairs of female rats (approximately 200 animals) using a multiple regression power analysis with an alpha of 0.001 (set low to ensure maximum sensitivity in this pilot assessment, as the true incidence in the colony was unknown) with 1 regressor and a rho^2 value of 0.1. This calculation resulted in a projection of the need for 208 rats to result in a power of 0.9183. To reduce the overall use of animals, the experimental subjects were divided into 2 cohorts (approximately 100 animals per cohort) to perform the initial analysis. It was planned to repeat the experiment with the second cohort if no significant findings were identified in the initial cohort, but this was determined to be unnecessary.

The P selectively bred rats were an average of 6 weeks of age when the study started (range of 4 to 8 weeks of age), produced on campus from P progenitors and housed with same-sex siblings. Only females were assessed in this study as this disorder occurs more frequently in females than in males. The rats were housed within the IU animal facilities in accordance with standard operating procedures, briefly summarized here. Rats were given unrestricted access to food (Teklad Diet 7001, Envigo, Indianapolis, IN, USA) and water. Room temperatures were maintained at 21.7 +/− 1 °C and humidity was at 55 +/− 5%. Lights were maintained on a 12:12-h light:dark cycle (lights on a 0700). The rats were pair or trio housed in individually ventilated caging systems (Lab Products, Seaford, DE, USA) using standard, clear polycarbonate shoebox cages with wire lids and filter tops. Contact bedding consisting of aspen chips (Sani-Chip, PJ Murphy Forest Products, Montville, NH, USA) and paper towel nesting materials were provided to each cage.

The colony was screened quarterly by using indirect sentinels. At the time of this study, the colony was free of the following pathogens: coronavirus (sialodacryoadenitis virus), parvoviruses (NS1, rat pneumonia virus, Kilham rat virus, H1 virus, rat minute virus), theliovirus, *Clostridium piliforme*, *Mycoplasma pulmonis*, pinworms (*Aspicularis tetraptera*, *Syphacia* spp.), and fur mites (*Radfordia ensifer*, *Orinthonyssus bacoti*).

2.3. Assessment

Grooming behavior was assessed with a spray test that has been previously described [50]. Briefly, rats were removed from their cages and placed in a clear, Plexiglass chamber. After a minimum of 5 min within the chamber without disturbance, each rat was gently sprayed with a single application mist of water, sufficient to lightly dampen the fur on the head and shoulders. Their behavior was recorded for 15 min on digital video. This spray test was performed once for each rat. The video was scored later by a single observer (AP), and the proportion of time engaged in three grooming behaviors during the 15 min following the spray was recorded (Table 1). The videos were scored continuously with a notation of each time that each behavior was initiated. The relative frequency of each behavior was calculated by dividing the number of times each behavior was initiated by 15 min. All behavioral testing was conducted between 1200 and 1700. All data were collected within a 3-month period. Rats were identified within the cage by applying a hash mark on the tail of the second rat of each pair (rat "B") or two hash marks on the tail of the third rat in a trio (rat "C") with a sharpie. Rat "A" was not marked.

Table 1. Description of grooming behaviors assessed.

Grooming Type	Description
Manual Grooming (MG)	Animal grooms its muzzle, face, and head using its forepaws.
Oral Grooming (OG)	Animal grooms its body by licking, usually beginning on the upper neck and back, then extending down to more caudal areas of the body, including the tail.
Scratching (SCR)	Animal uses its hindlimbs to scratch its head, neck, and back (very fast and of short duration).

All cages in the colony were assessed for hair loss and skin lesions by the research staff every two weeks (and daily by the animal care staff). Tail markings were refreshed, as needed, at this time. If animals developed severe dermatitis (defined as the presence of a single open or ulcerated lesion that exceeded 2 cm in diameter), the rat was euthanized for humane reasons before the end of the study. Additionally, an animal that developed a body condition score of 2 or lower [76] was humanely euthanized. Because the study was assessing the progression of hair loss and dermatitis, treatment of ulcerative dermatitis would confound the study and its exemption was approved in advance of the study. When the rats were approximately 8 months of age, they were euthanized by carbon dioxide. Carcasses were photographed for all animals to record the presence or absence of skin lesions—those euthanized at humane endpoints and at the end of the study.

2.4. Statistical Analysis

The relative frequency of each of the three grooming behaviors was calculated as described above. The normality of data was determined using the Anderson–Darling test. If the data set was normal, it was analyzed with a one-way ANOVA. If the data set was not normal, it was analyzed with the Kruskal–Wallis test.

To determine if the use of the sharpie to identify animals potentially confounded the study, a chi-square test was performed to test the association between being marked with the sharpie and the presence or absence of the lesions.

To calculate the odds ratio, the grooming behaviors for oral grooming and manual grooming were set at either greater or less than 50% of the time engaged in these behaviors. Because no rat engaged in scratching behavior greater than 33% of the time during the spray test, the odds ratio was set at either zero or greater than zero for the percentage of time spent engaged in scratching behavior. In all cases, the odds ratio included the presence or absence of lesions. The odds ratio was calculated first by dividing the number of animals with lesions who had engaged in the defined proportion of the behavior by the number of animals with lesions that did not engage in the defined proportion of the behavior. This number was divided by the number of animals without lesions who had engaged in the defined proportion behavior by the number of animals without lesions that did not engage in the defined proportion of the behavior.

3. Results

Two rats were excluded due to technical difficulties with the recording of video for scoring and associated data loss. Out of the remaining 110 rats, 19 developed dermatitis (19/110; 17.27%). The actual numbers of animals who developed lesions with the associated relative frequency of each behavior are presented in Table 2. The use of the sharpie to identify the rats did not correspond with the development of lesions (Chi-Square = 0.718, $p = 0.3967$).

Table 2. Relative frequency of each behavior and the number of animals who developed lesions associated with each behavior. Data used to calculate the reported odds ratios.

Behavior	Number of Animals with Lesions	Number of Animals without Lesions
MG relative frequency less than 0.50	11	46
MG relative frequency greater than 0.50	7	46
OG relative frequency less than 0.50	8	51
OG relative frequency greater than 0.50	10	41
SCR relative frequency of 0	3	6
SCR relative frequency great than 0	15	86

The data for the relative frequency of manual grooming and scratching behaviors were not normally distributed. The data for the relative frequency of oral grooming was normally distributed. There were no significant differences in the relative frequency of manual grooming ($p = 0.7147$) or scratching ($p = 0.1324$) in rats that developed lesions as compared to the rats that did not develop lesions using the Kruskal–Wallis test. The rats that developed lesions engaged in significantly higher relative frequencies of oral grooming as compared to those that did not develop lesions ($p = 0.0448$) using the one-way ANOVA. Data are presented in Table 3.

Table 3. Mean relative frequency of rats developing lesions versus those without developing lesions. Data presented as mean +/− standard deviation.

Behavior	Animals with Lesions	Animals without Lesions
Manual grooming (MG)	0.51 +/− 0.24	0.51 +/− 0.19
Oral grooming (OG)	0.60 +/− 0.27	0.47 +/− 0.23
Scratching (SCR)	0.04 +/− 0.10	0.01 +/− 0.03

The amount of time spent scratching was predictive of the development of future dermatitis in the female rat with an odds ratio of 2.87 (confidence interval of 0.65, 12.73). Assuming 50% of time spent engaged in oral grooming behavior to be predictive of the development of future dermatitis in the female P rat, the odds ratio was calculated to be 1.55 (confidence interval of 0.56, 4.3). Assuming 50% of time spent in manual grooming behavior to be predictive of the development of future dermatitis in the female P rat, the odds ratio was calculated to be 0.64 (confidence interval of 0.23, 1.79).

4. Discussion

The hypothesis of this study was that the assessment of grooming patterns following the application of the spray test would prove predictive of the development of dermatitis in the future. The relative frequency of time engaged in the oral grooming behaviors was found to be predictive of the development of dermatitis in this cohort of rats, suggesting that this screening test is a potentially valuable tool for the implementation of proactive studies. The behavior of scratching also had a high odds ratio, suggesting that the presence of this behavior is highly predictive of the development of future dermatitis, even if the relative frequency of this behavior is not significant between rats that developed dermatitis as compared to those who did not develop dermatitis. Of note, the incidence of dermatitis reported in the cohort of animals in this study is consistent with reported incidences of dermatitis in colonies of C57BL/6J mice [77–79].

The incidence of lesion development in our population of P rats was relatively high (over 17%) as compared to the reported incidence in humans (as high as 3%) [2], but the true frequency of the disorder in humans is likely underestimated as individuals work to hide the disorder to avoid chastisement or teasing [4,8–15]. Because this stock of rat has been selected to be prone to addictive behaviors [65,66], it is possible that the incidence is higher than in the general human population, but the incidence of almost 20% provides interesting avenues for potential future research directions.

Most animal model studies of hair loss and dermatitis are therapeutic studies, evaluating potential treatments of affected humans and animals. Although treatments have value for individual humans and other animals that are affected by these body-focused repetitive disorders, the ability to identify at-risk individuals and intervene before the disorders become established would be beneficial. The spray test has been implemented to identify at-risk C57BL/6J mice and to test nutritional interventions that may prevent the onset of these body-focused repetitive disorders [50]. However, as the C57BL/6J mouse is an inbred strain of mouse, the translatability of these findings to a heterogeneous population, such as humans, may be limited. The genetic diversity of the selectively bred P rat provides a better approximation of the genetic diversity of the human population, suggesting that preventative and therapeutic studies using this animal model will result in findings that are more translatable. For example, our laboratory recently reported on how the use of nail trims can improve outcomes for ulcerative dermatitis in the P rat [80], but completion of this study required waiting for animals to develop ulcerative dermatitis prior to enrollment in the study. To complete this study, it took over 5 years, due to the relatively low incidence in the colony. The ability to screen young rats who are at risk would allow the development of a case-control study with young females with similar behavioral profiles that can be used in a study comparing potential therapeutics.

Future directions for this model include the development of studies to test the effectiveness of therapeutic interventions (such as nutritional interventions, as has been suggested in mice [50]) in the prevention of the development of dermatitis in at-risk individuals. Genetic and epigenetic factors could be determined, providing insight into predisposing neurochemical or neurophysiological alterations observed in subjects that display or do not display TTM-like behaviors. For instance, the P rat has several neurochemical differences from their non-preferring (NP) counterparts, including serotonin and dopamine deficiencies, as well as that of associated receptors and transporters, in certain meso-corticolimbic nuclei (for a review of multiple neurotransmitter and receptor differences, see [55,65]). It is recognized that alcohol use disorders, OCD, and depression are often linked to imbalances in serotonin level and function [81,82].

5. Conclusions

The results of this study suggest that similar to the C57BL/6J mouse, the selectively bred P rat is a potential animal model for evaluating interventions to treat or prevent TTM behaviors. In addition to the P rat sharing clinical signs and behaviors consistent with what has been described for women affected by TTM, the spray test, and identification of rats who engage in increased scratching and oral grooming behaviors may allow for the identification of at-risk rats and the development of strategies for the prevention or early treatment of this disorder. Additional evaluation of behavioral and physiologic interactions will help to characterize the value of this model in the development of prevention and intervention strategies for TTM and other similar body-focused repetitive disorders.

Author Contributions: Conceptualization, D.H.; methodology, D.H.; validation, D.H. and A.P.; formal analysis, D.H.; resources, R.B. and D.H.; data curation, D.H.; writing—original draft preparation, A.P.; writing—review and editing, D.H., R.B. and A.P.; supervision, D.H.; project administration, D.H.; funding acquisition, R.B. and D.H. All authors have read and agreed to the published version of the manuscript.

Funding: This work was supported in part by the NIH/NIAAA P06AA007611 (Indiana Alcohol Research Center).

Institutional Review Board Statement: The study was conducted according to the applicable federal regulations in the United States of America, and approved by the Institutional Animal Care and Use Committee of Indiana University School of Medicine (study 10041, approved 27 September 2014).

Informed Consent Statement: Not applicable.

Data Availability Statement: The data presented in this study are available in Table 2 of this manuscript.

Acknowledgments: The authors would like to thank Melissa Swan and Ronald Wheeler for their assistance in the coordination of data collection. We would also like to thank the animal care staff for providing excellent support for the animals used in this study.

Conflicts of Interest: The authors declare no conflict of interest.

References

1. Association, A.P. *Diagnostic and Statistical Manual of Mental Disorders (DSM-V)*; American Psychiatric Publishing: Arlington, VA, USA, 2013.
2. Anwar, S.; Jafferany, M. Trichotillomania: A psychopathological perspective and the psychiatric comorbidity of hair pulling. *Acta Derm. Alp. Pannonica Adriat.* **2019**, *28*, 33–36. [CrossRef]
3. King, R.A.; Scahill, L.; Vitulano, L.A.; Schwab-Stone, M.; Tercyak, K.P., Jr.; Riddle, M.A. Childhood trichotillomania: Clinical phenomenology, comorbidity, and family genetics. *J. Am. Acad. Child Adolesc. Psychiatry* **1995**, *34*, 1451–1459. [CrossRef]
4. Pinhas, L.; Geist, R.; Katzman, D.K. Trichotillomania and anorexia nervosa in an adolescent female: A case study of comorbidity. *J. Adolesc. Health* **1996**, *19*, 141–144. [CrossRef]
5. Reeve, E.A.; Bernstein, G.A.; Christenson, G.A. Clinical characteristics and psychiatric comorbidity in children with trichotillomania. *J. Am. Acad. Child Adolesc. Psychiatry* **1992**, *31*, 132–138. [CrossRef] [PubMed]
6. Snorrason, I.; Belleau, E.L.; Woods, D.W. How related are hair pulling disorder (trichotillomania) and skin picking disorder? A review of evidence for comorbidity, similarities and shared etiology. *Clin. Psychol. Rev.* **2012**, *32*, 618–629. [PubMed]
7. Stein, D.J.; Grant, J.E.; Franklin, M.E.; Keuthen, N.; Lochner, C.; Singer, H.S.; Woods, D.W. Trichotillomania (hair pulling disorder), skin picking disorder, and stereotypic movement disorder: Toward DSM-V. *Depress. Anxiety* **2010**, *27*, 611–626. [CrossRef]
8. Walther, M.R.; Snorrason, I.; Flessner, C.A.; Franklin, M.E.; Burkel, R.; Woods, D.W. The trichotillomania impact project in young children (TIP-YC): Clinical characteristics, comorbidity, functional impairment and treatment utilization. *Child Psychiatry Hum. Dev.* **2014**, *45*, 24–31. [CrossRef] [PubMed]
9. Franklin, M.E.; Zagrabbe, K.; Benavides, K.L. Trichotillomania and its treatment: A review and recommendations. *Expert Rev. Neurother* **2011**, *11*, 1165–1174. [CrossRef]
10. Krishnan, K.R.; Davidson, J.R.; Guajardo, C. Trichotillomania—A review. *Compr. Psychiatry* **1985**, *26*, 123–128. [CrossRef]
11. Kuprevich, C.L.; Nagra, B.; Rosenbaum, R. Trichotillomania: A brief review. *Del. Med. J.* **2005**, *77*, 253–257.
12. Morales-Fuentes, B.; Camacho-Maya, U.; Coll-Clemente, F.L.; Vazquez-Minero, J.C. Trichotillomania, recurrent trichobezoar and Rapunzel syndrome: Case report and literature review. *Cir. Cir.* **2010**, *78*, 265–266.
13. Nuss, M.A.; Carlisle, D.; Hall, M.; Yerneni, S.C.; Kovach, R. Trichotillomania: A review and case report. *Cutis* **2003**, *72*, 191–196.
14. Odlaug, B.L.; Grant, J.E. Clinical characteristics and medical complications of pathologic skin picking. *Gen. Hosp. Psychiatry* **2008**, *30*, 61–66. [CrossRef]
15. Tay, Y.K.; Levy, M.L.; Metry, D.W. Trichotillomania in childhood: Case series and review. *Pediatrics* **2004**, *113*, e494–e498. [CrossRef]
16. Adam, B.S.; Kashani, J.H. Trichotillomania in children and adolescents: Review of the literature and case report. *Child Psychiatry Hum. Dev.* **1990**, *20*, 159–168. [CrossRef]
17. Bloch, M.H.; Landeros-Weisenberger, A.; Dombrowski, P.; Kelmendi, B.; Wegner, R.; Nudel, J.; Pittenger, C.; Leckman, J.F.; Coric, V. Systematic review: Pharmacological and behavioral treatment for trichotillomania. *Biol. Psychiatry* **2007**, *62*, 839–846. [CrossRef]
18. Chamberlain, S.; Ipser, J.; Stein, D.; Fineberg, N. Regarding "Systematic review: Pharmacological and behavioral treatment for trichotillomania". *Biol. Psychiatry* **2008**, *63*, e33, author reply e34–e35. [CrossRef] [PubMed]
19. Hamiel, D.; Yoffe, A.; Roe, D. Trichotillomania and the mourning process: A case report and review of the psychodynamics. *Isr J. Psychiatry Relat. Sci.* **1999**, *36*, 192–199; discussion 200–202.
20. Jones, G.C.; Coutinho, K.; Anjaria, D.; Hussain, N.; Dholakia, R. Treatment of recurrent Rapunzel syndrome and trichotillomania: Case report and literature review. *Psychosomatics* **2010**, *51*, 443–446. [CrossRef]
21. Kumar, D.; Singh, H.; Trivedi, J.K. Trichotillomania-a brief review and case report. *Indian J. Psychiatry* **1982**, *24*, 95–98. [PubMed]
22. Bordnick, P.S.; Thyer, B.A.; Ritchie, B.W. Feather picking disorder and trichotillomania: An avian model of human psychopathology. *J. Behav. Exp. Psychiatry* **1994**, *25*, 189–196. [CrossRef]
23. Feusner, J.D.; Hembacher, E.; Phillips, K.A. The mouse who couldn't stop washing: Pathologic grooming in animals and humans. *CNS Spectr.* **2009**, *14*, 503–513. [CrossRef]

24. Gaskins, L.A.; Hungerford, L. Nonmedical factors associated with feather picking in pet psittacine birds. *J. Avian Med. Surg.* **2014**, *28*, 109–117. [CrossRef] [PubMed]
25. Grindlinger, H.M.; Ramsay, E. Compulsive feather picking in birds. *Arch. Gen. Psychiatry* **1991**, *48*, 857. [CrossRef]
26. Jenkins, J.R. Feather picking and self-mutilation in psittacine birds. *Vet. Clin. N. Am. Exot. Anim. Pract.* **2001**, *4*, 651–667. [CrossRef]
27. Colton, S.; Fraley, G.S. The effects of environmental enrichment devices on feather picking in commercially housed Pekingducks. *Poult. Sci.* **2014**, *93*, 2143–2150. [CrossRef] [PubMed]
28. Riber, A.B.; Hinrichsen, L.K. Feather eating and its associations with plumage damage and feathers on the floor in commercial farms of laying hens. *Animal* **2016**, *10*, 1218–1224. [CrossRef]
29. Rubinstein, J.; Lightfoot, T. Feather loss and feather destructive behavior in pet birds. *Vet. Clin. N. Am. Exot. Anim. Pract.* **2014**, *17*, 77–101. [CrossRef]
30. Costa, P.; Macchi, E.; Valle, E.; De Marco, M.; Nucera, D.M., Gasco, L.; Schiavone, A. An association between feather damaging behavior and corticosterone metabolite excretion in captive African grey parrots (*Psittacus erithacus*). *PeerJ.* **2016**, *4*, e2462. [CrossRef]
31. Nasr, M.A.F.; Mohammed, H.; Hassan, R.A.; Swelum, A.A.; Saadeldin, I.M. Does light intensity affect the behavior, welfare, performance, meat quality, amino acid profile, and egg quality of Japanese quails? *Poult. Sci.* **2019**, *98*, 3093–3102. [CrossRef]
32. Mertens, P.A.; Torres, S.; Jessen, C. The effects of clomipramine hydrochloride in cats with psychogenic alopecia: A prospective study. *J. Am. Anim. Hosp. Assoc.* **2006**, *42*, 336–343. [CrossRef] [PubMed]
33. Sawyer, L.S.; Moon-Fanelli, A.A.; Dodman, N.H. Psychogenic alopecia in cats: 11 cases (1993–1996). *J. Am. Vet. Med. Assoc.* **1999**, *214*, 71–74. [PubMed]
34. Swanepoel, N.; Lee, E.; Stein, D.J. Psychogenic alopecia in a cat: Response to clomipramine. *J. S. Afr. Vet. Assoc.* **1998**, *69*, 22. [CrossRef] [PubMed]
35. Virga, V. Behavioral dermatology. *Vet. Clin. N. Am. Small Anim. Pract.* **2003**, *33*, 231–251. [CrossRef]
36. Waisglass, S.E.; Landsberg, G.M.; Yager, J.A.; Hall, J.A. Underlying medical conditions in cats with presumptive psychogenic alopecia. *J. Am. Vet. Med. Assoc.* **2006**, *228*, 1705–1709. [CrossRef]
37. Kramer, J.A.; Mansfield, K.G.; Simmons, J.H.; Bernstein, J.A. Psychogenic alopecia in rhesus macaques presenting as focally extensive alopecia of the distal limb. *Comp. Med.* **2011**, *61*, 263–268.
38. Novak, M.A.; Kelly, B.J.; Bayne, K.; Meyer, J.S. Behavioral Disorders of Nonhuman Primates. In *Nonhuman Primates in Biomedical Research: Biology and Management*, 2nd ed.; Abee, C.R., Mansfield, K., Tardif, S., Morris, T., Eds.; Elsevier: Waltham, MA, USA, 2012; Volume 1, pp. 177–196.
39. Novak, M.A.; Meyer, J.S. Alopecia: Possible causes and treatments, particularly in captive nonhuman primates. *Comp. Med.* **2009**, *59*, 18–26.
40. Dodman, N.H.; Shuster, L.; White, S.D.; Court, M.H.; Parker, D.; Dixon, R. Use of narcotic antagonists to modify stereotypic self-licking, self-chewing, and scratching behavior in dogs. *J. Am. Vet. Med. Assoc.* **1988**, *193*, 815–819.
41. Hewson, C.J.; Luescher, U.A.; Parent, J.M.; Conlon, P.D.; Ball, R.O. Efficacy of clomipramine in the treatment of canine compulsive disorder. *J. Am. Vet. Med. Assoc.* **1998**, *213*, 1760–1766.
42. Shumaker, A.K. Diagnosis and Treatment of Canine Acral Lick Dermatitis. *Vet. Clin. N. Am. Small Anim. Pract.* **2019**, *49*, 105–123. [CrossRef]
43. Dik, N.; Fermont, M.A.; Kuller, W.I.; van Leengoed, L.A. Ear injuries in young pigs. *Tijdschr. Diergeneeskd* **2011**, *136*, 870–873.
44. Taylor, N.R.; Main, D.C.; Mendl, M.; Edwards, S.A. Tail-biting: A new perspective. *Vet. J.* **2010**, *186*, 137–147. [CrossRef] [PubMed]
45. Sonoda, L.T.; Fels, M.; Oczak, M.; Vranken, E.; Ismayilova, G.; Guarino, M.; Viazzi, S.; Bahr, C.; Berckmans, D.; Hartung, J. Tail biting in pigs—Causes and management intervention strategies to reduce the behavioural disorder: A review. *Berl. Munch. Tierarztl. Wochenschr.* **2013**, *126*, 104–112. [PubMed]
46. Dodman, N.H.; Normile, J.A.; Shuster, L.; Rand, W. Equine self-mutilation syndrome (57 cases). *J. Am. Vet. Med. Assoc.* **1994**, *204*, 1219–1223.
47. Murray, M.J.; Crowell-Davis, S.L. Psychogenic colic in a horse. *J. Am. Vet. Med. Assoc.* **1985**, *186*, 381–383. [PubMed]
48. Osman, S.A.; Hanafy, A.; Amer, S.E. Clinical and therapeutic studies on mange in horses. *Vet. Parasitol.* **2006**, *141*, 191–195. [CrossRef]
49. Bechard, A.; Meagher, R.; Mason, G. Environmental enrichment reduces the likelihood of alopecia in adult C57BL/6J mice. *J. Am. Assoc. Lab. Anim. Sci.* **2011**, *50*, 171–174.
50. Dufour, B.D.; Adeola, O.; Cheng, H.W.; Donkin, S.S.; Klein, J.D.; Pajor, E.A.; Garner, J.P. Nutritional up-regulation of serotonin paradoxically induces compulsive behavior. *Nutr. Neurosci.* **2010**, *13*, 256–264. [CrossRef] [PubMed]
51. Garner, J.P.; Dufour, B.; Gregg, L.E.; Weisker, S.M.; Mench, J.A. Social and husbandry factors affecting the prevalence and severity of barbering ('whisker trimming') by laboratory mice. *Appl. Anim. Behav. Sci.* **2004**, *89*, 263–282. [CrossRef]
52. Kalueff, A.V.; Minasyan, A.; Keisala, T.; Shah, Z.H.; Tuohimaa, P. Hair barbering in mice: Implications for neurobehavioural research. *Behav. Processes* **2006**, *71*, 8–15. [CrossRef]
53. Sarna, J.R.; Dyck, R.H.; Whishaw, I.Q. The Dalila effect: C57BL6 mice barber whiskers by plucking. *Behav. Brain Res.* **2000**, *108*, 39–45. [CrossRef]
54. Nicholson, A.; Malcolm, R.D.; Russ, P.L.; Cough, K.; Touma, C.; Palme, R.; Wiles, M.V. The response of C57BL/6J and BALB/cJ mice to increased housing density. *J. Am. Assoc. Lab. Anim. Sci.* **2009**, *48*, 740–753. [PubMed]
55. Bell, R.L.; Sable, H.J.; Colombo, G.; Hyytia, P.; Rodd, Z.A.; Lumeng, L. Animal models for medications development targeting alcohol abuse using selectively bred rat lines: Neurobiological and pharmacological validity. *Pharm. Biochem. Behav.* **2012**, *103*, 119–155. [CrossRef]

56. Murphy, J.M.; Stewart, R.B.; Bell, R.L.; Badia-Elder, N.E.; Carr, L.G.; McBride, W.J.; Lumeng, L.; Li, T.K. Phenotypic and genotypic characterization of the Indiana University rat lines selectively bred for high and low alcohol preference. *Behav. Genet.* **2002**, *32*, 363–388. [CrossRef] [PubMed]
57. Hackam, D.G.; Redelmeier, D.A. Translation of research evidence from animals to humans. *JAMA* **2006**, *296*, 1731–1732. [CrossRef] [PubMed]
58. McGonigle, P.; Ruggeri, B. Animals models of human disease: Challenges in enabling translation. *Biochem. Pharm.* **2014**, *87*, 162–171. [CrossRef]
59. Sundberg, J.P.; Schofield, P.N. Living inside the box: Environmental effects on mouse models of human disease. *Dis. Model. Mech.* **2018**, *11*, dmm035360. [CrossRef]
60. Sunagar, R.; Kumar, S.; Namjoshi, P.; Rosa, S.J.; Hazlett, K.R.O.; Gosselin, E.J. Evaluation of an outbred mouse model for Francisella tularensis vaccine development and testing. *PLoS ONE* **2018**, *13*, e0207587. [CrossRef]
61. Frischknecht, H.R.; Siegfried, B.; Waser, P.G. Opioids and behavior: Genetic aspects. *Experientia* **1988**, *44*, 473–481. [CrossRef]
62. Tirelli, E.; Witkin, J.M. Verticalization of behavior elicited by dopaminergic mobilization is qualitatively different between C57BL/6J and DBA/2J mice. *Psychopharmacology* **1994**, *116*, 191–200. [CrossRef]
63. Warden, A.S.; DaCosta, A.; Mason, S.; Blednov, Y.A.; Mayfield, R.D.; Harris, R.A. Inbred substrain difference influence neuroimmune response and drinking behavior. *Alcohol. Clin. Exp. Res.* **2020**, *44*, 1760–1768. [CrossRef] [PubMed]
64. Lumeng, L.; Hawkins, T.D.; Li, T.-K. New strains of rats with alcohol preference and non-preference. In *Alcohol and Aldehyde Metabolizing Systems*; Thurman, R.G., Williamson, J.R., Drott, H., Chance, B., Eds.; Academic Press: New York, NY, USA, 1977; Volume 3, pp. 537–544.
65. Bell, R.L.; Hauser, S.; Rodd, Z.A.; Liang, T.; Sari, Y.; McClintick, J.; Rahman, S.; Engleman, E.A. A genetic animal model of alcoholism for screening medications to treat addiction. *Int. Rev. Neurobiol.* **2016**, *126*, 179–261. [PubMed]
66. Bell, R.L.; Hauser, S.R.; Liang, T.; Sari, Y.; Maldonado-Devincci, A.; Rodd, Z.A. Rat animal models for screening medications to treat alcohol use disorders. *Neuropharmacology* **2017**, *122*, 201–243. [CrossRef] [PubMed]
67. Lumeng, L.; Waller, M.B.; McBride, W.J.; Li, T.-K. Different sensitivities to ethanol in alcohol-preferring and –nonpreferring rats. *Pharm. Biochem. Behav.* **1982**, *16*, 125–130. [CrossRef]
68. Lumeng, L.; Li, T.-K. The development of metabolic tolerance in the alcohol-preferring P rats: Comparison of forced and free-choice drinking of ethanol. *Pharm. Biochem. Behav.* **1986**, *25*, 1013–1020. [CrossRef]
69. Kampov-Polevoy, A.B.; Matthews, D.B.; Gause, L.; Morrow, A.L.; Overstreet, D.H. P rats develop physical dependence on alcohol via voluntary drinking: Changes in seizure thresholds, anxiety, and patterns of alcohol drinking. *Alcohol. Clin. Exp. Res.* **2000**, *24*, 278–284. [CrossRef]
70. Rodd, Z.A.; Bell, R.L.; Sable, H.J.; Murphy, J.M.; McBride, W.J. Recent advances in animal models of alcohol craving and relapse. *Pharm. Biochem. Behav.* **2004**, *79*, 439–450. [CrossRef]
71. Bell, R.L.; Stewart, R.B.; Woods, J.E., II; Lumeng, L.; Li, T.-K.; Murphy, J.M.; McBride, W.J. Responsivity and development of tolerance to the motor impairing effects of moderate doses of ethanol in Alcohol- Preferring (P) and –Nonpreferring (NP) rat lines. *Alcohol. Clin. Exp. Res.* **2001**, *25*, 644–650. [CrossRef]
72. Stewart, R.B.; Kurtz, D.L.; Zweifel, M.; Li, T.-K.; Froehlich, J.C. Differences in the hypothermic response to ethanol in rats selectively bred for oral ethanol preference and nonpreference. *Psychopharmacology* **1992**, *106*, 169–174. [CrossRef]
73. Kurtz, D.L.; Stewart, R.B.; Zweifel, M.; Li, T.-K.; Froehlich, J.C. Genetic differences in tolerance and sensitization to the sedative/hypnotic effects of alcohol. *Pharm. Biochem. Behav.* **1996**, *53*, 585–591. [CrossRef]
74. Bell, R.L.; Rodd, Z.A.; Smith, R.J.; Toalston, J.E.; Franklin, K.M.; McBride, W.J. Modeling binge-like ethanol drinking by peri-adolescent and adult P rats. *Pharm. Biochem. Behav.* **2011**, *100*, 90–97. [CrossRef] [PubMed]
75. Bell, R.L.; Rodd, Z.A.; Engleman, E.A.; Toalston, J.E.; McBride, W.J. Scheduled access alcohol drinking by alcohol-preferring (P) and high alcohol-drinking (HAD) rats: Modeling adolescent and adult binge-like drinking. *Alcohol* **2014**, *48*, 225–234. [CrossRef] [PubMed]
76. Hickman, D.L.; Swan, M. Use of a body condition score technique to assess health status in a rat model of polycystic kidney disease. *J. Am. Assoc. Lab. Anim. Sci.* **2010**, *49*, 155–159. [PubMed]
77. Kastenmayer, R.J.; Fain, M.A.; Perdue, K.A. A retrospective study of idiopathic ulcerative dermatitis in mice with a C57BL/6 background. *J. Am. Assoc. Lab. Anim. Sci.* **2006**, *45*, 8–12. [PubMed]
78. Adams, S.C.; Garner, J.P.; Felt, S.A.; Geronimo, J.T.; Chu, D.K. A "Pedi" Cures All: Toenail Trimming and the Treatment of Ulcerative Dermatitis in Mice. *PLoS ONE* **2016**, *11*, e0144871. [CrossRef]
79. Alvarado, C.G.; Franklin, C.L.; Dixon, L.W. Retrospective Evaluation of Nail Trimming as a Conservative Treatment for Ulcerative Dermatitis in Laboratory Mice. *J. Am. Assoc. Lab. Anim. Sci.* **2016**, *55*, 462–466.
80. Skiles, B.A.; Boehm, C.A.; Peveler, J.P.; Hickman, D.L. Evaluation of treatment options for ulcerative dermatitis in the P rat. *J. Am. Assoc. Lab. Anim. Sci.* **2021**, *60*, 311–318. [CrossRef]
81. Lissemore, J.I.; Sookman, D.; Gravel, P.; Berney, A.; Barsoum, A.; Diksic, M.; Nordahl, T.E.; Pinard, G.; Sibon, I.; Cottraux, J.; et al. Brain serotonin synthesis capacity in obsessive-compulsive disorder: Effects of cognitive behavioral therapy and sertraline. *Transl. Psychiatry* **2018**, *8*, 82. [CrossRef]
82. Underwood, M.D.; Kassir, S.A.; Bakalian, M.J.; Galfalvy, H.; Dwork, A.J.; Mann, J.J.; Arango, V. Serotonin receptors and suicide, major depression, alcohol use disorder and reported early life adversity. *Transl. Psychiatry* **2018**, *8*, 279. [CrossRef]

veterinary sciences

MDPI

Article

Ischemia-Induced Cognitive Impairment Is Improved via Remyelination and Restoration of Synaptic Density in the Hippocampus after Treatment with COG-Up® in a Gerbil Model of Ischemic Stroke

Tae-Kyeong Lee [1,†], Junkee Hong [2,†], Ji-Won Lee [3], Sung-Su Kim [3], Hyejin Sim [4], Jae-Chul Lee [4], Dae Won Kim [5], Soon Sung Lim [1], Il Jun Kang [1,*] and Moo-Ho Won [4,*]

1 Department of Food Science and Nutrition, Hallym University, Chuncheon 24252, Korea; tk_lee@hallym.ac.kr (T.-K.L.); limss@hallym.ac.kr (S.S.L.)
2 Department of Global Innovative Drug, Chung-Ang University, Seoul 06974, Korea; jk.hong@famenity.com
3 Famenity Co., Ltd., Uiwang 16006, Korea; jiwon.lee@famenity.com (J.-W.L.); sungsu.kim@famenity.com (S.-S.K.)
4 Department of Neurobiology, School of Medicine, Kangwon National University, Chuncheon 24341, Korea; janny20@naver.com (H.S.); anajclee@kangwon.ac.kr (J.-C.L.)
5 Department of Biochemistry and Molecular Biology, Research Institute of Oral Sciences, College of Dentistry, Kangnung-Wonju National University, Gangneung 25457, Korea; kimdw@gwnu.ac.kr
* Correspondence: ijkang@hallym.ac.kr (I.J.K.); mhwon@kangwon.ac.kr (M.-H.W.); Tel.: +82-33-248-2135 (I.J.K.); +82-33-250-8891 (M.-H.W.); Fax: +82-33-255-4787 (I.J.K.); +82-33-256-1614 (M.-H.W.)
† Tae-Kyeong Lee and Junkee Hong have equally contributed to this work.

Citation: Lee, T.-K.; Hong, J.; Lee, J.-W.; Kim, S.-S.; Sim, H.; Lee, J.-C.; Kim, D.W.; Lim, S.S.; Kang, I.J.; Won, M.-H. Ischemia-Induced Cognitive Impairment Is Improved via Remyelination and Restoration of Synaptic Density in the Hippocampus after Treatment with COG-Up® in a Gerbil Model of Ischemic Stroke. *Vet. Sci.* **2021**, *8*, 321. https://doi.org/10.3390/vetsci8120321

Academic Editors: Ana Faustino and Paula A. Oliveira

Received: 18 November 2021
Accepted: 9 December 2021
Published: 10 December 2021

Abstract: Cerebrovascular disease such as ischemic stroke develops cognitive impairment due to brain tissue damage including neural loss, demyelination and decrease in synaptic density. In the present study, we developed transient ischemia in the forebrain of the gerbil and found cognitive impairment using the Barnes maze test and passive avoidance test for spatial memory and learning memory, respectively. In addition, neuronal loss/death was detected in the Cornu Ammonis 1 (CA1) region of the gerbil hippocampus after the ischemia by cresyl violet histochemistry, immunohistochemistry for neuronal nuclei and histofluorescence with Fluoro-Jade B. Furthermore, in the CA1 region following ischemia, myelin and vesicular synaptic density were significantly decreased using immunohistochemistry for myelin basic protein and vesicular glutamate transporter 1. In the gerbils, treatment with COG-up® (a combined extract of *Erigeron annuus* (L.) Pers. and *Brassica oleracea* Var.), which was rich in scutellarin and sinapic acid, after the ischemia, significantly improved ischemia-induced decline in memory function when compared with that shown in gerbils treated with vehicle after the ischemia. In the CA1 region of these gerbils, COG-up® treatment significantly promoted the remyelination visualized using immunohistochemistry myelin basic protein, increased oligodendrocytes visualized using a receptor-interacting protein, and restored the density of glutamatergic synapses visualized using double immunofluorescence for vesicular glutamate transporter 1 and microtubule-associated protein, although COG-up® treatment did not protect pyramidal cells (principal neurons) located in the CA1 region form the ischemic insult. Considering the current findings, a gerbil model of ischemic stroke apparently showed cognitive impairment accompanied by ischemic injury in the hippocampus; also, COG-up® can be employed for improving cognitive decline following ischemia-reperfusion injury in brains.

Keywords: Cornu Ammonis 1; glutaminergic synapse; ischemia-reperfusion injury; memory function; oligodendrocyte; pyramidal cells

1. Introduction

Ischemia/reperfusion injury in the brain is a common property of ischemic stroke, which involves a period of impaired blood circulation in the brain, followed by restoration of perfusion through medical intervention [1,2]. It was acknowledged that cognitive impairment is developed by brain ischemia/reperfusion injury which involves various neurological and behavioral changes such as paralysis in limbs, vertigo, confusion, cognitive dysfunction [3–5].

It is well accepted that the hippocampus plays a pivotal role in memory and cognitive function [6,7]. Therefore, ischemia/reperfusion in the hippocampus induces neuronal damage or death, which is accompanied by decreases in synaptic density, demyelination and axonal damage which can lead to memory and cognitive deficits [3,8–10]. It is well established that, in the hippocampus of the gerbil, delayed neuronal death is induced in the CA 1 region four to five days after in five-minute transient ischemia (TI) in the forebrain [8,11,12]. For this property, many studies have utilized this model to develop cognitive impairment followed by delayed neuronal death [3,8,13].

Many precedent studies show that the enhancement of remyelination after ischemic insults is important for the functional recovery of memory and cognition [3,8,14]. In particular, oligodendrocytes, a type of glial cells, form myelin sheath enveloping axons in the brain and spinal cord in order to accelerate neural transmission by saltatory conduction, and newly generated oligodendrocytes play an important role in remyelination [15–17]. In addition, accumulated clinical and preclinical data demonstrate that change in glutamate neurotransmission in the brain may be linked to cognitive impairment [18,19].

Numerous studies have reported the beneficial effects of medicinal herbs. *Erigeron annuus* (L.) Pers. (EALP) belongs to the Asteraceae family, and its extract shows anti-inflammatory effects in lipopolysaccharide-induced activated macrophage and carrageen-induced acute inflammation in rat paws [20]. In addition, the genus *Brassica* belonging to the Brassicaceae family has beneficial attributes. For instance, the extract of *Brassica oleracea* Var. *Italica* attenuates amyloid $\beta_{1\text{–}42}$-induced learning and memory impairment in mice [21]. However, to the best of our knowledge, experiments on the effects of the extracts from *Erigeron annuus* (L.) Pers. and *Brassica oleracea* Var. on the improvement of cognitive impairment following TI in rodent brains have rarely been conducted. Therefore, herewith we developed transient ischemia in the forebrain using gerbils, analyzed COG-up® (a combined extract of *Erigeron annuus* (L.) Pers. and *Brassica oleracea* Var.) and investigated whether COG-up® improve cognitive impairment induced by TI in the hippocampus. In addition, we examined the effects of COG-up® on remyelination and restoration of synaptic density in the damaged hippocampus.

2. Materials and Methods

2.1. Experimental Animals

Eighty-four male gerbils at the age of six months (85 ± 5 g of body weight) were provided by Experimental Animal Center of Kangwon National University (Chuncheon, Gangwon, Korea). The gerbils were housed in conventional room with optimum conditions (24 ± 1 °C of room temperature; 50 ± 5% of relative humidity). A steady cycle of light and dark was controlled every 12 h, and pellet feed (DBL Co. Ltd., Chungbuk, Korea) and water were freely accessible.

All experimental processes were according to the guidelines described in the "Current International Laws and Policies", a part of the "Guide for the Care and Use of Laboratory Animals". Approval for the experimental protocol was sanctioned by Institutional Animal Care and Use Committee of Kangwon National University (Chuncheon, Korea) on 18 February 2020 (approval no., KW-200113-1).

2.2. Experimental Groups

In this study, four groups were used: (1) sham+vehicle group (n = 21) which was given sham surgery and treated with vehicle (saline); (2) TI+vehicle group (n = 21) which was given TI surgery and treated with vehicle; (3) sham+COG-up® group (n = 21) which was undergone sham operation and treated with 100 mg/kg COG-up®; and (4) TI+COG-up® group (n = 21) which was undergone TI operation and treated with 100 mg/kg COG-up®.

2.3. Qualitative Analysis of COG-Up®

COG-up® (a combined extract of *Erigeron annuus* (L.) Pers. and *Brassica oleracea* Var.) was provided by Famenity (Uiwang, Korea). Scutellarin (Glentham, Corsham, UK) standard sample and test sample (COG-up®) were precisely weighed and dissolved. These samples (5 μL, respectively) were subjected to HPLC (Agilent 1260 Infinity II Prime LC System) (Agilent Technologies, Inc., Waldbronn, Germany) using Discovery C18 column (diameter of 4.6 mm, length of 250 mm) (Sigma-Aldrich Co., St. Louis, MO, USA), which was filled with octadecylsilyl silica gel (diameter of 5 μm) at 1.0 mL/min of flow rate. Optimum HPLC separation was achieved at 30 °C. UV wavelength was 335 nm for the scutellarin. Phosphoric acid (Junsei Chemical Co., Ltd., Tokyo, Japan) was mixed with distilled water and methanol (JT Baker, Phillipsburg, NJ, USA) which were used as a mobile phase. Mobile phase condition for the scutellarin using A (0.5% phosphoric acid in water) and B (methanol) was as follows: 0–5 min (25% B), 5–25 min (25–55% B), 25–30 min (55–100% B), 30–32 min (100% B), 32–33 min (100–25% B), 33–40 min (25% B).

For high analytical performance, pretreatment using alkaline solvent was required for sinapic acid. Sinapic acid (Sigma-Aldrich Co., St. Louis, MO, USA) standard sample and the test sample (COG-up®) were precisely weighed and dissolved. These samples (5 μL) were subjected to HPLC (Agilent 1260 infinity II, Agilent Technologies, Inc., Santa Clara, CA, USA) using the same column and flow rate. Optimum HPLC separation was achieved at 30 °C. UV wavelength was 320 nm for sinapic acid. The mobile phase condition for sinapic acid using the same solution A and B was as follows: 0–15 min (10–30% B), 15–25 min (30% B), 25–26 min (30–50% B), 26–27 min (50–90% B), 27–30 min (90% B), 30–32 min (90–10% B) and 32–37 min (10% B).

2.4. Induction of TI and Treatment with COG-Up®

The gerbils underwent TI in the forebrain to develop cognitive impairment due to ischemic damage [3]. As described previously [22], the animals were adequately anesthetized with 2.5% isoflurane (Hana Pharmaceutical Co. Ltd., Seoul, Korea) (in mixture of 67% N_2O and 33% O_2) using an inhaler. Under the anesthesia, an incision was made on the midline of the ventral neck. Bilateral common carotid arteries (BCCA) were freed from the vagus nerve and occluded with aneurysm clips (0.69 N) (Yasargil FE 723K; Aesculap, Tuttlingen, Germany) for five minutes. For the complete occlusion of the BCCA, the stop of blood flow was verified through observing in right and left central arteries (branches of the internal carotid arteries) located at both retinae using ophthalmoscope (HEINE K180®) (Heine Optotechnik; Herrsching, Germany). The aneurysm clips were removed at five minutes after ligating the BCCA, and the skin was sutured with 3–0 suture silk (Ethicon Inc, Somerville, NJ, USA). In this study, sham TI operation was performed like the identical surgical procedure, excepting BCCA ligation. In particular, body temperature was controlled at normothermia (37 ± 0.2 °C) during the surgical procedure using rectal temperature probe (TR-100) (Fine Science Tools Inc.; Foster City, CA, USA).

In this study, vehicle and COG-up® were orally administrated once a day five days to 30 days after TI.

2.5. Barnes Maze Test (BMT)

To examine spatial memory, BMT was daily conducted 26 to 30 days after TI (Figure 1). According to a published method [3], visual signs were located around the maze at a level that would be perceptible to the gerbils. Lights and a stereo speaker were installed beneath the ceiling in order to maintain brightness (220 lx) and background sound (85 dB). The gerbils were adapted to a refuge for two minutes on the first day of training. Each gerbil was given training three times per day with 15 min of intervals per day for four successive days. When the gerbil failed to find the refuge within 180 s, we escorted the gerbil toward the escape hole and let the gerbil stay for 35 s in the refuge. Thereafter, the gerbil was placed onto the center of the maze to explore an escape hole that is linked to the refuge. Each trial was finished when the gerbil had entered the refuge and stayed for 30 s. The substantial test was carried out one day after the final training. The refuge was removed, and, when the gerbil went to the entry area where the refuge had been previously located, the latency time was recorded within 90 s.

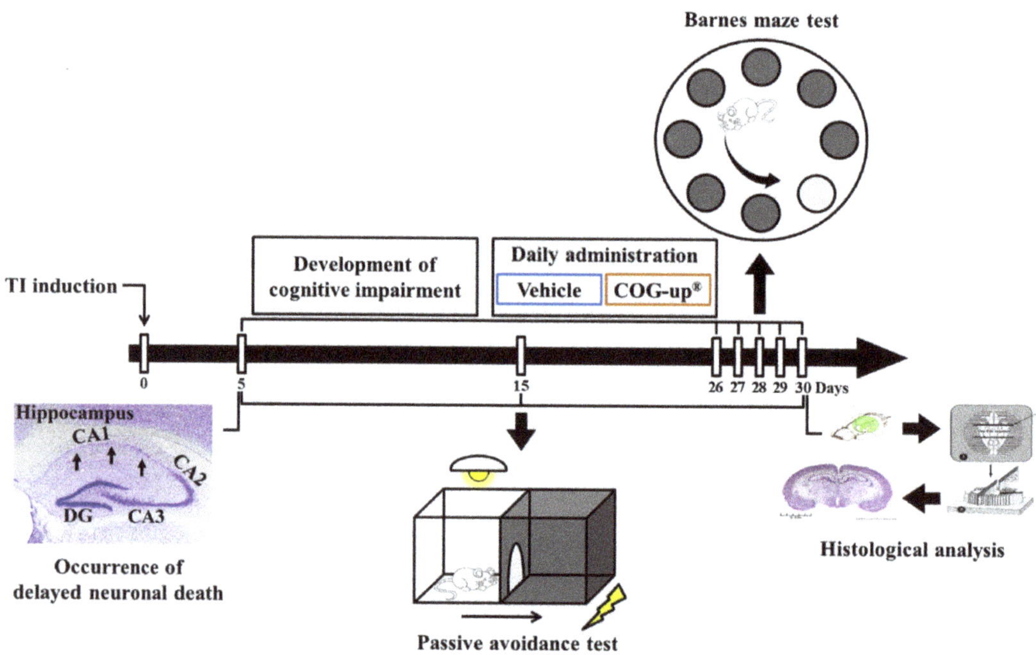

Figure 1. Experimental timeline. Cognitive impairment is developed for 25 days from five days after TI operation. Gerbils are daily treated with vehicle or COG-up® from five days to 30 days after TI. Delayed neuronal death is examined five days after TI. Passive avoidance test is conducted zero, five, 15 and 30 days after TI, and Barnes maze test is carried out from 26 days to 30 days after TI. Histological examination is performed five, 15 and 30 days after TI.

2.6. Passive Avoidance Test (PAT)

In accordance with some references [3,8], PAT was performed to investigate learning memory on day 5, day 15 and day 30 after TI (Figure 1). We used Gemini Avoidance System (GEM 392) (San Diego Instruments Inc., San Diego, CA, USA) for PAT. This apparatus consisted of a dark and light compartment that was connected through a vertical gate between the compartments. The experimental procedure was separated into two sections: training section and substantial trial section. The training was performed one day before each substantial trial. For the training, the gerbil was allowed to freely explore the dark and light compartment for 60 s. Thereafter, the vertical door was shut as soon as the gerbil entered the dark compartment, and the animals received an inescapable foot-shock of 0.5 mA for five seconds. For the substantial trial, each animal was put in the light compartment, and the latency time to go into the dark compartment was evaluated within 180 s.

2.7. Preparation of Brain Sections

The brain tissue sections containing the hippocampus were prepared to carry out histological analyses in the four groups five days (n = 7 in each group), 15 days (n = 7 in each group) and 30 days (n = 7 in each group) after TI. According to previously published methods [3,23], the gerbils of the four groups (n = 21, respectively) were profoundly anesthetized with 200 mg/kg pentobarbital sodium (intraperitoneal injection) (JW Pharmaceutical Co., Ltd., Seoul, South Korea) [22]. Under anesthesia, the gerbils were perfused (flow rate of 6 mL/min, total perfused volume of 60 mL) with 100 mM phosphate-buffered saline (PBS, pH 7.4) and subsequently fixed with paraformaldehyde solution (4% *w/v*; in 100 mM phosphate buffer (PB), pH 7.4) with same flow rate and total perfused volume via the ascending aorta. When their brains were fixed, the brains were harvested and individually immersed in the same fixative for post-fixation for six hours at room temperature. Thereafter, these brains were infiltrated with 30% *w/v* sucrose (in 100 mM PB) to be cryoprotected for 24 h at room temperature. The brain tissues were serially cut into 30-μm coronal sections using sliding microtome (SM2020 R) (Leica, Nussloch, Germany) equipped with freezing stage (BFS-40MP) (Physitemp Instruments Inc., Clifton, NJ, USA), Representative sections were selected at antero-posterior of −1.8 mm to 2.7 mm from the bregma with a reference of the "Brain Atlas of the Mongolian Gerbil (*Meriones unguiculatus*)" [24].

2.8. Cresyl Violet (CV) Staining

CV staining was carried out to examine the change in the distribution of hippocampal cells. As described in a previous paper [25], the brain sections were mounted onto the slide glasses coated with gelatin. After confirming their attachment, the prepared brain sections were immersed in 0.1% *w/v* CV acetate (Sigma-Aldrich Co., St. Louis, MO, USA) for 20 min. Thereafter these stained sections were briefly washed followed by decolorized in 50% ethanol for two minutes, subsequently dehydrated through successive incubation in 70%, 80%, 90%, 95% and 100% ethanol and then cleared in xylene. These stained sections were mounted with cover glasses.

To observe the changes of the CV-stained cells in gerbil hippocampus, images of the cells stained with CV were taken using microscope (BX53) (Olympus, Tokyo, Japan).

2.9. Fluoro-Jade B (FJB) Histofluorescence

FJB histofluorescence was performed to examine neuronal death (loss) in the hippocampus with reference to some previous reports [26,27]. The brains sections were put onto the slide glasses, which were coated with gelatin. These sections were soaked in 0.06% $KMnO_4$ (Sigma-Aldrich Co., St. Louis, MO, USA) for 20 min and briefly washed. After washing, these sections were soaked in 0.0004% FJB (Histo-chem Inc., Jefferson, AR, USA) for 30 min and rinsed. Thereafter, these sections were placed onto slide warmer until they were completely dried. The reacted sections were finally cleared in xylene and coverslipped with dibutyl phthalate polystyrene xylene (Sigma-Aldrich Co., St. Louis, MO, USA).

In order to count numbers of FJB positive cells, five sections per gerbil were chosen. According to a paper [28] with some modification, the stained sections were observed using epifluorescent microscope (Carl Zeiss, Göttingen, Germany) with blue excitation fluorescent filter (wavelength of 450–490 nm). Images of FJ B positive cells, which underwent degeneration (bright fluoresce) when compared with the background [26], were captured and counted in 250 × 250 μm at the middle of the CA1 region. Finally, the mean number of FJB positive cells was calculated using NIH Image 1.59 software (NIH, Bethesda, Rockville, MD, USA).

2.10. Immunohistochemistry

To examine changes in neurons, myelin and oligodendrocytes in the hippocampus, immunohistochemistry was carried out using avidin–biotin complex (ABC) method. According to previous studies [25,29] with minor modifications, the prepared sections were washed with PBS (pH 7.4) and soaked in 0.3% H_2O_2 (in 100 mM PBS, pH 7.4) for 30 min to block endogenous peroxidase activity. Subsequently, to block non-specific immunoreaction, these sections were incubated in five percent goat serum or horse (in 100 mM PBS, pH 7.4) for 30 min. Next, the sections were immunoreacted with primary antibodies (Table 1) for

24 h at 4 °C. Thereafter, the immunoreacted sections were incubated in each secondary antibody (Table 1) for two hours at room temperature followed by ABC (diluted 1:300) (Vector Laboratories, Burlingame, CA, USA). After briefly washing, these sections were reacted with 0.06% 3, 3′-diaminobenzidine tetrahydrochloride (Sigma-Aldrich Co, St Louis, MO, USA) (in 100 mM PBS containing 0.1% H_2O), washed with 100 mM PBS (pH 7.4), mounted onto the microscopic slides, dehydrated in 70%, 80%, 90%, 95% and 100% ethanol and cleared in xylene. Lastly, these stained sections were coverslipped with Canada balsam (Kanto Chemical Co., Inc., Tokyo, Japan).

Table 1. Primary and secondary antibodies for immunohistochemical staining.

Primary Antibodies	Dilution	Suppliers
Mouse anti-neuronal nuclei (NeuN)	1:1000	Chemicon, Temecula, CA, USA
Rabbit anti-myelin basic protein (MBP)	1:200	Abcam, Cambridge, UK
Mouse anti-receptor interacting protein (RIP)	1:200	Santa Cruz Biotechnology, Santa Cruz, CA, USA
Secondary Antibodies	**Dilution**	**Suppliers**
Biotinylated horse anti-mouse IgG	1:250	Vector Laboratories Inc., Burlingame, CA, USA
Biotinylated goat anti-rabbit IgG	1:250	Vector Laboratories Inc., Burlingame, CA, USA

For negative controls, adjacent tissue sections were immersed in pre-immune serum without each primary antibody. After testing, immunoreactive structures were not shown in the sections (data not shown).

In order to evaluate changes in neurons and oligodendrocytes, the numbers of NeuN immunoreactive neurons and RIP immunoreactive oligodendrocytes, five sections/gerbil were selected and analyzed using microscope (BX53) (Olympus, Tokyo, Japan) in the same way described above (in Section 2.8).

To evaluate changes in myelin, the optical density of MBP immunoreactive structure was presented. In accordance with a previous study [30], five sections/gerbils were chosen, and the image of MBP immunoreactive structure was taken using microscope (BX53) equipped with cellSens Standard software (Olympus, Tokyo, Japan). The captured image was converted to eight bits of grey scale (range, 0–255; from black to white) to measure grey scale intensity. Average density of MBP immunoreactive structure was computed using Image J software (version 1.46) (National Institutes of Health, Bethesda, Rockville, MD, USA). Lastly, the density of the MBP immunoreactive structure was presented as relative optical density (ROD) as percentage of the Sham+vehicle group (100%).

2.11. Double Immunofluorescence

Double immunofluorescence was performed to examine synaptic formation by examining the co-localization of vesicular glutamate transporter 1 (VGLUT-1; a marker for excitatory synapse) and microtubule-associated protein 2 (MAP2; a marker for apical dendrites) immunoreactive structures. As described previously [3], in brief, rabbit anti-VGLUT-1 (diluted 1:500) (Synaptic Systems GmbH, Göttingen, Germany) and mouse anti-MAP2 (diluted 1:400) (Chemicon International Inc., Temecula, CA, USA) were used as primary antibodies. The sections were reacted with secondary antibody—mixture of Alexa Fluor® 488-conjugated donkey anti-mouse IgG (diluted 1:500) (Invitrogen, Waltham, MA, USA) and Alexa Fluor® 546-conjugated goat anti-rabbit IgG (diluted 1:500) (Invitrogen, Waltham, MA, USA).

In accordance with our previous study [3], in each group, five sections at each time were observed using epifluorescent microscope (Carl Zeiss, Göttingen, Germany) with blue (450–490 nm of wavelength; for observing VGLUT-1 immunoreactive structure) and green (510–560 nm of wavelength; for observing MAP2 immunoreactive structure) excitation fluorescent filters. Using cellSens Standard software (Olympus, Tokyo, Japan), each VGLUT-

1 and MAP2 immunoreactive structure was captured and merged. To quantitatively analyze the co-localized VGLUT-1/MAP2 immunoreactive structure, the merged image was converted into grey scale with 0–255 of range from black and white. Using Image J software (version 1.46) (National Institutes of Health, Bethesda, Rockville, MD, USA), the density of the co-localized VGLUT-1/MAP2 immunoreactive structure was presented as ROD as %: the ROD of the sham+vehicle group was designated as 100%.

2.12. Statistical Analysis

To perform all statistical analyses, in this study, SPSS software (version 15.0) (SPSS Inc., Chicago, IL, USA) was used. In addition, Kolmogorov and Smirnov test to evaluate normal distributions and Bartlett test to evaluate identical standard error of the mean (SEM) were used. Moreover, all presented data were taken for the normality test. The statistical significances of the mean among the experimental groups were determined by two-way analysis of variance followed by post hoc Tukey's test for all pairwise multiple comparisons. All presented data were shown as the mean ± SEM, and statistical significance was designated when *p*-value was less than 0.05.

3. Results

3.1. Major Ingredients of COG-Up®

Scutellarin standard was 19.803 min at the retention time, as shown in Figure 2A, which was also detected in COG-up® (19.729 min of retention time (Figure 2B). The retention time of sinapic acid standard was 24.098 min (Figure 2C), which was also shown in COG-up® (24.004 min of retention time) (Figure 2D). Accordingly, scutellarin and sinapic acid were revealed as major ingredients of the COG-up®.

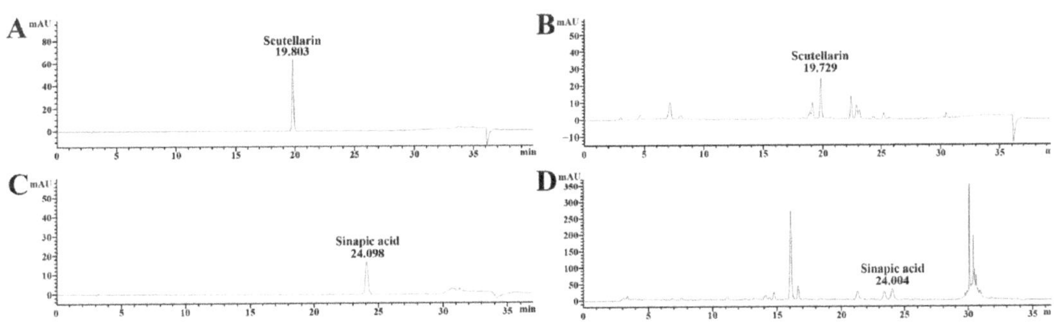

Figure 2. Qualitative analysis of COG-up® via high-performance liquid chromatography (**A**,**B**) are produced under the same mobile phase condition for the analysis of scutellarin. (**C**,**D**) were produced under the same mobile phase condition for the analysis of sinapic acid. The retention time of the standard scutellarin and sinapic acid was 19.803 and 24.098 min, respectively. The retention time of scutellarin and sinapic acid in COG- up® was 19.729 and 24.004 min, respectively.

3.2. Cognitive Function

3.2.1. Spatial Memory by BMT

As shown in Figure 3A, in all four groups, the latency time to find the escape hole evaluated from 26 days to 30 days after TI was gradually shortened. Latency time in the two sham groups was not different from each other and latency time in the TI+vehicle group was significantly longer than that shown in the sham groups. In the TI+COG-up® group, latency time was also significantly longer than that shown in the sham groups 26, 27 and 28 days after TI. However, in this group, latency time on day 28, day 29 and day 30 after TI was significantly shortened when as compared to that shown in the TI+vehicle group.

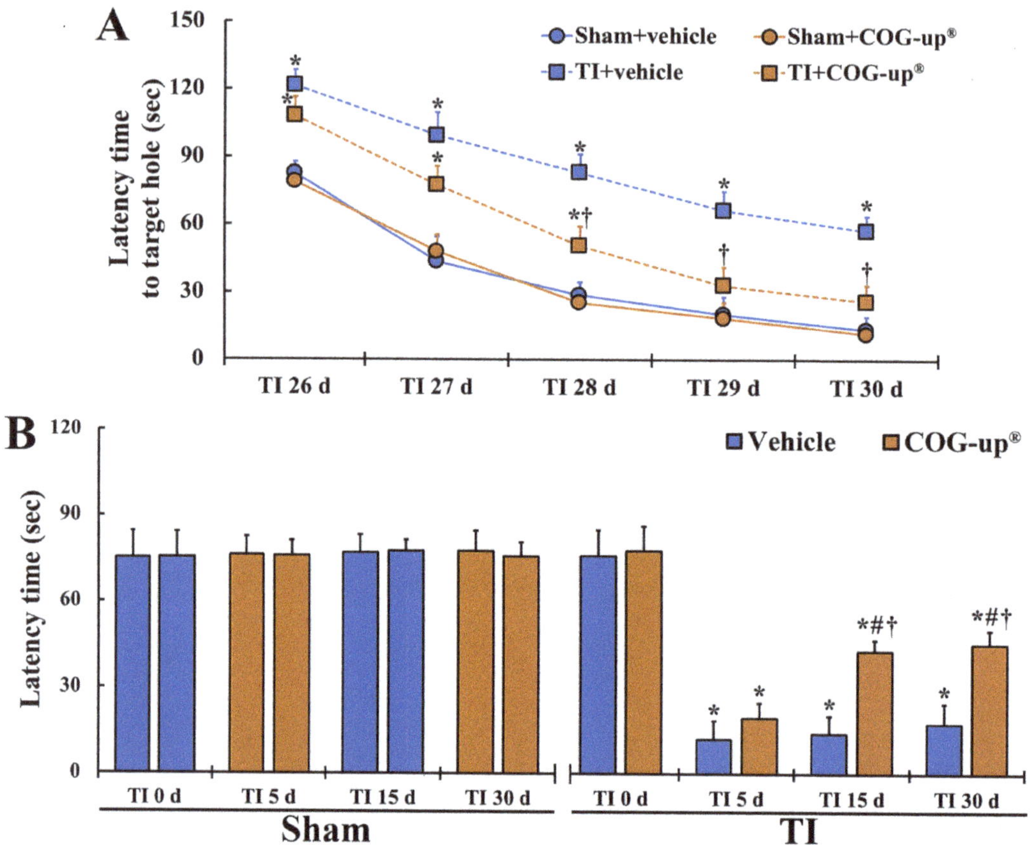

Figure 3. (**A**) Spatial memory by BMT. In the TI+COG-up® group, latency time in the BMT is significantly shortened from 28 days after TI as compared with the TI+vehicle group. The bars indicate mean ± SEM ($n = 7$ at each time in each group; * $p < 0.05$ versus sham+vehicle group and † $p < 0.05$ versus corresponding time of TI+vehicle group). (**B**) Learning memory by PAT. In the TI+COG-up® group, latency time 15 and 30 days after TI is significantly increased as compared with the TI+vehicle group. The bars indicate mean ± SEM ($n = 7$ at each time in each group; * $p < 0.05$ versus sham+vehicle group, # $p < 0.05$ versus prior time of each group, and † $p < 0.05$ versus corresponding time of TI+vehicle group).

3.2.2. Learning Memory by PAT

In all four groups, latency time evaluated at zero-day after TI did not exhibit differences (Figure 3B). In both sham groups, latency time shown at every time point after sham TI operation was similar to that found at zero-day (Figure 3B). After TI, latency time in the TI+vehicle group was significantly short as compared with that shown in the sham groups, although the latency time was gradually increased with time after TI (Figure 3B). In the TI+COG-up® group, latency time shown on day 5 after TI was increased as compared to the TI+vehicle group, but significant difference in the latency time was not found between the two groups (Figure 3B). However, latency time at 15 and 30 days after TI was significantly longer as compared to that found in the TI+vehicle group (Figure 3B).

3.3. Cellular Change in the Hippocampus

To examine change in cells in the gerbil hippocampus following TI, CV staining was performed. Cells stained by CV, in all sham groups, were easily identified in the hippocampus (Figure 4A,E). In particular, CV-stained cells formed the stratum pyramidale

(SP): these cells are principal cells and called pyramidal cells (Figure 4A,E). Five days after TI, in the TI+vehicle and TI+COG-up® groups, the CV stainability was decreased in the SP of the CA1 region, not CA2 and 3 regions (Figure 4B,F). This finding means that delayed neuronal damage/death occurs only in the CA1 region following TI. Thereafter, in these two groups, the distribution pattern of CV-stained cells was not altered until 30 days after TI (Figure 4C,D,G,H).

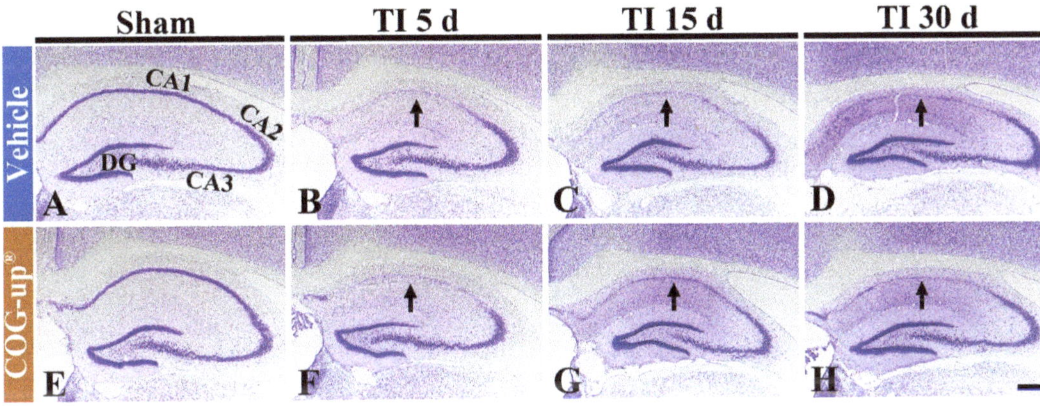

Figure 4. CV staining in the hippocampus of the sham+vehicle (**A**), sham+COG-up® (**E**), TI+vehicle and TI+COG-up® groups five (**B**,**F**), 15 (**C**,**G**) and 30 (**D**,**H**) days after TI. In the TI+vehicle and TI+COG-up® groups, CV stainability is decreased in the stratum pyramidale (arrows) of the CA1 region from five days after TI: the distribution pattern of CV-stained cells is not changed until 30 days after TI. Scale bar = 400 μm. DG, dentate gyrus.

3.4. TI-Induced Neuronal Death (Loss) in the CA1 Region

3.4.1. Findings by NeuN Immunohistochemistry

In all sham groups, numerous NeuN immunoreactive pyramidal cells stained with NeuN were located in the SP of the CA1 region (about 84 cells/250 μm^2) (Figure 5(Aa,Ae)). In the TI+vehicle and TI-COG-up® groups, NeuN immunoreactive cells were rarely detected in the SP five days after TI, showing that their number was about nine cells/250 μm^2 (Figure 5(Ab,Af,B)). In these two groups, at 15 and 30 days after TI, the number of NeuN immunoreactive pyramidal cells was not significantly different from that shown five days after TI (Figure 5(Ac,Ad,Ag,Ah,B)). This finding means that treatment with COG-up® does not influence TI-induced delayed neuronal death.

3.4.2. Findings by FJB Histofluorescence

In both sham+vehicle and sham+COG-up® groups, FJB positive cells were not detected in the CA1 region (Figure 5(Ca,Ce)). In the TI+vehicle and TI+COG-up® groups, a plenty number of FJB positive cells (about 74 cells/250 μm^2) were observed in the SP five days after TI (Figure 5(Cb,Cf,D)). In these groups, the numbers of FJB positive cells found 15 and 30 days after TI were not different from those found five days after TI (Figure 5(Cc,Cd,Cg,Ch,D)). Definitely, this finding shows that treatment with COG-up® does save CA1 pyramidal cells from TI-induced death.

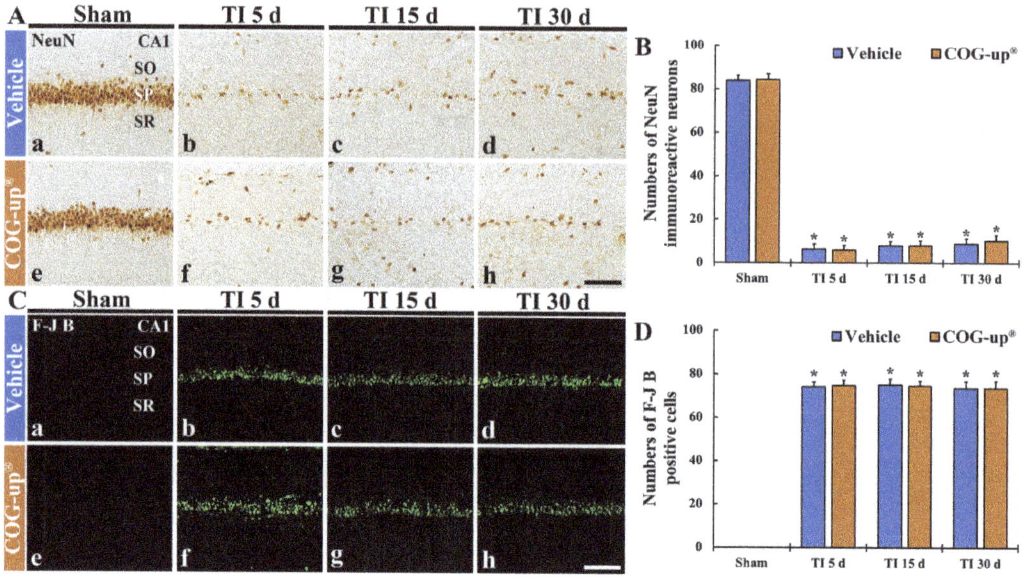

Figure 5. (**A**,**B**) Immunohistochemistry for NeuN (**A**) and FJB histofluorescence (**C**) in the CA1 region of the sham+vehicle (**Aa**,**Ca**), sham+COG-up® (**Ae**,**Ce**), TI+vehicle and TI+COG-up® groups five (**Ab**,**Af**,**Cb**,**Cf**), 15 (**Ac**,**Ag**,**Cc**,**Cg**) and 30 (**Ad**,**Ah**,**Cd**,**Ch**) days after TI. In both TI+vehicle and TI+COG-up® groups, NeuN immunoreactive cells are rarely detected, and numerous FJB positive cells are shown. Scale bar = 100 μm. (**B**,**D**) Mean numbers of NeuN immunoreactive cells (**B**) and mean numbers of FJB positive cells (**D**). SO, stratum oriens; SP, stratum pyramidale; SR, stratum radiatum. The bars indicate mean ± SEM ($n = 7$, respectively; * $p < 0.05$ versus sham+vehicle group).

3.5. Myelin Using MBP Immunohistochemistry

In all sham groups, MBP immunoreactive structures, as myelinated nerve fibers, were distributed throughout all layers in the CA1 region (Figure 6(Aa,Ad)). In the TI+vehicle group, the density of MBP immunoreactive structures was significantly decreased (about 9% and 12% on day 15 and day 30 after TI, respectively, versus sham+vehicle group) as compared with that shown in the sham+vehicle group (Figure 6(Ab,6Ac,B)). In contrast, in the TI+COG-up® group, the density of MBP immunoreactive structure was significantly higher (about 46% and 55% at 15 and 30 days, respectively, after TI versus sham+vehicle group) than that evaluated in the TI+vehicle group (Figure 6(Ae,Af,B)).

3.6. Oligodendrocytes Using RIP Immunohistochemistry

In both sham+vehicle and sham+COG-up® groups, RIP immunoreactive structures, as oligodendrocytes, were clearly shown in all layers of the CA1 region (Figure 6(Ca,Cd)). In the TI+vehicle group, the number of RIP immunoreactive oligodendrocytes was significantly increased (about 20 cells/250 μm^2 and 28 cells/250 μm^2 at 15 and 30 days, respectively, after TI) as compared to that evaluated in the sham+vehicle group (Figure 6(Cb,Cc,D)). In the TI+COG-up® group, however, a significantly increased number of RIP immunoreactive oligodendrocytes was detected (about 25 cells/250 μm^2 and 37 cells/250 μm^2 on day 15 and day 30 after TI, respectively) when compared with that evaluated in the TI+vehicle group (Figure 6(Ce,Cf,D)).

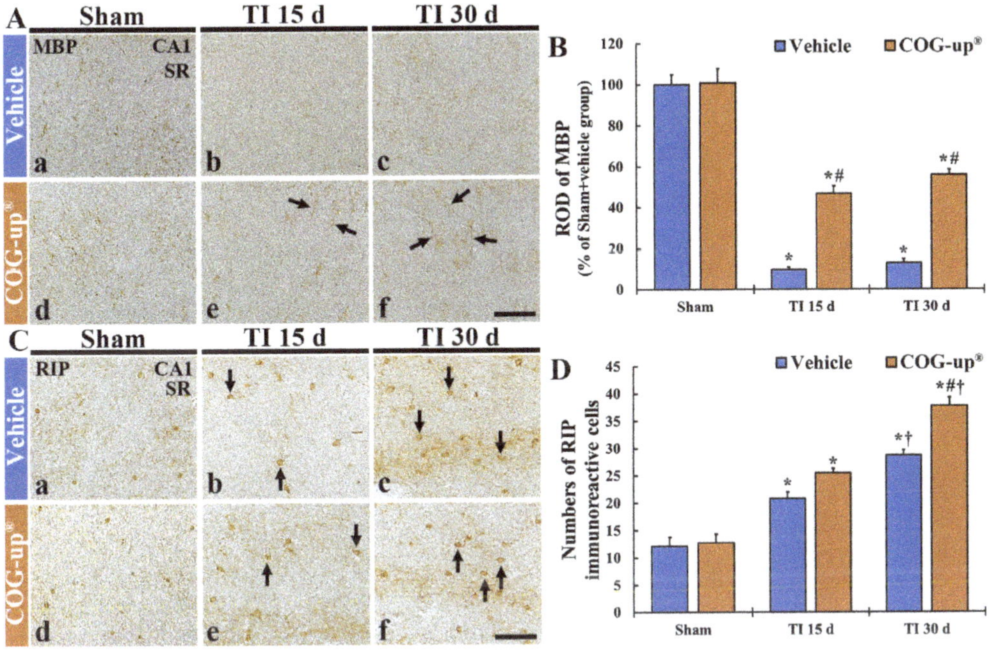

Figure 6. (**A**,**B**) Immunohistochemistry for MBP (**A**) and RIP (**C**) in the CA1 region of the sham+vehicle (**Aa**,**Ca**), sham+COG-up® (**Ae**,**Ce**), TI+vehicle and TI+COG-up® groups at five (**Ab**,**Af**,**Cb**,**Cf**), 15 (**Ac**,**Cc**) and 30 (**Ad**,**Cd**) days after TI. In the TI+vehicle group, MBP immunoreactive nerve fibers are decreased at 15 and 30 days after TI. However, in the TI+COG-up® group, MBP immunoreactive structures (arrows) are significantly increased at 15 and 30 days after TI when compared to those shown in the TI+vehicle group. RIP immunoreactive cells (arrows) are increased in both TI+vehicle and TI+COG-up® group, at 15 and 30 days after TI; however, RIP immunoreactive oligodendrocytes in the TI+ COG-up® group are more than those shown in the TI+vehicle group. SO, stratum oriens; SP, stratum pyramidale; SR, stratum radiatum. Scale bar = 100 μm. (**B**,**D**) ROD of MBP immunoreactive structure (**B**) and mean numbers of RIP immunoreactive oligodendrocytes (**D**). The bars indicate mean ± SEM ($n = 7$ at each time in each group; * $p < 0.05$ versus sham+vehicle group, # $p < 0.05$ versus prior time point of each group, and † $p < 0.05$ versus corresponding time of TI+vehicle group).

3.7. Synaptic Density Using Double Immunofluorescence for VGLUT-1/MAP2

In all sham groups, VGLUT-1 immunoreactive structures, as glutamate transporter in the membrane of synaptic vesicles, were predominately distributed in the stratum oriens and stratum radiatum (Figure 7(Aa,Ba)). The VGLUT-1 immunoreactive structures were co-localized with MAP2 immunoreactive structures, as neuronal dendritic extensions (main dendrites), in the CA1 region (Figure 7(Ab,Ac,Bb,Bc)). In the TI+vehicle group, when VGLUT-1/MAP2 immunoreactive structures were observed at 30 days after TI, they were significantly decreased (about 2% versus sham+vehicle group) as compared to the sham+vehicle group (Figure 7(Ad–Af),C). However, in the TI+COG-up® group at 30 days after TI, a significant increase in VGLUT-1/MAP2 immunoreactive structures were found (about 25% versus TI+vehicle group) when compared with the TI+vehicle group (Figure 7(Bd–Bf),C).

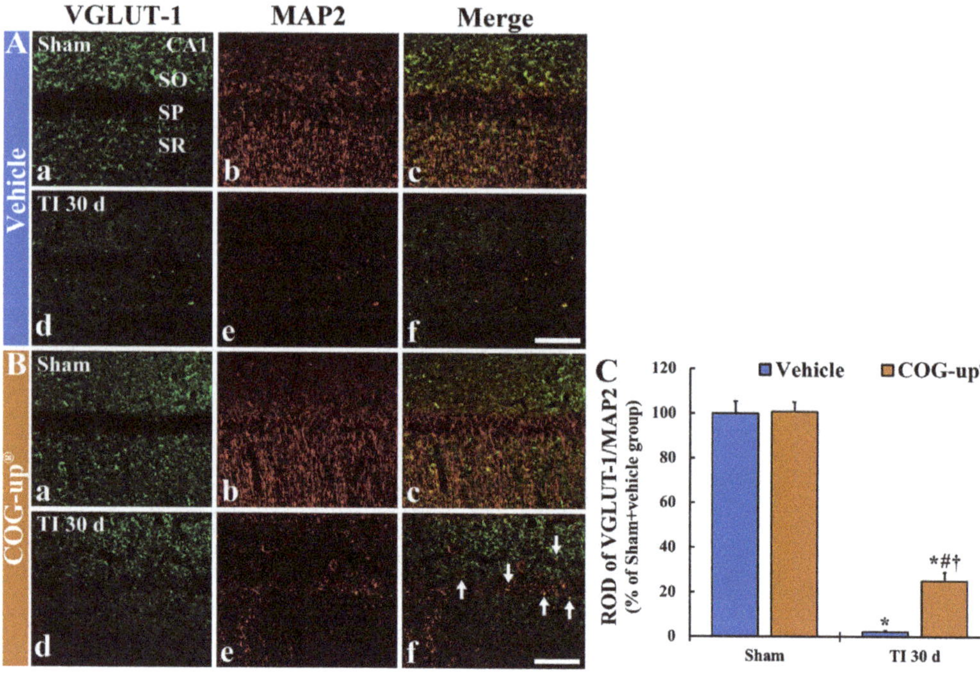

Figure 7. (**A**,**B**) Double immunofluorescence for VGLUT-1 (**a**,**d**; green)/MAP2 (**b**,**e**; red), and merged images (**c**,**f**) in the CA1 region of the sham+vehicle (**Aa–Ac**), sham+COG-up® (**Ba–Bc**), TI+vehicle (**Ad–Af**) and TI+COG-up® (**Bd–Bf**) groups at 30 days after TI. In the TI+vehicle group, VGLUT-1/MAP2 immunoreactive structures are hardly found. However, in the TI+COG-up® group, co-localized VGLUT-1/MAP2 immunoreactive structures are easily shown (arrows). SO, stratum oriens; SP, stratum pyramidale; SR, stratum radiatum. Scale bar = 100 μm. (**C**) ROD of VGLUT-1/MAP2 fluorescent structure. The bars indicate mean ± SEM (n = 7 at each time in each group; * $p < 0.05$ versus sham+vehicle group, # $p < 0.05$ versus prior time point of each group, and † $p < 0.05$ versus corresponding time of TI+vehicle group).

4. Discussion

In the present study, cognitive impairment was apparently induced in a gerbil model of TI which was accompanied by the death of pyramidal cells (neurons as principal cells) of the hippocampal CA1 region as shown by CV staining, NeuN immunohistochemistry and FJB histofluorescence. In addition, in the CA1 region following ischemia, myelin and vesicular synaptic density were significantly decreased as shown by immunohistochemistry for MBP and VGLUT-1. Therefore, we had tried to investigate whether COG-up® treatment improved microenvironmental damages such as demyelination (loss of myelin) and decreased synaptic density in the hippocampus.

Accumulating experimental data show that herbal medicines were used thanks to their beneficial attributes such as anti-inflammatory and antioxidant activities [12,20,31,32]. Furthermore, many studies have investigated active ingredients of herbal medicines. For instance, the therapeutic effect of *Angelica gigas* Nakai (Umbelliferae family) root extract containing decursin (a coumarin derivate compound) was shown in the ischemic hippocampus via protecting blood–brain barrier leakage and astrocyte endfeet damage in a gerbil model of TI [33]. Additionally, it was reported that, in a gerbil model of TI, pretreated with YES-10®, a combined extract of EALP and *Clematis mandshurica* RUPR. (Ranunculaceae family), containing scutellarin and chlorogenic acid showed strong neuroprotective effect in the hippocampus after TI [12]. We, in this study, investigated the major ingredients of COG-up®; and, as shown in the HPLC result, we found scutellarin and sinapic acid as active ingredients of COG-up®. A precedent study reported that scutellarin, as a flavonoid glycoside compound, ameliorated learning and memory deficit induced by chronic cerebral hypoperfusion in rats [34]. In addition, Kim et al. (2011) reported that synaptic acid treatment attenuated memory impairment in a rat model of global cerebral ischemia induced by ligation of four vessels [35].

Recently, studies developed combined extracts originating natural resources and reported that they improved cognitive dysfunctions following dementia. For example, Shenmayizhi decoction, a Chinese herbal prescription, consists of four herbal extracts—*Panax ginseng* C. A. Mey. (Araliaceae family), *Gastrodia elata* Bl. (Orchidaceae family), *Ligusticum chuanxiong* Hort. (Apiaceae family) and *Euonymus alatus* Sieb. (Celastraceae family)—ameliorate declined memory and learning function in a rat model of vascular cognitive impairment induced by BCCA occlusion [36]. Based on these previous studies, in this experiment, we treated COG-up® after TI in gerbils and found that TI-induced cognitive impairment was significantly improved when we performed BMT and PAT for spatial memory and learning memory, respectively.

In the present study, although COG-up® treatment after TI showed an improvement of cognitive impairment in the gerbils, the COG-up® treatment failed to protect CA1 pyramidal cells from TI when we performed CV histochemistry, NeuN immunohistochemistry and FJB histofluorescence. Therefore, we had tried to investigate whether COG-up® treatment improved microenvironmental damages such as demyelination (loss of myelin) and decreased synaptic density in the hippocampus.

As reviewed by Baaklini et al. (2019), demyelination attributed to various pathological conditions retards axonal conduction and may bring a failure to transmit neural information, passing by demyelinated segments [37]. Demyelinated axons due to injuries undergo remyelination for the functional recovery of neural activities, and this process is facilitated by newly generated oligodendrocytes [37,38]. It was demonstrated that the improvement of cognitive impairment induced by ischemic stroke is accompanied by remyelination and proliferation of oligodendrocytes. For example, Chen et al. (2018) reported that treatment with *N*-acetyl-5-methoxytryptamine (melatonin), a lipophilic hormone synthesized by the pineal gland, after ischemic stroke in gerbils, excellently improved memory impairment and showed remyelination in the hippocampus [3]. Furthermore, Qu et al. (2014) demonstrated that quercetin, a flavonoid abundant in diverse plants, improved cognitive deficit induced by hypoxia-ischemia via promoting remyelination in neonatal rats [14]. With correspondence to these precedent data, in this study, COG-up® treatment after TI in the gerbils increased MBP immunoreactive structure (myelin) and RIP immunoreactive cells (oligodendrocytes) in the ischemic CA1 region.

In our current study, COG-up® treatment after TI in the gerbils increased the density of glutamatergic synapses in the ischemic CA1 region using double immunofluorescence for VGLUT-1/MAP2. Recently, we reported that treatment with melatonin after TI in gerbils improved the expression level of VGLUT-1 (higher synaptic density) when compared with that evaluated in ischemic gerbils treated with vehicle [3]. Glutamate, as a major excitatory neurotransmitter in the brain, contributes to aspects of higher intellectual function [39]. It is acknowledged that glutamatergic neurotransmission is mainly achieved by VGLUT-1 [40]. A study described that VGLUT-1-knockout mice displayed a decline in hippocampal long-term potentiation which was in company with impairment of spatial memory [41]. In addition, Cao et al. (2016) showed that the expression level of VGLUT-1 was decreased in the hippocampal CA1–3 regions accompanied with cognitive impairment after chronic cerebral ischemia induced by BCCA occlusion in rats [42]. Moreover, a case-control autopsy study by Kirvell et al. (2010) showed that, in patients with stroke, the preservation of glutamatergic synapses in the frontal cortex against the temporal cortex might play a role in maintaining cognitive function against dementia following a stroke [39].

5. Conclusions

Our present data showed that COG-up® contained scutellarin and sinapic acid as major ingredients. Treatment with COG-up® after TI in gerbils improved cognitive impairment (decline in spatial and learning memory) induced by TI. However, COG-up® treatment did not protect against the death of pyramidal cells (principal cells) located in the hippocampal CA1 region following TI. Instead, we found that treatment with COG-up® improved remyelination and restored the density of glutamatergic synapses in the ischemic CA1 region. Based on this result, we suggest that follow-up studies such as mechanisms and optimization improving cognitive function need to be conducted; through further studies, COG-up® can be employed for improving cognitive decline following ischemic stroke via commercializing as health/functional foods and medicines. Ultimately, it can enhance national health.

Author Contributions: Conceptualization, M.-H.W. and I.J.K.; Methodology, J.-C.L., D.W.K. and S.S.L.; Software, H.S.; Validation, D.W.K., S.-S.K. and J.-C.L.; Investigation, T.-K.L. and H.S.; Resources, J.H., J.-W.L. and S.-S.K.; Data Curation, T.-K.L. and J.-W.L.; Writing—original draft preparation, T.-K.L. and J.H.; Writing—review and editing M.-H.W.; Supervision, I.J.K. and M.-H.W.; Project Administration, M.-H.W. and I.J.K.; Funding Acquisition, S.-S.K., I.J.K. and S.S.L. All authors have read and agreed to the published version of the manuscript.

Funding: This work was supported by the Korean Institute of Planning and Evaluation for Technology in Food, Agriculture, Forestry (IPET) through the High Value-added Food Technology Development Program (or Project) funded by the Ministry of Agriculture, Food and Rural Affairs (MAFRA) (117055-03), by Basic Science Research Program through the National Research Foundation (NRF) of Korea funded by the Ministry of Education (NRF-2020R1F1A1071973), and by the Brain Korea 21 (BK21) Fostering Outstanding Universities for Research (FOUR, 4220200913807) funded by the NRF of Korea.

Institutional Review Board Statement: All experimental processes adhered to the guidelines described in the "Current International Laws and Policies", a part of the "Guide for the Care and Use of Laboratory Animals". Approval for the experimental protocol was sanctioned by the Institutional Animal Care and Use Committee of Kangwon National University (Chuncheon, Korea) on 18, February 2020 (approval no., KW-200113-1).

Informed Consent Statement: Not applicable.

Data Availability Statement: The data presented in this study are available on request from the corresponding author.

Acknowledgments: The authors would like to thank Seung Uk Lee and Hyun Sook Kim for their technical help in this work.

Conflicts of Interest: The authors have declared that there are no conflict of interest relating to this article.

Abbreviations

ABC: avidin-biotin complex; BCCA, bilateral common carotid arteries; BMT, Barnes maze test; CA1, Cornu Ammonis 1 region; CV, cresyl violet; FJB, Fluoro-Jade B; MAP2, microtubule-associated protein 2; MBP, myelin basic protein; NeuN, neuronal nuclei; PAT, Passive avoidance test; ROD, relative optical density; SO, stratum oriens; SP, stratum pyramidale; SR, stratum radiatum; RIP, receptor-interacting protein; TI, transient ischemia; VGLUT-1, vesicular glutamate transporter 1.

References

1. Lin, L.; Wang, X.; Yu, Z. Ischemia-reperfusion injury in the brain: Mechanisms and potential therapeutic strategies. *Biochem. Pharmacol. Open Access* **2016**, *5*, 213.
2. Xu, M.S.; Yin, L.M.; Cheng, A.F.; Zhang, Y.J.; Zhang, D.; Tao, M.M.; Deng, Y.Y.; Ge, L.B.; Shan, C.L. Cerebral ischemia-reperfusion is associated with upregulation of cofilin-1 in the motor cortex. *Front. Cell Dev. Biol.* **2021**, *9*, 634347. [CrossRef] [PubMed]
3. Chen, B.H.; Park, J.H.; Lee, Y.L.; Kang, I.J.; Kim, D.W.; Hwang, I.K.; Lee, C.H.; Yan, B.C.; Kim, Y.M.; Lee, T.K.; et al. Melatonin improves vascular cognitive impairment induced by ischemic stroke by remyelination via activation of ERK1/2 signaling and restoration of glutamatergic synapses in the gerbil hippocampus. *Biomed. Pharmacother.* **2018**, *108*, 687–697. [CrossRef] [PubMed]
4. Della-Morte, D.; Rundek, T. Dizziness and vertigo. *Front. Neurol. Neurosci.* **2012**, *30*, 22–25. [PubMed]

5. Liao, L.Y.; Lau, B.W.; Sanchez-Vidana, D.I.; Gao, Q. Exogenous neural stem cell transplantation for cerebral ischemia. *Neural Regen. Res.* **2019**, *14*, 1129–1137. [PubMed]
6. Sweatt, J.D. Hippocampal function in cognition. *Psychopharmacology* **2004**, *174*, 99–110. [CrossRef] [PubMed]
7. Bird, C.M.; Burgess, N. The hippocampus and memory: Insights from spatial processing. *Nat. Rev. Neurosci.* **2008**, *9*, 182–194. [CrossRef]
8. Ahn, J.H.; Chen, B.H.; Shin, B.N.; Cho, J.H.; Kim, I.H.; Park, J.H.; Lee, J.C.; Tae, H.J.; Lee, Y.L.; Lee, J.; et al. Intravenously infused F3.Olig2 improves memory deficits via restoring myelination in the aged hippocampus following experimental ischemic stroke. *Cell Transplant.* **2016**, *25*, 2129–2144. [CrossRef]
9. Lee, J.C.; Park, J.H.; Yan, B.C.; Kim, I.H.; Cho, G.S.; Jeoung, D.; Kwon, Y.G.; Kim, Y.M.; Lee, Y.L.; Shin, H.C.; et al. Effects of transient cerebral ischemia on the expression of DNA methyltransferase 1 in the gerbil hippocampal ca1 region. *Neurochem. Res.* **2013**, *38*, 74–81. [CrossRef] [PubMed]
10. Yan, B.C.; Park, J.H.; Ahn, J.H.; Lee, J.C.; Won, M.H.; Kang, I.J. Postsynaptic density protein (PSD)-95 expression is markedly decreased in the hippocampal ca1 region after experimental ischemia-reperfusion injury. *J. Neurol. Sci.* **2013**, *330*, 111–116. [CrossRef]
11. Kirino, T. Delayed neuronal death in the gerbil hippocampus following ischemia. *Brain Res.* **1982**, *239*, 57–69. [CrossRef]
12. Park, Y.E.; Noh, Y.; Kim, D.W.; Lee, T.K.; Ahn, J.H.; Kim, B.; Lee, J.C.; Park, C.W.; Park, J.H.; Kim, J.D.; et al. Experimental pretreatment with YES-10®, a plant extract rich in scutellarin and chlorogenic acid, protects hippocampal neurons from ischemia/reperfusion injury via antioxidant role. *Exp. Ther. Med.* **2021**, *21*, 183. [CrossRef] [PubMed]
13. Chen, W.; Zhang, J.; Wang, J.; Li, Y.; Liu, W.; Xia, J. Lycopene supplementation protects vascular dementia gerbils against the impairment of learning and memory. *Folia Neuropathol.* **2021**, *59*, 161–173. [CrossRef]
14. Qu, X.; Qi, D.; Dong, F.; Wang, B.; Guo, R.; Luo, M.; Yao, R. Quercetin improves hypoxia-ischemia induced cognitive deficits via promoting remyelination in neonatal rat. *Brain Res.* **2014**, *1553*, 31–40. [CrossRef]
15. Nave, K.A.; Werner, H.B. Myelination of the nervous system: Mechanisms and functions. *Annu. Rev. Cell Dev. Biol.* **2014**, *30*, 503–533. [CrossRef]
16. Domingues, H.S.; Portugal, C.C.; Socodato, R.; Relvas, J.B. Oligodendrocyte, astrocyte, and microglia crosstalk in myelin development, damage, and repair. *Front. Cell Dev. Biol.* **2016**, *4*, 71. [PubMed]
17. Duncan, I.D.; Radcliff, A.B.; Heidari, M.; Kidd, G.; August, B.K.; Wierenga, L.A. The adult oligodendrocyte can participate in remyelination. *Proc. Natl. Acad. Sci. USA* **2018**, *115*, E11807–E11816. [CrossRef]
18. Kugaya, A.; Sanacora, G. Beyond monoamines: Glutamatergic function in mood disorders. *CNS Spectr.* **2005**, *10*, 808–819. [CrossRef] [PubMed]
19. Robbins, T.W.; Murphy, E.R. Behavioural pharmacology: 40+ years of progress, with a focus on glutamate receptors and cognition. *Trends Pharmacol. Sci.* **2006**, *27*, 141–148. [CrossRef] [PubMed]
20. Jo, M.J.; Lee, J.R.; Cho, I.J.; Kim, Y.W.; Kim, S.C. Roots of erigeron annuus attenuate acute inflammation as mediated with the inhibition of NF-kappa B-associated nitric oxide and prostaglandin E2 production. *Evid.-Based Complement. Altern. Med.* **2013**, *2013*, 297427. [CrossRef]
21. Park, S.K.; Ha, J.S.; Kim, J.M.; Kang, J.Y.; Lee du, S.; Guo, T.J.; Lee, U.; Kim, D.O.; Heo, H.J. Antiamnesic effect of broccoli (Brassica oleracea var. Italica) leaves on amyloid beta (Aβ)1-42-induced learning and memory impairment. *J. Agric. Food Chem.* **2016**, *64*, 3353–3361. [CrossRef]
22. Carpenter, J.W.; Marion, C. *Exotic Animal Formulary-E-Book*; Elsevier Health Sciences: Amsterdam, The Netherlands, 2017.
23. Mahar, I.; Albuquerque, M.S.; Mondragon-Rodriguez, S.; Cavanagh, C.; Davoli, M.A.; Chabot, J.G.; Williams, S.; Mechawar, N.; Quirion, R.; Krantic, S. Phenotypic alterations in hippocampal NPY- and PV-expressing interneurons in a presymptomatic transgenic mouse model of Alzheimer's disease. *Front. Aging Neurosci.* **2016**, *8*, 327. [CrossRef]
24. Radtke-Schuller, S.; Schuller, G.; Angenstein, F.; Grosser, O.S.; Goldschmidt, J.; Budinger, E. Brain atlas of the Mongolian gerbil (meriones unguiculatus) in CT/MRI-aided stereotaxic coordinates. *Brain Struct. Funct.* **2016**, *221* (Suppl. 1), 1–272. [CrossRef]
25. Zhu, Y.; Liu, F.; Zou, X.; Torbey, M. Comparison of unbiased estimation of neuronal number in the rat hippocampus with different staining methods. *J. Neurosci. Methods* **2015**, *254*, 73–79. [CrossRef]
26. Schmued, L.C.; Hopkins, K.J. Fluoro-Jade B: A high affinity fluorescent marker for the localization of neuronal degeneration. *Brain Res.* **2000**, *874*, 123–130. [CrossRef]
27. Anderson, K.J.; Miller, K.M.; Fugaccia, I.; Scheff, S.W. Regional distribution of Fluoro-Jade B staining in the hippocampus following traumatic brain injury. *Exp. Neurol.* **2005**, *193*, 125–130. [CrossRef] [PubMed]
28. Sharma, S.S.; Dhar, A.; Kaundal, R.K. Fetpps protects against global cerebral ischemic-reperfusion injury in gerbils. *Pharmacol. Res.* **2007**, *55*, 335–342. [CrossRef] [PubMed]
29. Meurer, R.T.; Martins, D.T.; Hilbig, A.; Ribeiro Mde, C.; Roehe, A.V.; Barbosa-Coutinho, L.M.; Fernandes Mda, C. Immunohistochemical expression of markers Ki-67, NeuN, synaptophysin, p53 and HER2 in medulloblastoma and its correlation with clinicopathological parameters. *Arq. De Neuro-Psiquiatr.* **2008**, *66*, 385–390. [CrossRef] [PubMed]
30. Paizs, M.; Engelhardt, J.I.; Siklos, L. Quantitative assessment of relative changes of immunohistochemical staining by light microscopy in specified anatomical regions. *J. Microsc.* **2009**, *234*, 103–112. [CrossRef]
31. Cho, J.H.; Kwon, J.E.; Cho, Y.; Kim, I.; Kang, S.C. Anti-inflammatory effect of Angelica gigas via heme oxygenase (HO)-1 expression. *Nutrients* **2015**, *7*, 4862–4874. [CrossRef]

32. Lee, C.H.; Park, J.H.; Ahn, J.H.; Kim, J.D.; Cho, J.H.; Lee, T.K.; Won, M.H. Stronger antioxidant enzyme immunoreactivity of Populus tomentiglandulosa extract than ascorbic acid in rat liver and kidney. *Iran. J. Basic Med. Sci.* **2019**, *22*, 963–967. [PubMed]
33. Lee, T.K.; Kang, I.J.; Sim, H.; Lee, J.C.; Ahn, J.H.; Kim, D.W.; Park, J.H.; Lee, C.H.; Kim, J.D.; Won, M.H.; et al. Therapeutic effects of decursin and Angelica gigas nakai root extract in gerbil brain after transient ischemia via protecting BBB leakage and astrocyte endfeet damage. *Molecules* **2021**, *26*, 2161. [CrossRef]
34. Shin, J.W.; Kweon, K.J.; Kim, D.K.; Kim, P.; Jeon, T.D.; Maeng, S.; Sohn, N.W. Scutellarin ameliorates learning and memory deficit via suppressing beta-amyloid formation and microglial activation in rats with chronic cerebral hypoperfusion. *Am. J. Chin. Med.* **2018**, *46*, 1203–1223. [CrossRef] [PubMed]
35. Kim, Y.O.; Lee, S.W.; Oh, M.S.; Lee, H.J. Effects of sinapic acid of 4 vessel occlusion model-induced ischemia and cognitive impairments in the rat. *Clin. Psychopharmacol. Neurosci. Off. Sci. J. Korean Coll. Neuropsychopharmacol.* **2011**, *9*, 86–90. [CrossRef] [PubMed]
36. Sun, C.; Liu, M.; Liu, J.; Zhang, T.; Zhang, L.; Li, H.; Luo, Z. Shenmayizhi decoction improves the mitochondrial structure in the brain and ameliorates cognitive impairment in VCI rats via the AMPK/UCP2 signaling pathway. *Neuropsychiatr. Dis. Treat.* **2021**, *17*, 1937–1951. [CrossRef]
37. Baaklini, C.S.; Rawji, K.S.; Duncan, G.J.; Ho, M.F.S.; Plemel, J.R. Central nervous system remyelination: Roles of glia and innate immune cells. *Front. Mol. Neurosci.* **2019**, *12*, 225. [CrossRef] [PubMed]
38. Jia, W.; Kamen, Y.; Pivonkova, H.; Karadottir, R.T. Neuronal activity-dependent myelin repair after stroke. *Neurosci. Lett.* **2019**, *703*, 139–144. [CrossRef]
39. Kirvell, S.L.; Elliott, M.S.; Kalaria, R.N.; Hortobagyi, T.; Ballard, C.G.; Francis, P.T. Vesicular glutamate transporter and cognition in stroke: A case-control autopsy study. *Neurology* **2010**, *75*, 1803–1809. [CrossRef]
40. Du, X.; Li, J.; Li, M.; Yang, X.; Qi, Z.; Xu, B.; Liu, W.; Xu, Z.; Deng, Y. Research progress on the role of type i vesicular glutamate transporter (VGLUT1) in nervous system diseases. *Cell Biosci.* **2020**, *10*, 26. [CrossRef]
41. Balschun, D.; Moechars, D.; Callaerts-Vegh, Z.; Vermaercke, B.; Van Acker, N.; Andries, L.; D'Hooge, R. Vesicular glutamate transporter vglut1 has a role in hippocampal long-term potentiation and spatial reversal learning. *Cereb. Cortex* **2010**, *20*, 684–693. [CrossRef]
42. Cao, Y.; Gou, Z.; Du, Y.; Fan, Y.; Liang, L.; Yan, Y.; Lin, P.; Jin, M.; Du, Y. Glutamatergic and central cholinergic dysfunction in the CA1, CA2 and CA3 fields on spatial learning and memory in chronic cerebral ischemia-induced vascular dementia of rats. *Neurosci. Lett.* **2016**, *620*, 169–176. [CrossRef] [PubMed]

veterinary sciences

MDPI

Article

Locomotor Behavior Analysis in Spinal Cord Injured *Macaca radiata* after Predegenerated Peripheral Nerve Grafting—A Preliminary Evidence

Anand Paramasivam [1,2,3,*], Suresh Mickymaray [4], Saikarthik Jayakumar [1,3], Mathew Jeraud [2,5], Periasamy Perumal [2,6], Abdullah Alassaf [7], Abdullah Abdulrahman Aljabr [3], Sridevi Dasarathy [8] and Suresh Babu Rangasamy [2,8]

1 Department of Basic Medical Sciences, College of Dentistry, Majmaah University, Al-Majmaah 11952, Saudi Arabia; s.jaya@mu.edu.sa
2 Department of Physiology, Dr. ALM PGIBMS, University of Madras, Chennai 600113, India; mathewjeraud@gmail.com (M.J.); periasamysumi@yahoo.co.in (P.P.); SureshBabu_Rangasamy@rush.edu (S.B.R.)
3 Department of Medical Education, College of Dentistry, Majmaah University, Majmaah 11952, Saudi Arabia; aa.jabr@mu.edu.sa
4 Department of Biology, College of Science, Majmaah University, Majmaah 11952, Saudi Arabia; s.maray@mu.edu.sa
5 Department of Physiology, Ibn Sina National College for Medical Studies, Jeddah 22421, Saudi Arabia
6 Department of Physiology, Vellore Medical College, Vellore 632002, India
7 Department of Preventive Dental Sciences, College of Dentistry, Majmaah University, Majmaah 11952, Saudi Arabia; am.assaf@mu.edu.sa
8 Department of Neurological Sciences, Rush University Medical Centre, Chicago, IL 60612, USA; anandsivam@gmail.com
* Correspondence: anand.p@mu.edu.sa

Citation: Paramasivam, A.; Mickymaray, S.; Jayakumar, S.; Jeraud, M.; Perumal, P.; Alassaf, A.; Aljabr, A.A.; Dasarathy, S.; Rangasamy, S.B. Locomotor Behavior Analysis in Spinal Cord Injured *Macaca radiata* after Predegenerated Peripheral Nerve Grafting—A Preliminary Evidence. *Vet. Sci.* **2021**, *8*, 288. https://doi.org/10.3390/vetsci8120288

Academic Editors: Ana Faustino and Paula A. Oliveira

Received: 28 September 2021
Accepted: 19 November 2021
Published: 23 November 2021

Abstract: Introduction: Primate animal models are being utilized to explore novel therapies for spinal cord injuries. This study aimed to evaluate the efficiency of the transplantation of predegenerated nerve segments in unilateral spinal cord-hemisected bonnet monkeys' (*Macaca radiata*) locomotor functions using the complex runways. Materials and Methods: The bonnet monkeys were initially trained to walk in a bipedal motion on grid and staircase runways. In one group of trained monkeys, surgical hemisection was made in the spinal cord at the T12-L1 level. In the other group, hemisection was induced in the spinal cord, and the ulnar nerve was also transected at the same time (transplant group). After one week, the hemisected cavity was reopened and implanted with predegenerated ulnar nerve segments obtained from the same animal of the transplant group. Results: All the operated monkeys showed significant deficits in locomotion on runways at the early postoperative period. The walking ability of operated monkeys was found to be gradually improved, and they recovered nearer to preoperative level at the fourth postoperative month, and there were no marked differences. Conclusion: The results demonstrate that there were no significant improvements in the locomotion of monkeys on runways after the delayed grafting of nerve segments until one year later. The failure of the predegenerated nerve graft as a possible therapeutic strategy to improve the locomotion of monkeys may be due to a number of factors set in motion by trauma, which could possibly prevent the qualities of regeneration. The exact reason for this ineffectiveness of predegenerated nerve segments and their underlying mechanism is not known.

Keywords: spinal cord hemisection; *Macaca radiata*; predegenerated nerve transplantation; locomotor recovery

1. Introduction

Impairment/injury in the spinal cord (SC) in mammals usually causes temporary or permanent paraplegia based on the severity of the injury. Several transplants, such as embryonic neural tissue [1,2], peripheral nerve tissue [3,4], and Schwann cells [5,6] are commonly used as a therapeutic approach to repair SC injuries (SCI), which differ by their nature, origin, and mode of action. However, many scientists in SC injury research were specifically attracted to the use of the peripheral nerve graft (PNG) as a possible means for promoting SC repair. This may be due to the well-known fact that the use of embryonic neural tissue for transplantation is not only difficult to obtain but has also inherent ethical problems to use in patients. Therefore, the benefit of other tissues, viz., peripheral nerve tissue taken from the same animal as a donor material, has been investigated to elucidate their capacity to repair the damaged SC [7]. Additionally, the use of PNG prepared from the same animal is more appropriate, as it overcomes the immunological problems of the rejection of transplanted tissue from the lesioned area of SC [8]. Furthermore, all types of transplants can stimulate the growth of axons from host nerve fibers, but these actions are far more important and conspicuous with PNG [9,10], which provides not only a favorable atmosphere for the growth of axons but also provides guidance by directing the regrowth of fibers to a specific target [11].

Many studies in neuroscience research have demonstrated the interaction of CNS neurons to sustain the outgrowth of their axons with PNG [12]. Using the retrograde and anterograde tracing techniques, some investigators have demonstrated that neurons in SC, brain stem, and dorsal root ganglia can revive as intraspinal grafts of peripheral nerve segments [13]. To produce a good surgical reconstruction, Kao and his colleagues demonstrated a "delayed reconstruction" method with an improved microsurgical technique by using PNG on the SC cavities. The implantation of nerve graft into transected SC of dog was performed after a delay of about 1 week after the lesion and good reconstruction of regenerated axons within the graft after 1 month or later times following surgical reconstruction was observed. Additionally, the results of the delayed nerve graft exhibited no additional cord autotomy, and the grafted nerve had adhered to both cord stumps after 1 month with no cavitation [14].

The usefulness of a degenerated versus fresh nerve graft was well documented. Earlier investigators demonstrated that predegenerated PNG may be as effective as fresh PNG or more efficient in promoting speedy regeneration of peripheral nerves [15,16]. Predegenerated PNG may also lessen the initial time lapse before regenerating nerve fibers enter the graft without affecting or enhancing the regeneration rate and maturation of fibers [17,18]. Senoo et al., [19] examined the effect of prelesioned PNG versus fresh PNG in the dorsal funiculi of rat SC and observed the number of regenerating axons was around 10-fold greater in the prelesioned graft than the untreated grafts. Similarly, the usefulness of predegenerated PNG on axonal regeneration of retinal ganglion cells after optic nerve pre-lesions was also demonstrated in adult hamsters [19]. To date, most of the studies on the effect of predegenerated PNG as a possible source to treat SCIs have been performed in rodents [3]. To analyze specifically whether the predegenerated PNG promotes axonal elongation in the damaged SC of higher-order mammals, we resorted to studying the usefulness of predegenerated PNG in the recovery of locomotor functions of bonnet monkeys after inducing unilateral SC hemisection using the complex runways. Such a wide knowledge about the usefulness of predegenerated nerve grafting in the SC in the primate model will provide valuable information for SCI patients in future attempts.

2. Materials and Methods

2.1. Animal Selection

The selected bonnet monkeys (*Macaca radiata*) of either sex with bodyweights ranging between 4–5 kg (3–4 years of age) were used for this study. Necessary ethical approval was obtained from the Animal Care and Institutional Ethical Committee of the University of Madras (PGIBMS/PHY/DRRR/2021/0702021). The selected animals were fit in general health and limb activities with no external injuries. All experimental procedures were conducted as per the Committee for the Purpose of Control and Supervision of Experiments on Animals (CPCSEA) guidelines, India. Monkeys were grouped into unilateral

SC-hemisected control monkeys (G_{SC}) (n = 6) and unilateral SC-hemisected as well as predegenerated nerve-transplanted monkeys ($G_{SC+PDNT}$) (n = 6). The lighting operated on a 12 h on-and-off schedule in the primate facility house, and monkeys were fed with freshly prepared vegetables with boiled rice, commercial food pellets (Gold Mohur, India), and water provided ad libitum. Selection of non-human primate models has benefits of genetic similarities, spinal cord length and physiological and biochemical responses to SCI which are similar to those in humans [20,21].

2.2. Experimental Design

To begin with, the monkeys were accustomed to our primate facility house for 3 to 4 months and were housed one per cage. This type of habituation helped the animals develop quietness and kept them confined for easy handling during the training period. On completion of the habituation period, all monkeys were gently trained (30 min for each animal; 2 sessions/day) to perform the locomotion in a bipedal fashion (i.e., the hindlimbs participate in locomotion while forelimbs are restrained behind its back) on the two complex runways preoperatively, viz. staircase and grid runway. The locomotor training of these monkeys was continued until they achieved acceptable presentation (1 to 2 months). At the end of the completion of the training period, a quantitative preoperative assessment of the ability of animals on these runways was tested for multiple days by two independent observers to ensure the consistency of the animals' maximum competencies. Following the preoperative training and assessment, the spinal cord was hemisected on the right side at the level of T12 and L1 vertebral level in G_{SC}. In addition to the SC hemisection, the ulnar nerve was exposed and severed (either on the right or left side) in the $G_{SC+PDNT}$ group. Subsequently, a predegenerated segment of the ulnar nerve was placed in the hemisected site of the $G_{SC+PDNT}$ group. (Detailed in Sections 2.4–2.8). After the postoperative recovery period of 4 weeks, locomotor functional assessment was performed weekly for a period of one year. At the end of the postoperative observatory period (pop), the animals were sacrificed and the spinal cord tissue was collected from the operated site in both groups for histological evaluation and 3D graphical reconstruction (Illustrated in Figure 1).

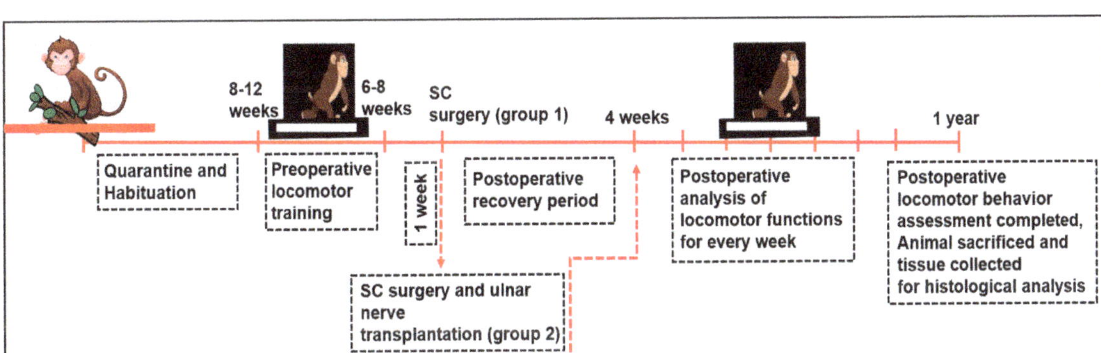

Figure 1. Illustrative of the experimental design.

2.3. The Locomotor Tests

Fresh fruits (small pieces of apples, guava, bananas, grapes) were provided as food rewards at either end of the runways during the training period. Additional fresh fruits, food pellets, and boiled rice with vegetables were given if any weight loss was noted during the training period. The bipedal locomotor training was conducted between 8 a.m. and 5 p.m. by a modified locomotor behavior test for primates originally designed to study the recovery of motor functions after SCI in rodents [22]. All images were captured using Pentax K1000 Manual (Pentax, Tokyo, Japan) focus with 50 mm lens. The details of each runway in this study are as follows.

2.4. Staircase Runway

The habituated monkeys were preoperatively trained to perform bipedal walking in the ascent and descent of a staircase. The staircase was built with a smooth wooden material fixed with adjustable support on the wall. This runway consisted of 10 identical steps, measuring 45 × 15 × 15 cm, which were fitted at a 25-degree inclination. Attractive food rewards were supplied at both corners of the staircase during the training sessions (Figure 2A).

Figure 2. Locomotor behavior of trained Bonnet monkeys in (**A**) staircase runways and (**B**) grid runway. In staircase and grid runways, trained monkeys performed bipedal locomotion to receive the food rewards placed on either end of runways.

2.5. Grid Runway

Monkeys were trained to perform bipedal locomotion on four iron grids that have fixed parallel bars with different inter bar intervals (4 cm, 5 cm, 6 cm, 7 cm; 1.5 m × 30 cm each). Food rewards were provided at both corners of the grid during the training sessions (Figure 2B). The errors in locomotor behavior over the grid runway were observed as the misplacement of an animal's foot in such a way that it falls between fixed parallel bars rather than being placed onto the rungs.

At each end of the training, the animals were freed to access a routine normal pellet feed in a limited quantity. All the observations were done quantitatively in relation to the time and number of steps for the evaluation of gait function recovery. During the pop, the functional status and progressive modification in the locomotion of monkeys on the grid or staircase runway was analyzed by using the 10-point locomotor scoring technique at weekly intervals [22]. The points would be maximized if the animal was normal, and the reverse would be true in a completely paralyzed animal (Table 1). The score ranges from 0 points to 10 points for a completely paralyzed monkey and a normal monkey (i.e., an unoperated animal), respectively.

2.6. Spinal Cord Lesions

Following the presurgical locomotor training, the monkeys were anesthetized with Thiopentane (28 mg/kg/body weight) by intraperitoneal route. On the backside of the animals, the hairs were shaved, and the skin was disinfected with 1% povidone -iodine. A rectal thermometer was used to measure the animals' body temperature and the temperature was maintained with a heating pad. The cardiac rhythm was monitored throughout surgery. The level of laminectomy was marked by palpating the posterior spinous process. A laminectomy was performed at T12-L1 vertebral level and the SC with intact dura was gently exposed with the help of dissecting microscope. The dura was slit with a number

11 scalpel blade and unilateral hemisection was produced with sterilized micro scissors followed by aspiration on the right side of SC (Figure 3).

Table 1. Score for Grid runway and Staircase runway.

Score	Behavior in Staircase Runway	Behavior in Grid Runway
0 grade/10 points	No attempt to stand, walk, no weight bearing	No attempt to stand, walk, no weight bearing
1 grade/8 points	Weak and/or delayed attempt to stand, no attempt to walk on the runway (initiated one or two steps)	Weak and/or delayed attempt to stand, no attempt to walk on the runway (initiated one or two steps)
2 grade/6 points	Good attempt to support body weight, weak attempt to walk on the runway with frequent slipping/errors seen	Good attempt to support body weight, weak attempt to walk on the runway with frequent errors seen
3 grade/4 points	Good attempt to stand and walk on the runway with few errors, significant change in time to cross the runway	Good attempt to stand and walk on the runway with few errors, significant change in time to cross the runway
4 grade/2 points	Good attempt to stand and walk on the runway with only mild deficits, no slipping, no significant change in time taken to cross the runway	Good attempt to stand and walk on the runway with only mild deficits, no misplacement, no significant change in time taken to cross the runway
5 grade/0 points	Good attempt to walk on the runway, no significant change in time compared to control animals	Good attempt to walk on the runway, no significant change in time compared to control animals

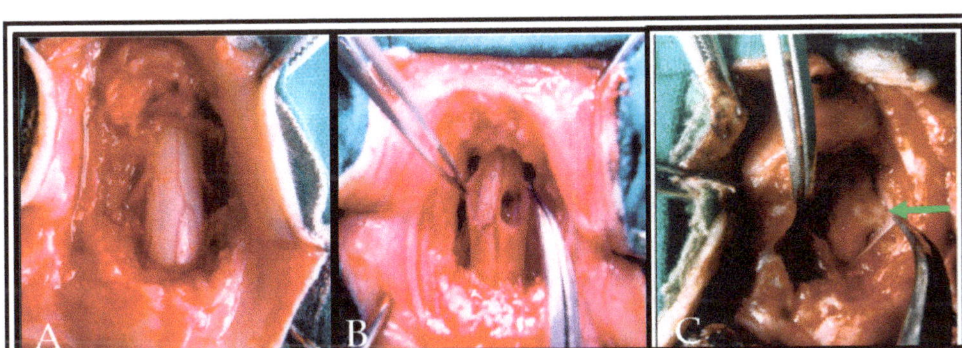

Figure 3. Photograph showing the completion of laminectomy and exposure of the dorsal and lateral surface of normal spinal cord segments in situ with intact meningeal coverings, neighboring blood vessels and dorsal rootlets (**A**). In (**B**), the dura was slit and a hemisection lesion cavity (approximately 5 mm) was created unilaterally. Care has been taken to not disrupt the neighboring blood vessels in the left side of spinal cord or along the dorsal nerve rootlets. In (**C**), the hemisected lesion cavity was filled with the donor tissue and plasma clot obtained from the same animal. An arrow indicates the transplanted tissue in lesioned cavity of the spinal cord.

Using durafilm (Durafilm, Codman Shurtleff, Inc., Randolph, MA, USA), the site of hemisection was covered and sutured with 4/0 silk thread. The muscles and skin were sutured in layers with the help of 1/0 silk thread in order to cover the laminectomized area, and the operated animal was kept in a thermostatic bed until the anesthetic drowsiness effect wore off (Including injuries that occur during the experiments). The operated animals were not injured and survived till the end of the study with zero mortality rate. Animals showed hemiplegic signs due to unilateral spinal cord hemisection but managed to move up and down within the cage. This movement prevented bedsore formation in the operated animals during the post-operative recovery period.

2.7. Preparation and Collection of Donor Tissue for Transplantation

The ulnar nerve was exposed and severed (either from the right or left arm) and the wound was closed and sutured, in addition to the induction of unilateral SC hemisection in the same animal. One week later, the wound at the elbow was reopened under anesthesia and an autologous nerve graft was obtained from the predegenerative nerve segments of the ulnar nerve, and 3–4 fascicles were separated to fill the hemisectioned cavity of SC. The predegenerated ulnar nerve segment was placed longitudinally at the site of hemisection in spinal cord, and analogous plasma clot was added in order to maintain the transected nerve fibres in situ [23] (Figure 4).

2.8. Transplantation Animal Model

Before the implantation of predegenerated nerve grafting, unilateral hemisection was performed on the right side of SC of the trained monkey. After one week, the hemisectioned cavity of SC was reopened, and the accumulated glial scar was aspirated at the lesioned area. Subsequently, the prepared predegenerated segment of the ulnar nerve (3–4 fascicles) obtained from the same animal was placed in the reopened hemisectioned cavity by using the sterile pointed micro forceps along with plasma clot. The operated animal was left in an undisturbed position for 15–20 min for retention of transplanted predegenerated tissue in the lesioned cavity (Figure 4F). In all cases, bleeding was arrested before transplantation of donor tissue. The dura was again closed serially by continuous suture using 4/0 silk thread and covered with dura film to prevent the transplanted tissue from getting dislodged from the lesioned site. The skin was closed over the wound with 6/0 silk thread, and the animal was taken to postoperative care. Perfect closure of dura is essential for better graft survival, as this impedes the epidural scar from penetrating the zone, in which the transplanted tissue has been deposited [24]. Recovery of the animal was assessed by observing the amelioration of neurologic deficits at different postoperative intervals in locomotor behavior on runways.

2.9. Histological Evaluation

At the end of the pop, the animals were deeply anesthetized with Pentathol Sodium by the intraperitoneal route and sacrificed using the method of transcardial perfusion with 10% buffered formalin. Freshly prepared normal saline was allowed to flow into the left ventricle through a cannula fitted with a needle. As the right atrium starts bulging, a nick was made on its wall for the drainage of infusing fluids. After brief washing of the vascular systems with normal saline, two liters of 10% formal saline were infused at room temperature. After a period from 12 to 24 h, the entire spinal cord and brain were dissected in situ. The operated spinal cord segment of animals was confirmed by counting the spinal nerves from the cervical level. After a secondary fixation in buffered 10% formalin, the spinal cord segment of animals was processed for routine paraffin wax block techniques. Serial sections 10 microns thick were taken using a rotary microtome. The sections containing the lesioned site/transplanted tissue of the SC segment were practically saved without any wastage. Subsequently, the well-prepared thin sections were stained with Cresyl fast violet (E.Merck, Darmstadt, Germany) for the microscopical examination of the extent of the lesion and identification of transplanted nerve fascicles in the SC segment of operated animals. Subsequently, the remnants (cadavers) of the animals were incinerated.

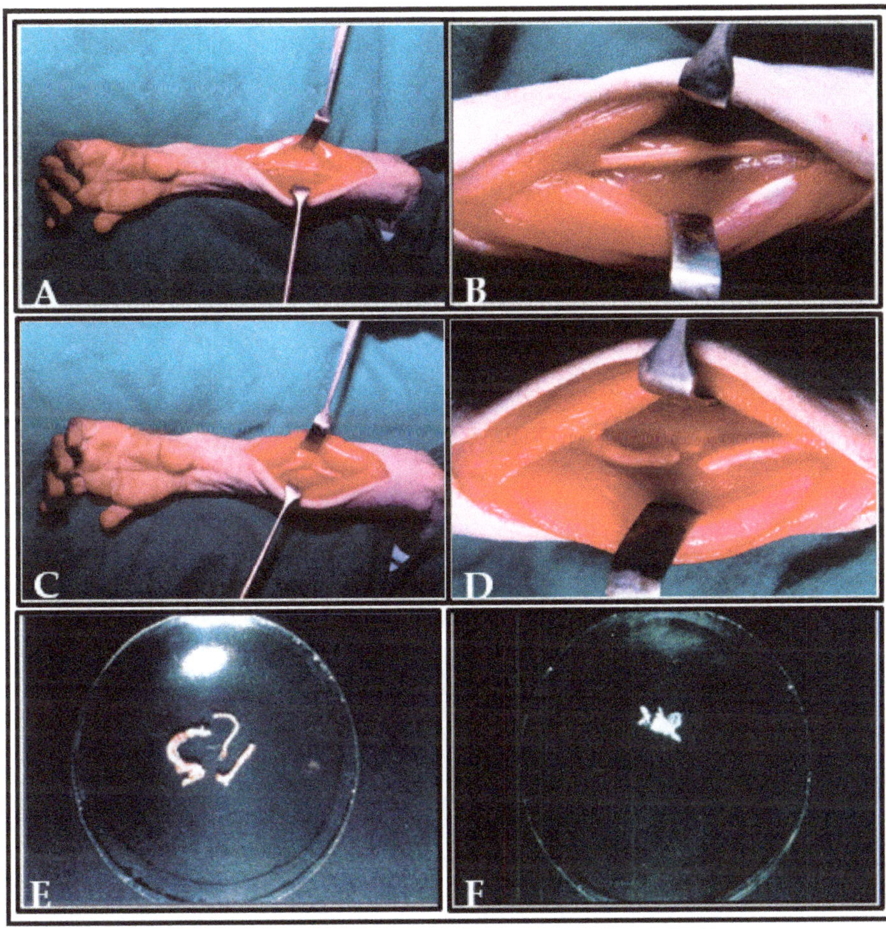

Figure 4. Photographs showing the preparation of ulnar nerve as donor tissue for transplantation. (**A,B**) Picture showing the incision made and the exposure of ulnar nerve. (**C,D**) Picture showing the ulnar nerve ligation. (**E**) Picture showing the ulnar nerve fascicles in dish. (**F**) Picture showing the transplanted predegenerated nerve fascicles in plasma clot at the hemisected cavity of spinal cord (arrow indicates the transplanted predegenerated nerve fascicles at the lesioned area of spinal cord).

2.10. Three (3-D) Dimensional Graphic Reconstruction

A three-dimensional graphic representation of the spinal cord showing lesion cavity and transplant was done for animals where not many spinal cord sections are lost during sectioning or staining. Reconstruction was done on a centimeter graph paper using a Nikon draw tube fitted to a Nikon labophot binocular light microscope with a magnification of 20×. This was kept constant throughout all the serials. Since every tenth-stained section ten microns thick was used for reconstruction, the total thickness was 100 microns, and 20 times magnification of this tube was 2000 microns, which is equivalent to 2.00 mm. To avoid overlapping, a distance of 2.00 cm has been provided between the adjacent drawings. A straight line passing through the central canal as the central axis (third axis) of the section, a horizontal line touching the dorsal or ventral limit of the section, and a ventral line passing through the ventral median fissure were used as a reference point. Then the outline of the lesioned cavity was plotted and colored. A similar method was followed for the reconstruction of the transplanted segment of the spinal cord in another group of animals.

2.11. Statistical Analysis

The data obtained were analyzed by one-way analysis of variance (ANOVA) and post hoc test was performed by Newman–Keuls multiple comparison test. $p < 0.05$ was considered as statistically significant.

3. Results

In the first 4 weeks of pop, the behavioral status of the animals was not a good indicator of the subsequent outcomes, as the operated animals showed paralysis of hindlimb on the operated side, i.e., right side of SC hemisected/transplanted group and failed to respond normally to behavioral tests. The recovery signs were noticeable after the fourth pop week. Hence, the analysis of locomotor functions on grid and staircase runways was carried out from the fifth pop week onwards.

3.1. Staircase Runway

All SC-hemisected animals failed to traverse the staircase runway successfully until the seventh pop week (4.50 ± 0.22; 5.55 ± 0.31 steps, grade 1). By the eight pop week (12.50 ± 0.22 steps; 14.33 ± 0.21 s and 13.43 ± 0.22 steps; 14.63 ± 0.16 s, grade 2), the operated animals had the ability to walk up and down on the runway and completed the task with a longer time. The locomotor performance of the SC-hemisected animals when climbing up and down the runway gradually improved and gained the preoperative values by 14th and 13th pop week (5.50 ± 0.22 steps; 2.67 ± 0.21 s and 6.38 ± 0.24 steps; 3.65 ± 0.13 s, grade 4), respectively. Upon follow-up observation, the animals did not show any further improvements for one year.

The walking pattern of predegenerated peripheral nerve-transplanted animals to cross the staircase runway was also similar to SC-hemisected animals. All SC-transplanted animals took a longer time to climb up and down the runway and completed the task successfully by the end of the ninth pop week (13.02 ± 0.29 steps; 13.96 ± 0.11 s and 14.40 ± 0.15 steps; 14.65 ± 0.13 s, grade 2), which was 1 week delayed compared to the recovery of SC-hemisected animals. However, in subsequent periods, all the predegenerated peripheral nerve-transplanted animals showed an improvement in traversing the runway and attained near normal values by the end of 16th and 17th pop week (6.58 ± 0.16 steps; 3.68 ± 0.04 s and 5.75 ± 0.11 steps; 3.57 ± 0.13 s, grade 4), respectively, which did not improve further (Figures 5 and 6).

3.2. Grid Runways

During the initial pop, the SC hemisected/transplanted group could not complete the walking task on the runways. Due to the increased inter grid distances, the SC-hemisected animals did not complete the task during the early POP. However, the SC-hemisected animals were able to traverse the inter grid distance relatively easily for 4 cm (7.16 ± 0.30 steps; 2.00 ± 0.25 s, grade 4) or 5 cm (7.16 ± 0.30 steps; 2.50 ± 0.22 s, grade 4) indicating almost complete functional recovery by 10th and 12th pop week, respectively. The recovery of functions at 6 cm inter grid distance was similar to the standard score only by the 12th pop week (7.83 ± 0.30 steps; 2.66 ± 0.21 s, grade 4) and for 7 cm inter grid distance by the 14th pop week (10.00 ± 0.25 steps; 3.83 ± 0.30 s, grade 4).

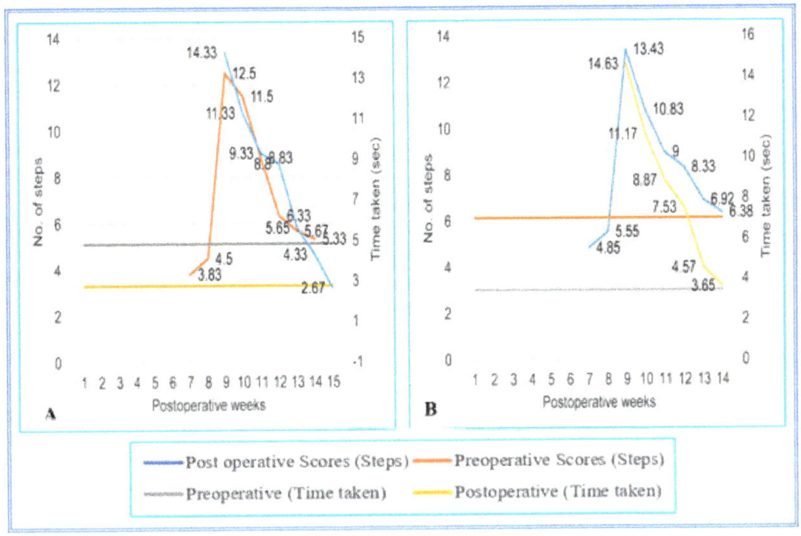

Figure 5. (**A**,**B**) Evaluation of locomotor behavior in staircase test after SPC hemisection in Bonnet monkey (*Macaca radiata*) on staircase runway. X-axis represents duration of recovery of locomotor functions in terms of weeks; primary Y-axis and secondary Y-axis represent the number of steps and time taken by the spinal cord unilateral hemisection plus predegenerated peripheral nerve transplantation monkeys to climb up/down the staircase.

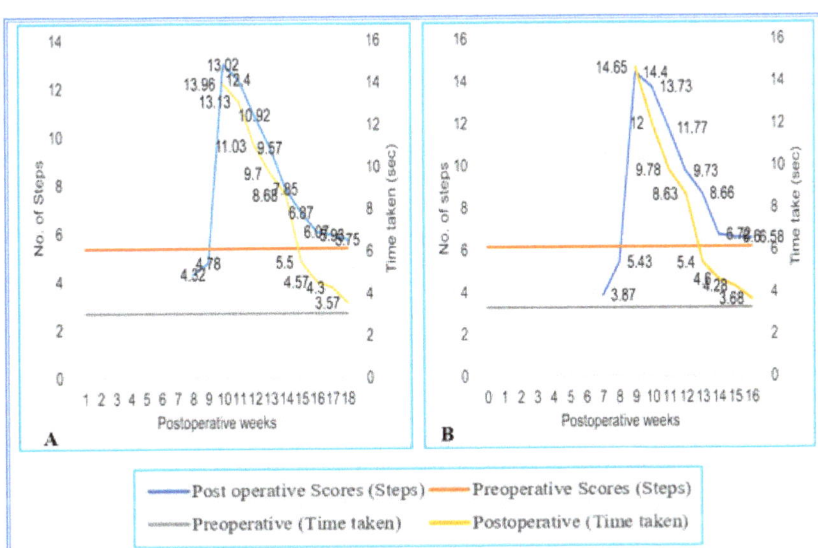

Figure 6. (**A**,**B**) Evaluation of locomotor behavior in staircase test after SPC hemisection plus predegenerated peripheral nerve transplantation in Bonnet monkey (*Macaca radiata*) on staircase runway. Primary X-axis represents duration of recovery of locomotor functions in terms of weeks; primary Y-axis and secondary Y-axis represent the number of steps and time taken by the spinal cord unilateral hemisection plus predegenerated peripheral nerve transplantation monkeys to climb up/down the staircase.

On the 4 cm inter grid runway, the performance of transplanted animals was identical to that of the SC-hemisected animals, but in the 5 cm inter grid interval, the animals showed a maximal recovery of locomotor function on the 11th pop week itself, i.e., 1 week earlier than SC-hemisected animals (7.20 ± 0.27 steps; 3.27 ± 0.32 s, grade 4). A similar type of earlier recovery was also seen in 6 cm grid interval runway, whereas in the 7 cm grid interval, the locomotor performances of transplanted and SC-hemisected animals were similar by the 14th pop week (Figures 7 and 8).

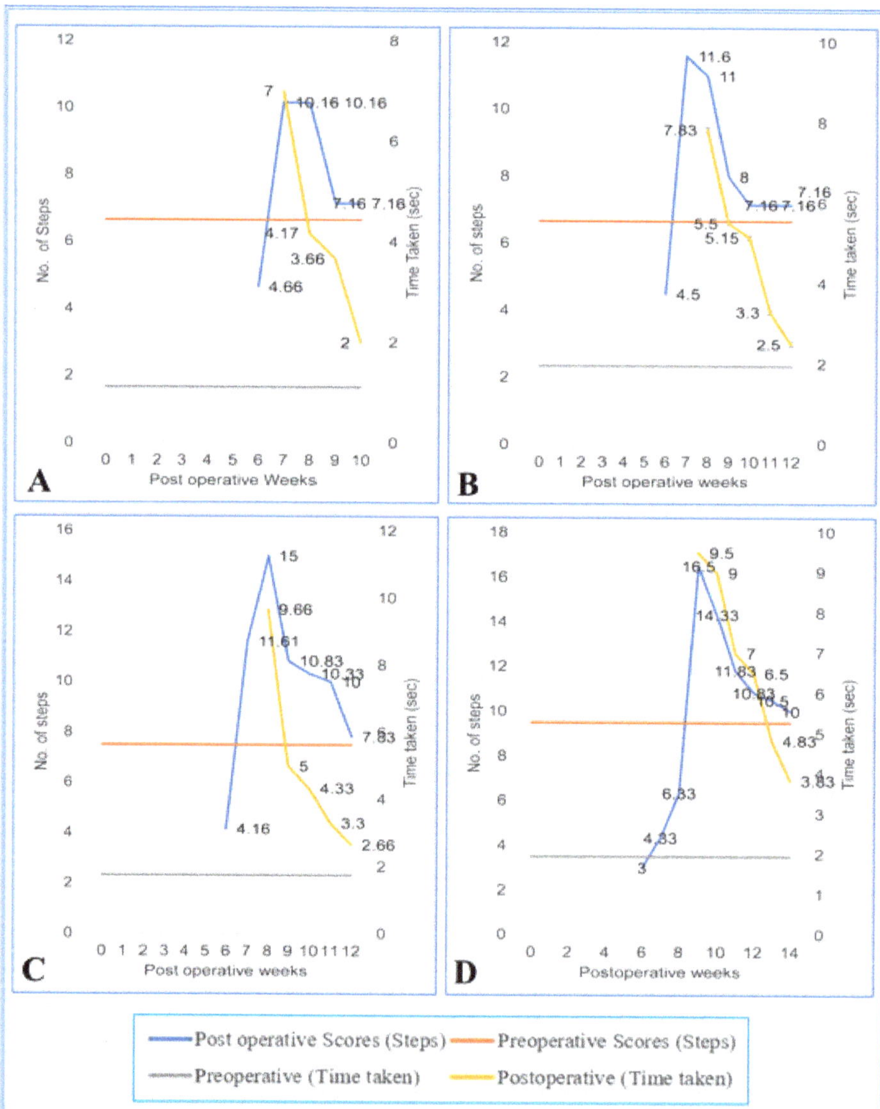

Figure 7. (**A–D**) Evaluation of locomotor behavior in grid runways after SPC hemisection in Bonnet monkey (*Macaca radiata*) on grid runways of 4, 5, 6, and and 7 cm intervals, respectively. Primary X-axis represents duration of recovery of locomotor functions in terms of weeks; primary Y-axis and secondary Y-axis represent the number of steps and time taken by the spinal cord unilateral hemisected monkeys to cross the grid runways.

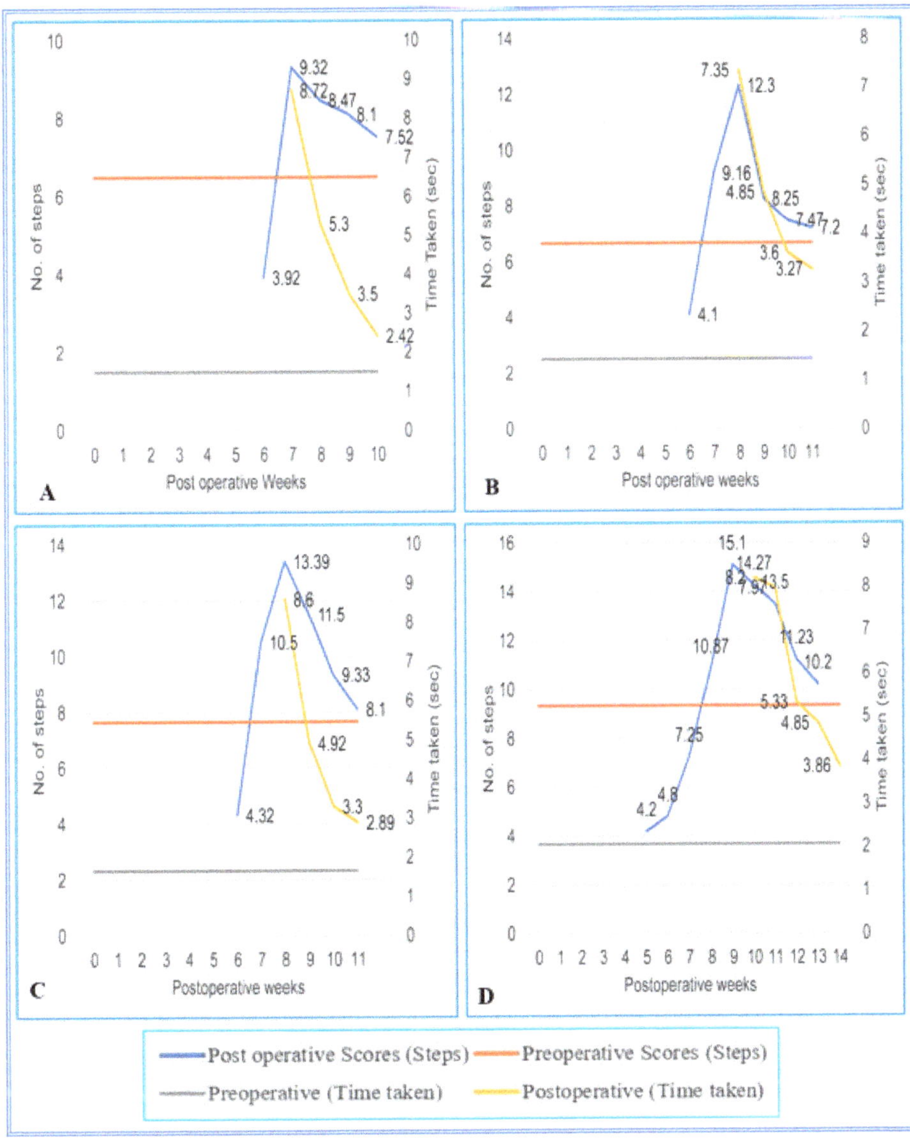

Figure 8. (**A–D**) Evaluation of locomotor behavior in grid runways after SPC hemisection as well as predegenerated peripheral nerve transplantation in Bonnet monkey (*Macaca radiata*) on Grid runways of 4, 5, 6, and 7 cm intervals respectively. Primary X-axis represents duration of recovery of locomotor functions in terms of weeks; primary Y-axis and secondary Y-axis represent the number of steps and time taken by the spinal cord unilateral hemisection as well as predegenerated peripheral nerve transplantation monkeys to cross the grid runways.

In addition to the specific observation above regarding the locomotor performance in G_{SC} and $G_{SC+PDNT}$ monkeys on both staircase and grid runways, we noticed slipping on the staircase runway and misplacement of hindlimb on the grid runway which could not be quantified but is of relevance in assessing the recovery of function after SCI. All the SCI animals sat on the middle of the runway without completing the task during the initial period of observations, as seen in the other runways. Though there was successful completion in traversing the runway by 14th pop week in SC-hemisected animals, by the 17th pop week, there was extensive slipping in the stairs by predegenerated peripheral nerve-transplanted animals, especially while climbing down the stairs. Additionally, the SC-injured animals frequently jumped instead of using normal stepping to traverse the runway, which did not improve until the end of the one-year observatory period. Even though the animals were able to walk successfully across the runway, changes were noted in its gait. The animals were able to grasp the grid bar firmly with their unoperated side of the left hindlimb and supported their maximum bodyweight towards the left hindlimb. In addition to this, as the distance between the grid bars increased to greater than the length of the animal's foot, slipping on the grid increased and was more frequent at 6 cm and 7 cm grid intervals.

3.3. Histological and 3D Graphic Reconstruction Evaluation

Histological and 3D graphical reconstruction of the spinal cord was done to supplement the preliminary findings of the functional outcome of this study. Histological examination revealed clear differentiation of gray and white matter bilaterally in the normal spinal cord. The central canal of the spinal cord was found to be prominent without any alterations in shape (Figure 9A). Following the unilateral spinal cord hemisection, we found deformations and irregularities in both gray and white matter around the area of hemisection. Furthermore, we noticed a few neuronal cell bodies found scattered away from the damaged site of gray matter on the right side of the spinal cord section. This section was analyzed at the maximal extent of the lesion. The central canal of the spinal cord was also found widely opened, and significant alterations in shape were noticed. However, the unlesioned left side was found to be intact, and no serious changes in the arrangement of gray and white matter were noted (Figure 9B). In Figure 9C, the gray and white matter in the right side of the spinal cord section in the transplanted animal was found to be disturbed and interconnected with predegenerated nerve fascicles along with scar on the dorsal side of spinal cord, and no serious changes in the unlesioned side of spinal cord were noted.

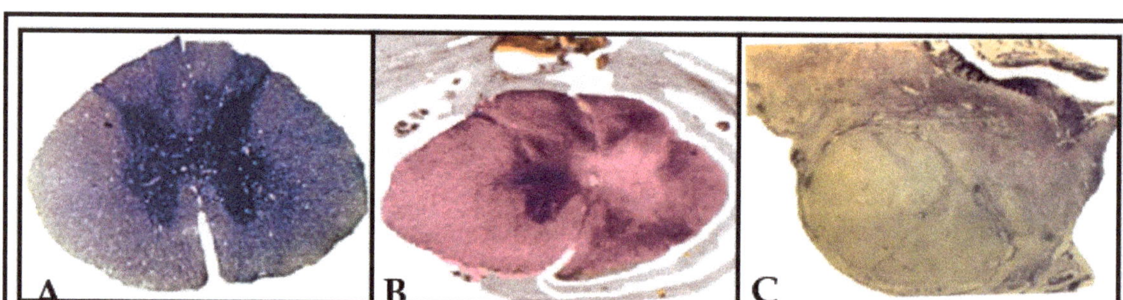

Figure 9. Photographs showing the Nissl (cresyl fast violet)-stained sections of spinal cord of the (**A**) normal animal, (**B**) unilateral spinal-hemisected animal (pop 360 days) and (**C**) peripheral nerve-transplanted animal (pop 360 days). In (**A**), the arrangement of gray and white matter is clearly differentiated in both right and left side of normal spinal cord section. In (**B**), a typical hemisection lesion cavity is viewed in the right side of spinal cord section. The lesion included the total damage of gray and white matter in right side of spinal cord section. The hemisection lesion caused the section to be deformed into an irregular shape of surrounding tissue in right side (indicated by an arrow). This section was at the maximal extent of lesion. The unlesioned left side was found intact, and no serious changes in the arrangement of gray and white matter were noted. In (**C**), the transplanted predegenerated nerve fascicles were found to be interlinked with the unlesioned left side of spinal cord. A black arrow indicates the transplanted predegenerated tissue in right side of lesioned spinal cord.

Lesion studies are generally helpful to convey the information on functional aspects about the effects of lesion on different region of brain or spinal cord for particular behaviors. Three-dimensional graphic representation allows the use of the image and enables a more suitable and effective understanding about the extent of induced lesion or damage occurred (Figure 10). Hence, we created a serial image-reconstructive representation in our study to reveal the accuracy and extent of lesion so as to understand where the induced lesion cavity starts (rostral) and ends (caudal). This information helps us analyze and measure the lesion volume and verify the reproducibility for constant level of lesion and so, as a preliminary data for this study, we have shown these reconstructive images in Figure 10.

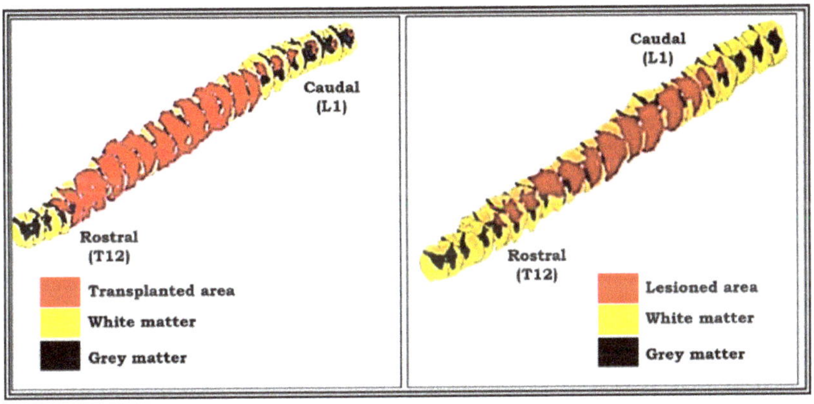

Figure 10. Pictures showing three-dimensional graphic representation of the unilateral hemisected segment of spinal cord (pop 360 days) and transplanted segment of spinal cord (pop 360 days). Reconstruction was done on a centimeter graph paper using Nikon draw tube fitted to a Nikon labophot binocular light microscope with a magnification of 20×. This was kept constant throughout for all the serials. Red-colored part shows the lesioned area, and yellow-colored part shows the intact area of spinal cord segment (see the procedure details for three-dimensional graphic reconstruction in materials and methods).

4. Discussion

The present study demonstrates that transplantation of predegenerated peripheral nerve segments as a therapeutic aid was found to be ineffective in improving the functional recovery of bipedal locomotion assessed using the two complex runways in unilaterally SC-hemisected monkeys. To our knowledge the present study is the first of its kind in the field of neuroscience to use complex runways in non-human primates (*Macaca radiata*) for the assessment of bipedal locomotor behavior after SC damage.

To appraise the valuable effect of transplants and the mechanisms underlying the recovery of motor functions, it is essential to use behavioral analysis [25]. This helps to differentiate between various kinds of SC injuries or to segregate the effects of several therapeutic interventions. Although most of the investigations concerning the assessment of behavioral functions after SCI have been carried out in rodents, only a handful of studies have been extended to higher species. Primate and large animal models are being increasingly used and tested to explore the possible novel therapies for spinal cord injuries and to advance the ideal methods for human translation.

The PNG may depend on several factors for effective SC "reconstruction". The most common factor is the secondary necrosis that occurs in stumps of surgically transected SC, which results in the loss of additional SC tissue and appearance of cavities during the first week of transection [26]. It was thus indicated that immediate grafting of peripheral nerve grafts to SC transection was found to be innervated but separated with some degree of cyst cavities from the SC stumps [27]. The PNG can be collected instantly after nerve transection (fresh), or the nerve can be cut at its proximal end and left in the animal to undergo Wallerian degeneration and collected later (predegenerated). Available evidence from the earlier investigations revealed that the most common period of predegeneration is 7 days; however, a positive effect on axon growth can be seen with predegeneration times as short as 3 days and as long as 35 days [28]. Nevertheless, Decherchi and Gauthier [17] reported that the most effective period of predegeneration was found to occur between 3–5 days for the rapid axonal growth of peripheral nerves. Oudega et al. [15], Zhao and Kerns [29], Gordon [30], and Zhou and Notterpek [31], in their studies, have shown a mild to moderate rise in the rate and extent of axon growth within predegenerated as opposed to fresh PNG. This effect is likely due to an axotomy-induced increase in Schwann cell mitotic activity and trophic factors production. Studies have shown that a predegenerated peripheral nerve graft reduced the initial lapse before regenerating nerve fibers enter the graft without affecting the regeneration rate and was more effective in promoting rapid regeneration of transected fibers of SC [32,33]. However, the effectiveness of the predegenerated versus fresh nerve graft remains controversial. In some studies, the predegenerated peripheral nerve graft either did not result in enhanced axonal growth compared to fresh grafts [34] or the increased regeneration was only evident at early time points with no difference at longer survival times [28,35].

Remarkably, few variances were observed between "fresh" transplants and nerves previously denervated in vivo; the latter were more densely innervated by fibers. From the above observation, two conclusions were derived (1) in a peripheral nerve microenvironment, the CNS neuron can regenerate fibers and (2) this may be due to the synthesis of chemotrophic and neurotrophic factors by peripheral Schwann cells in response to denervation, a phenomenon which hypothetically would be absent in CNS tissue [14,26]. Furthermore, the degenerated nerves excluded the possibility of early intervention of infiltrating macrophages during the storage phase and their consequent influence both on Schwann cell activity and on the removal of debris. Some authors believe that the premature removal of debris within predegenerated peripheral nerves could be described as facilitating regeneration by reducing the initial stage of regeneration, whereas others believe that the presence of debris has little effect, specifically in the initial stage while the advancing axons are so thin that they can easily pass through debris [36,37]. In our study, the transplantation of predegenerated peripheral nerve segments into SC-hemisected monkeys showed no improvement in the recovery of locomotor functions using the complex runways. This is contrary to the previous literature on rats [29,31,38] and cats [23], which have shown moderate improvement in SC function following transplantation with PNG. However, these studies did not focus on the behavioral parameters but mostly focused on the histological changes in SC segments following transplantation. The failure of this technique using predegenerated peripheral nerve grafts as a donor material in the repair of lost functions in SC-hemisected monkeys may be due to a number of factors set in motion by trauma, which could considerably prevent the qualities of regeneration, some of which are discussed below.

Many investigations in CNS injury research have delineated the benefit of PNGs as [39] a favorable and growth-promoting environment to a variety of lesioned axons. However, the effectiveness of the peripheral nerve environment depends on the level of injury and may not be suitable for the regrowth of long axons into these conduits. For instance, the axonal growth into PNG is limited from injured retinal ganglion cells or rubrospinal neurons when the axons are injured a long distance away from the neuronal cell body [40–42]. On the other hand, most CNS axons are capable of lengthy growth within the permissive substrates (e.g., PNG) following axotomy close to the cell body. The biochemical changes involved in axon regeneration are correlated with a neuronal cell body response that includes significant upregulation of regeneration-associated-gens (RAGs) viz., GAP-43, Tα1-tubulin, and c-Jun [43–46]. Available literature proved that infusion of neurotrophic factors or transplantation of small peripheral nerve segments in close proximity to the neuronal cell body has not only increased RAG expression but also enhanced the extension of injured CNS axons into PNG. In our study, the failure in the regeneration of PNG to form suitable functional connections between the rostral and caudal end of SC on the recovery of bipedal locomotion is probably due to the distant location of the transplanted segment of SC from the neuronal cell body of medulla.

Experimental studies in animals have shown that a small number of regenerating axonal fibers was capable of leaving the PNG and re-entering the CNS. These structural rearrangements may be seen due to formation of glial scar at the PNS-CNS boundary, which mostly includes reactive astrocytes, fibroblasts, and microglia. Reactive astrocytes at the boundary were found to inhibit the growth of axons by producing a physiological [47] stop signal within the growing axonal fibers [48,49]. Furthermore, reactive astrocytes present in glial scar express significant levels of inhibitory molecules such as tenascin and chondroitin sulphate proteoglycans [50]. Another important factor is the glial scarring that prevents the diffusion process between cells at the injury site. For instance, using the evaluation of diffusion co-efficient with tetramethyl ammonium-sensitive electrodes, Roitback and Sykova [51] observed that the diffusion process decreased with astrocytic hypertrophy and increased chondroitin sulphate. From the available reports, it is evident that formation of glial scar can express higher levels of molecular inhibitors and also act as a barricade to prevent the diffusion of growth-promoting molecules. Another study by Stichel et al. [52] demonstrated that the impermeable nature of glial scar at the injured site is due to its basal membrane. Moreover, in the lesioned fornix, the reduction in collagen type IV resulted in axonal migration in glial scar region despite continued expression of proteoglycans and glial activation. Nevertheless, this approach did not influence the significant growth of the injured corticospinal tract [47].

There is also convincing evidence that some axons re-entering the CNS was found to grow only short distances, approximately 1–2 mm within the adult mammalian nervous system. Another strategy to facilitate the growth of CNS axons after injury is to neutralize the molecules that act as putative inhibitors of axonal growth [53]. In 1988, some investigators demonstrated the presence of actual growth inhibitors in the CNS of mammals. They reported that that the inhibitory nature of cultured oligodendrocytes could be blocked by the monoclonal antibody. IN-1. In addition, the gene for this inhibitor was also sequenced and named nogo [54,55]. Several other factors that prevent the axon growth have also been identified, including the glycoproteins and proteoglycans [56–59].

Several investigators have reported that a number of inhibitory factors found in immature and adult brains may be re-expressed after the injury. For example, the chemorepellent semaphorin III/collapsing I is re-expressed in scar tissue after an injury in adult CNS [60–62]. Similarly, the other growth inhibitors such as chondroitin sulphate and keratin are also re-expressed in mature brains [63] and the SC after injury [64–67]. Even though some of the earlier investigations showed encouraging results with the use of artificial materials in the form of bridges across the SC lesions, there are also conflicting results seen in relation to achieve the proper regeneration. For example, the injection of Schwann cell in the form of suspension or as a "Schwann cell cable" in a guidance channel of semipermeable membrane was observed to enhance the regeneration of SC fibers. Schreyer and Jones [68] identified the growth of newborn fibers in laminin-coated nitrocellulose implants in the SC of neonatal rats. On the other hand, some scientists reported poor regeneration of corticospinal fibers in adult rats in laminin-coated nitrocellulose bridges [69]. Paino et al. [38] reported negative results in the regrowth of axons with the use of collagen in lesioned adult rat SC. Conversely, Khan et al. [70] observed minimal growth of corticospinal fibers across carbon filament implants in transected rat SCs. In some cases, many axons were observed to grow in SC-filled guidance channels, but they failed to exit the graft and do not re-enter the CNS environment [71].

In the current study, we carefully tested the possible recovery of bipedal locomotor functions in unilateral SC-hemisected monkeys transplanted with predegenerated PNG. In SCI research, behavioral analysis is an indispensable tool to evaluate the functional recovery. The results of bipedal locomotor functional analysis in primates might be more reliable to relate with the results to humans. In SCI research, the use of behavioral analysis is indispensable to evaluate the appreciated outcome of transplants and their underlying mechanisms on the recovery of motor functions in animals. Forerunners in this field of research have emphasized the importance of behavioral tests to analyze the recovery of motor functions in animals after SCI for the following reasons: (1) the behavioral analysis may postulate hidden deficits of the undamaged part from the damaged; (2) deficit nature can be quantified, and quantification might be valuable in diagnosing the types of abnormalities and (3) multiple tests of behavioral analysis might minimize the experimental artifact of a single test which significantly alters the results. However, several scientists in SCI research have mostly used behavioral tests for analyzing the recovery of general motor functions in animals after inducing SCI.

Many factors lead to the promotion or inhibition of axonal growth and functional improvement after SCI. For instance, the role of regeneration-associated genes (RAGs) such as GAP-43, Tα1-tubulin, and c-Jun may not be amply expressed or weak enough to support the promotion of axonal regeneration following predegenerated peripheral nerve transplantation. The failure of RAGs on the regeneration of axonal fibers may be ineffective to initiate the connection between the transplanted tissue and host fibers. Similarly, there are also certain growth-promoting molecules that influence axonal regeneration in adult CNS. Notable exceptions include laminin 1 and polysialic acids, which play a role in axonal regeneration in an SC-lesioned animal. Other inhibitory factors identified, viz., chemorepellent semaphorin III/collapsing I, chondroitin sulphate, and proteoglycan NG2, may be re-expressed in scar tissue or axons or glial cells after injury in adult CNS. These molecular inhibitors present in the scar at the damaged site may act as a simple barrier to the diffusion of growth-promoting molecules and do not provide sufficient improvement in SC repair and restitution of motor functions. Considering these problems can be resolved, there is a significant possibility of making a suitable rewiring of ingrowth axonal fibers between transplants and host tissue, which helps ameliorate the recovery of locomotor functions following the predegenerated peripheral nerve transplantation in SC injured animals.

Strength and Limitations

The present study analyzed the bipedal locomotor recovery in SCI monkeys. The animal model used is a major strength to this study, as it closely resembles human responses in SCI. The study analyzed the effects of SCI in monkeys longitudinally for a period of 1 year, and hence, the results are comprehensive.

However, the study has its own limitations. This study primarily focused on the functional motor recovery using a behavioral test and not on other parameters, such as histology, immunohistochemistry and electromyography studies.

To overcome these limitations, our experimental group planned to study the advantages of different axonal growth-promoting molecules and neurotrophic factors in conjunction with glial scar-modifying or myelin-inhibiting blocking factors.

Author Contributions: A.P., M.J., S.B.R. and P.P. engaged in the analytical experimental part. A.P., S.B.R., S.M., S.J. and S.D. were involved in literature collection and manuscript preparation. A.P., S.B.R., S.M. and S.J. performed the statistical analysis. A.P., S.B.R., A.A. and A.A.A. coordinated the manuscript preparation and revision. All authors have read and agreed to the published version of the manuscript.

Funding: This research received no external funding.

Institutional Review Board Statement: The ethical approval was obtained from the Animal Care and Institutional Ethical Committee of the University of Madras (PGIBMS/PHY/DRRR/2021/0702021).

Informed Consent Statement: The informed consent was obtained from the Zoological Park, Chennai, Tamilnadu, India.

Data Availability Statement: The data presented in this study are available in the manuscript.

Acknowledgments: The authors would like to thank the Deanship of Scientific Research, Majmaah University, Kingdom of Saudi Arabia for research support under project number: R-2021-268. And the authors also would like to thank ALM PGIBMS, University of Madras, Chennai 600113, Tamil Nadu, India for the research facilities and academic support.

Conflicts of Interest: The authors declare no conflict of interest.

References

1. Reier, P.; Bregman, B.; Wujek, J. Intraspinal transplantation of embyronic spinal cord tissue in neonatal and adult rats. *J. Comp. Neurol.* **1986**, *247*, 275–296. [CrossRef]
2. Lin, C.-C.; Lai, S.-R.; Shao, Y.-H.; Chen, C.-L.; Lee, K.-Z. The therapeutic effectiveness of delayed fetal spinal cord tissue transplantation on respiratory function following mid-cervical spinal cord injury. *Neurotherapeutics* **2017**, *14*, 792–809. [CrossRef]
3. Côté, M.P.; Amin, A.A.; Tom, V.J.; Houle, J.D. Peripheral nerve grafts support regeneration after spinal cord injury. *Neurotherapeutics* **2011**, *8*, 294–303. [CrossRef] [PubMed]
4. DePaul, M.A.; Lin, C.-Y.; Silver, J.; Lee, Y.-S. Peripheral nerve transplantation combined with acidic fibroblast growth factor and chondroitinase induces regeneration and improves urinary function in complete spinal cord transected adult mice. *PLoS ONE* **2015**, *10*, e0139335. [CrossRef]
5. Kanno, H.; Pearse, D.D.; Ozawa, H.; Itoi, E.; Bunge, M.B. Schwann cell transplantation for spinal cord injury repair: Its significant therapeutic potential and prospectus. *Rev. Neurosci.* **2015**, *26*, 121–128. [CrossRef] [PubMed]
6. De la Garza-Castro, O.; Martínez-Rodríguez, H.G.; Sánchez-González, S.G.; Vidal-Torres, O.; Arreola-Romero, A.; de la Garza-Pineda, O.; Ancer-Arellano, A.G.; Guzmán-López, S.; Elizondo-Omaña, R.E. Schwann cell precursor transplant in a rat spinal cord injury model. *Rev. Investig. Clín.* **2018**, *70*, 88–95. [CrossRef]
7. Levi, A.D.; Dancausse, H.; Li, X.; Duncan, S.; Horkey, L.; Oliviera, M. Peripheral nerve grafts promoting central nervous system regeneration after spinal cord injury in the primate. *J. Neurosurg. Spine* **2002**, *96*, 197–205. [CrossRef] [PubMed]
8. Gao, Y.; Wang, Y.-L.; Kong, D.; Qu, B.; Su, X.-J.; Li, H.; Pi, H.-Y. Nerve autografts and tissue-engineered materials for the repair of peripheral nerve injuries: A 5-year bibliometric analysis. *Neural Regen. Res.* **2015**, *10*, 1003.
9. Villegas-Perez, M.P.; Vidal-Sanz, M.; Bray, G.M.; Aguayo, A.J. Influences of peripheral nerve grafts on the survival and regrowth of axotomized retinal ganglion cells in adult rats. *J. Neurosci.* **1988**, *8*, 265–280. [CrossRef]
10. Quan, M.-Z.; Kosaka, J.; Watanabe, M.; Wakabayashi, T.; Fukuda, Y. Survival of Axotomized Retinal Ganglion Cells in Peripheral Nerve–Grafted Ferrets. *Investig. Ophthalmol. Vis. Sci.* **1999**, *40*, 2360–2366.
11. Richardson, P.; Issa, V.; Aguayo, A. Regeneration of long spinal axons in the rat. *J. Neurocytol.* **1984**, *13*, 165–182. [CrossRef]
12. El Seblani, N.; Welleford, A.S.; Quintero, J.E.; van Horne, C.G.; Gerhardt, G.A. Invited review: Utilizing peripheral nerve regenerative elements to repair damage in the CNS. *J. Neurosci. Methods* **2020**, *335*, 108623. [CrossRef] [PubMed]
13. Ramón-Cueto, A.; Plant, G.W.; Avila, J.; Bunge, M.B. Long-distance axonal regeneration in the transected adult rat spinal cord is promoted by olfactory ensheathing glia transplants. *J. Neurosci.* **1998**, *18*, 3803–3815. [CrossRef] [PubMed]
14. Kao, C.C.; Chang, L.W.; Bloodworth, J., Jr. Axonal regeneration across transected mammalian spinal cords: An electron microscopic study of delayed microsurgical nerve grafting. *Exp. Neurol.* **1977**, *54*, 591–615. [CrossRef]
15. Oudega, M.; Varon, S.; Hagg, T. Regeneration of adult rat sensory axons into intraspinal nerve grafts: Promoting effects of conditioning lesion and graft predegeneration. *Exp. Neurol.* **1994**, *129*, 194–206. [CrossRef] [PubMed]
16. Dam-Hieu, P.; Liu, S.; Choudhri, T.; Said, G.; Tadié, M. Regeneration of primary sensory axons into the adult rat spinal cord via a peripheral nerve graft bridging the lumbar dorsal roots to the dorsal column. *J. Neurosci. Res.* **2002**, *68*, 293–304. [CrossRef] [PubMed]
17. Decherchi, P.; Gauthier, P. In vitro pre-degenerated nerve autografts support CNS axonal regeneration. *Brain Res.* **1996**, *726*, 181–188. [CrossRef]
18. Gauthier, P.; Rega, P.; Lammari-Barreault, N.; Polentes, J. Functional reconnections established by central respiratory neurons regenerating axons into a nerve graft bridging the respiratory centers to the cervical spinal cord. *J. Neurosci. Res.* **2002**, *70*, 65–81. [CrossRef]
19. Senoo, E.; Tamaki, N.; Fujimoto, E.; Ide, C. Effects of prelesioned peripheral nerve graft on nerve regeneration in the rat spinal cord. *Neurosurgery* **1998**, *42*, 1347–1356. [CrossRef]
20. Nout, Y.S.; Rosenzweig, E.S.; Brock, J.H.; Strand, S.C.; Moseanko, R.; Hawbecker, S.; Zdunowski, S.; Nielson, J.L.; Roy, R.R.; Courtine, G.; et al. Animal models of neurologic disorders: A nonhuman primate model of spinal cord injury. *Neurotherapeutics* **2012**, *9*, 380–392. [CrossRef] [PubMed]
21. Nardone, R.; Florea, C.; Höller, Y.; Brigo, F.; Versace, V.; Lochner, P.; Golaszewski, S.; Trinka, E. Rodent, large animal and non-human primate models of spinal cord injury. *Zoology* **2017**, *123*, 101–114. [CrossRef]
22. Kerasidis, H.; Wrathall, J.R.; Gale, K. Behavioral assessment of functional deficit in rats with contusive spinal cord injury. *J. Neurosci. Methods* **1987**, *20*, 167–179. [CrossRef]
23. Wrathall, J.; Rigamonti, D.; Braford, M.; Kao, C. Reconstruction of the contused cat spinal cord by the delayed nerve graft technique and cultured peripheral non-neuronal cells. *Acta Neuropathol.* **1982**, *57*, 59–69. [CrossRef]
24. Fernandez, E.; Pallini, R.; Maira, G.; Rossi, G. Peripheral nerve autografts to the injured spinal cord of the rat: An experimental model for the study of spinal cord regeneration. *Acta Neurochir.* **1985**, *78*, 57–64. [CrossRef]
25. Suresh Babu, R.; Anand, P.; Jeraud, M.; Periasamy, P.; Namasivayam, A. Bipedal locomotion of bonnet macaques after spinal cord injury. *Mot. Control* **2007**, *11*, 322–347. [CrossRef]

26. Kao, C.C.; Chang, L.W.; Bloodworth, J.M. The mechanism of spinal cord cavitation following spinal cord transection: Part 2: Electron microscopic observations. *J. Neurosurg.* **1977**, *46*, 745–756. [CrossRef] [PubMed]
27. Richardson, P.; McGuinness, U.; Aguayo, A. Peripheral nerve autografts to the rat spinal cord: Studies with axonal tracing methods. *Brain Res.* **1982**, *237*, 147–162. [CrossRef]
28. Lewin-Kowalik, J.; Sieron, A.L.; Krause, M.; Barski, J.-J.; Górka, D. Time-dependent regenerative influence of predegenerated nerve grafts on hippocampus. *Brain Res. Bull.* **1992**, *29*, 831–835. [CrossRef]
29. Zhao, Q.; Kerns, J.M. Effects of predegeneration on nerve regeneration through silicone Y-chambers. *Brain Res.* **1994**, *633*, 97–104. [CrossRef]
30. Gordon, T. The role of neurotrophic factors in nerve regeneration. *Neurosurg. Focus* **2009**, *26*, E3. [CrossRef] [PubMed]
31. Zhou, Y.; Notterpek, L. Promoting peripheral myelin repair. *Exp. Neurol.* **2016**, *283*, 573–580. [CrossRef] [PubMed]
32. Danielsen, N.; Kerns, J.M.; Holmquist, B.; Zhao, Q.; Lundborg, G.; Kanje, M. Pre-degenerated nerve grafts enhance regeneration by shortening the initial delay period. *Brain Res.* **1994**, *666*, 250–254. [CrossRef]
33. Danielsen, N.; Kerns, J.M.; Holmquist, B.; Zhao, Q.; Lundborg, G.; Kanje, M. Predegeneration enhances regeneration into acellular nerve grafts. *Brain Res.* **1995**, *681*, 105–108. [CrossRef]
34. Ellis, J.C.; McCaffrey, T.V. Nerve grafting: Functional results after primary vs delayed repair. *Arch. Otolaryngol.* **1985**, *111*, 781–785. [CrossRef]
35. Lewin-Kowalik, J.; Górka, D.; Larysz-Brysz, M.; Goka, B.; Swiech-Sabuda, E.; Małecka-Tendera, E.; Krause, M. Short-time predegenerated peripheral nerve grafts promote regrowth of injured hippocampal neurites. *Acta Physiol. Hung.* **1997**, *85*, 259–268.
36. Ochi, M.; Wakasa, M.; Ikuta, Y.; Kwong, W. Nerve regeneration in predegenerated basal lamina graft: The effect of duration of predegeneration on axonal extension. *Exp. Neurol.* **1994**, *128*, 216–225. [CrossRef] [PubMed]
37. Hall, S. Axonal regeneration through acellular muscle grafts. *J. Anat.* **1997**, *190*, 57–71. [CrossRef] [PubMed]
38. Paino, C.; Fernandez-Valle, C.; Bates, M.; Bunge, M. Regrowth of axons in lesioned adult rat spinal cord: Promotion by implants of cultured Schwann cells. *J. Neurocytol.* **1994**, *23*, 433–452. [CrossRef]
39. Oudega, M.; Hagg, T. Nerve growth factor promotes regeneration of sensory axons into adult rat spinal cord. *Exp. Neurol.* **1996**, *140*, 218–229. [CrossRef]
40. Richardson, P.; Issa, V.; Shemie, S. Regeneration and retrograde degeneration of axons in the rat optic nerve. *J. Neurocytol.* **1982**, *11*, 949–966. [CrossRef]
41. Kobayashi, N.R.; Fan, D.-P.; Giehl, K.M.; Bedard, A.M.; Wiegand, S.J.; Tetzlaff, W. BDNF and NT-4/5 prevent atrophy of rat rubrospinal neurons after cervical axotomy, stimulate GAP-43 and Tα1-tubulin mRNA expression, and promote axonal regeneration. *J. Neurosci.* **1997**, *17*, 9583–9595. [CrossRef]
42. Kwon, B.K.; Liu, J.; Oschipok, L.; Teh, J.; Liu, Z.W.; Tetzlaff, W. Rubrospinal neurons fail to respond to brain-derived neurotrophic factor applied to the spinal cord injury site 2 months after cervical axotomy. *Exp. Neurol.* **2004**, *189*, 45–57. [CrossRef]
43. Tetzlaff, W.; Alexander, S.W.; Miller, F.D.; Bisby, M.A. Response of facial and rubrospinal neurons to axotomy: Changes in mRNA expression for cytoskeletal proteins and GAP-43. *J. Neurosci.* **1991**, *11*, 2528–2544. [CrossRef] [PubMed]
44. Fernandes, K.J.; Fan, D.P.; Tsui, B.; Cassar, S.; Tetzlaff, W. Influence of the axotomy to cell body distance in rat rubrospinal and spinal motoneurons: Differential regulation of GAP-43, tubulins, and neurofilament-M. *J. Comp. Neurol.* **1999**, *414*, 495–510. [CrossRef]
45. Jenkins, R.; Tetzlaff, W.; Hunt, S. Differential expression of immediate early genes in rubrospinal neurons following axotomy in rat. *Eur. J. Neurosci.* **1993**, *5*, 203–209. [CrossRef] [PubMed]
46. Broude, E.; McAtee, M.; Kelley, M.; Bregman, B. c-Jun expression in adult rat dorsal root ganglion neurons: Differential response after central or peripheral axotomy. *Exp. Neurol.* **1997**, *148*, 367–377. [CrossRef]
47. Weidner, N.; Blesch, A.; Grill, R.; Tuszynski, M. Nerve growth factor–hypersecreting Schwann cell grafts augment and guide spinal cord axonal growth and remyelinate central nervous system axons in a phenotypically appropriate manner that correlates with expression of L1. *J. Comp. Neurol.* **1999**, *413*, 495–506. [CrossRef]
48. Liuzzi, F.J.; Tedeschi, B. Axo-glial interactions at the dorsal root transitional zone regulate neurofilament protein synthesis in axotomized sensory neurons. *J. Neurosci.* **1992**, *12*, 4783–4792. [CrossRef]
49. Fraher, J.P. The transitional zone and CNS regeneration. *J. Anat.* **2000**, *196*, 137–158. [CrossRef] [PubMed]
50. Pindzola, R.; Doller, C.; Silver, J. Putative inhibitory extracellular matrix molecules at the dorsal root entry zone of the spinal cord during development and after root and sciatic nerve lesions. *Dev. Biol.* **1993**, *156*, 34–48. [CrossRef] [PubMed]
51. Roitbak, T.; Syková, E. Diffusion barriers evoked in the rat cortex by reactive astrogliosis. *Glia* **1999**, *28*, 40–48. [CrossRef]
52. Stichel, C.C.; Hermanns, S.; Luhmann, H.J.; Lausberg, F.; Niermann, H.; D'Urso, D.; Servos, G.; Hartwig, H.G.; Müller, H.W. Inhibition of collagen IV deposition promotes regeneration of injured CNS axons. *Eur. J. Neurosci.* **1999**, *11*, 632–646. [CrossRef] [PubMed]
53. Keirstead, H.S.; Dyer, J.K.; Sholomenko, G.N.; McGraw, J.; Delaney, K.; Steeves, J. Axonal regeneration and physiological activity following transection and immunological disruption of myelin within the hatchling chick spinal cord. *J. Neurosci.* **1995**, *15*, 6963–6974. [CrossRef] [PubMed]
54. Chen, M.S.; Huber, A.B.; van der Haar, M.E.; Frank, M.; Schnell, L.; Spillmann, A.A.; Christ, F.; Schwab, M.E. Nogo-A is a myelin-associated neurite outgrowth inhibitor and an antigen for monoclonal antibody IN-1. *Nature* **2000**, *403*, 434–439. [CrossRef] [PubMed]

55. Oertle, T.; Van Der Haar, M.E.; Bandtlow, C.E.; Robeva, A.; Burfeind, P.; Buss, A.; Huber, A.B.; Simonen, M.; Schnell, L.; Brösamle, C.; et al. Nogo-A inhibits neurite outgrowth and cell spreading with three discrete regions. *J. Neurosci.* **2003**, *23*, 5393–5406. [CrossRef] [PubMed]
56. Fitch, M.T.; Silver, J. Glial cell extracellular matrix: Boundaries for axon growth in development and regeneration. *Cell Tissue Res.* **1997**, *290*, 379–384. [CrossRef] [PubMed]
57. Fawcett, J.W.; Asher, R.A. The glial scar and central nervous system repair. *Brain Res. Bull.* **1999**, *49*, 377–391. [CrossRef]
58. Tang, X.; Davies, J.E.; Davies, S.J. Changes in distribution, cell associations, and protein expression levels of NG2, neurocan, phosphacan, brevican, versican V2, and tenascin-C during acute to chronic maturation of spinal cord scar tissue. *J. Neurosci. Res.* **2003**, *71*, 427–444. [CrossRef] [PubMed]
59. Buss, A.; Pech, K.; Kakulas, B.A.; Martin, D.; Schoenen, J.; Noth, J.; Brook, G.A. NG2 and phosphacan are present in the astroglial scar after human traumatic spinal cord injury. *BMC Neurol.* **2009**, *9*, 32. [CrossRef] [PubMed]
60. Pasterkamp, R.J.; Giger, R.J.; Ruitenberg, M.J.; Holtmaat, A.; De Wit, J.; De Winter, F.; Verhaagen, J. Expression of the gene encoding the chemorepellent semaphorin III is induced in the fibroblast component of neural scar tissue formed following injuries of adult but not neonatal CNS. *Mol. Cell. Neurosci.* **1999**, *13*, 143–166. [CrossRef]
61. De Winter, F.; Oudega, M.; Lankhorst, A.; Hamers, F.; Blits, B.; Ruitenberg, M.J.; Pasterkamp, R.J.; Gispen, W.H.; Verhaagen, J. Injury-induced class 3 semaphorin expression in the rat spinal cord. *Exp. Neurol.* **2002**, *175*, 61–75. [CrossRef] [PubMed]
62. Hashimoto, M.; Ino, H.; Koda, M.; Murakami, M.; Yoshinaga, K.; Yamazaki, M.; Moriya, H. Regulation of semaphorin 3A expression in neurons of the rat spinal cord and cerebral cortex after transection injury. *Acta Neuropathol.* **2004**, *107*, 250–256. [CrossRef]
63. Haas, C.A.; Rauch, U.; Thon, N.; Merten, T.; Deller, T. Entorhinal cortex lesion in adult rats induces the expression of the neuronal chondroitin sulfate proteoglycan neurocan in reactive astrocytes. *J. Neurosci.* **1999**, *19*, 9953–9963. [CrossRef] [PubMed]
64. Lemons, M.L.; Howland, D.R.; Anderson, D.K. Chondroitin sulfate proteoglycan immunoreactivity increases following spinal cord injury and transplantation. *Exp. Neurol.* **1999**, *160*, 51–65. [CrossRef] [PubMed]
65. Ikegami, T.; Nakamura, M.; Yamane, J.; Katoh, H.; Okada, S.; Iwanami, A.; Watanabe, K.; Ishii, K.; Kato, F.; Fujita, H. Chondroitinase ABC combined with neural stem/progenitor cell transplantation enhances graft cell migration and outgrowth of growth-associated protein-43-positive fibers after rat spinal cord injury. *Eur. J. Neurosci.* **2005**, *22*, 3036–3046. [CrossRef] [PubMed]
66. Iseda, T.; Okuda, T.; Kane-Goldsmith, N.; Mathew, M.; Ahmed, S.; Chang, Y.-W.; Young, W.; Grumet, M. Single, high-dose intraspinal injection of chondroitinase reduces glycosaminoglycans in injured spinal cord and promotes corticospinal axonal regrowth after hemisection but not contusion. *J. Neurotrauma* **2008**, *25*, 334–349. [CrossRef] [PubMed]
67. Wang, Z.; Winsor, K.; Nienhaus, C.; Hess, E.; Blackmore, M.G. Combined chondroitinase and KLF7 expression reduce net retraction of sensory and CST axons from sites of spinal injury. *Neurobiol. Dis.* **2017**, *99*, 24–35. [CrossRef] [PubMed]
68. Schreyer, D.; Jones, E. Growth of corticospinal axons on prosthetic substrates introduced into the spinal cord of neonatal rats. *Dev. Brain Res.* **1987**, *35*, 291–299. [CrossRef]
69. Schnell, L.; Schwab, M. Sprouting and regeneration of lesioned corticospinal tract fibres in the adult rat spinal cord. *Eur. J. Neurosci.* **1993**, *5*, 1156–1171. [CrossRef] [PubMed]
70. Khan, T.; Dauzvardis, M.; Sayers, S. Carbon filament implants promote axonal growth across the transected rat spinal cord. *Brain Res.* **1991**, *541*, 139–145. [CrossRef]
71. Xu, X.M.; Chen, A.; Guenard, V.; Kleitman, N.; Bunge, M.B. Bridging Schwann cell transplants promote axonal regeneration from both the rostral and caudal stumps of transected adult rat spinal cord. *J. Neurocytol.* **1997**, *26*, 1–16. [CrossRef] [PubMed]

veterinary sciences

MDPI

Article

The Infectious Bronchitis Coronavirus Pneumonia Model Presenting a Novel Insight for the SARS-CoV-2 Dissemination Route

Ekaterina Nefedova [1], Vyacheslav Koptev [1], Anna S. Bobikova [1,2], Viktoria Cherepushkina [1], Tatyana Mironova [1,2], Vasily Afonyushkin [1,3], Nikolai Shkil [1], Nikolai Donchenko [1], Yulia Kozlova [3], Natalia Sigareva [2], Natalia Davidova [1], Nina Bogdanchikova [4,*], Alexey Pestryakov [5] and Yanis Toledano-Magaña [6,*]

[1] Siberian Federal Scientific Center of Agro-BioTechnologies of Russian Academy of Sciences, Novosibirsk Region, Novosibirsk District, 630501 Krasnoobsk, Russia; filll555@mail.ru (E.N.); kastrolog@mail.ru (V.K.); bobikova.anna97@gmail.com (A.S.B.); vicky88@bk.ru (V.C.); mironova.tanya1994@mail.ru (T.M.); lisocim@mail.ru (V.A.); nicola07@mail.ru (N.S.); tbc2009@yandex.ru (N.D.); ulona79@mail.ru (N.D.)

[2] Molecular Biology Department, Federal State Budgetary Educational Institution Higher Education Novosibirsk State Agrarian University, 630090 Novosibirsk, Russia; natalias72@mail.ru

[3] Institute of Chemical Biology and Fundamental Medicine SB RAS, 630090 Novosibirsk, Russia; ramira_@bk.ru

[4] Centro de Nanociencias y Nanotecnología, Universidad Nacional Autónoma de México, Ensenada 22860, BC, Mexico

[5] Research School of Chemistry & Applied Biomedical Sciences, National Research Tomsk Polytechnic University, 634050 Tomsk, Russia; pestryakov2005@yandex.ru

[6] Escuela de Ciencias de la Salud Unidad Valle Dorado, Universidad Autónoma de Baja California, Ensenada 22890, BC, Mexico

* Correspondence: nina@cnyn.unam.mx (N.B.); yanis.toledano@uabc.edu.mx (Y.T.-M.)

Citation: Nefedova, E.; Koptev, V.; Bobikova, A.S.; Cherepushkina, V.; Mironova, T.; Afonyushkin, V.; Shkil, N.; Donchenko, N.; Kozlova, Y.; Sigareva, N.; et al. The Infectious Bronchitis Coronavirus Pneumonia Model Presenting a Novel Insight for the SARS-CoV-2 Dissemination Route. *Vet. Sci.* **2021**, *8*, 239. https://doi.org/10.3390/vetsci8100239

Academic Editors: Ana Faustino and Paula A. Oliveira

Received: 7 September 2021
Accepted: 12 October 2021
Published: 18 October 2021

Abstract: Infectious bronchitis (IB) of chickens is a highly contagious disease characterized by damage of the respiratory system and reproductive organs in young animals caused by a virus of the genus Gamma coronavirus. The condition of the respiratory system caused by the IB virus in chickens has many similarities with the pathology of the respiratory system caused by SARS-CoV-2 in humans. The effectiveness of virucidal drugs (Argovit, Triviron, Ecocid, and lauric acid monoglyceride) was tested on chickens inoculated with a tenfold dose of a vaccine strain based on the attenuated virus H120 against IB of chickens. On the 6th day after inoculation, inflammatory changes in the intestines, lungs, and thymus were observed in the control group. The experimental groups were characterized by less pronounced inflammatory reactions and a lower proportion of thymus and lung probes containing genomic IB virus RNA. Since the virucidal activity of four orally administrated formulations was possible only in the intestine, the experimental data indirectly confirmed the hypothesis of the possibility of the predominant accumulation of coronaviruses in the intestine and subsequent lung damage due to the hematogenous redistribution of viral particles and IBV antigens. It was suggested that other coronaviruses including SARS-CoV-2 can implement a similar mechanism.

Keywords: virucidal drugs; IBV; chickens; Coronaviridae; silver nanoparticles; artificial ribonuclease; monoglyceride lauryl acid

1. Introduction

Infectious bronchitis (IB) of chickens is a highly contagious disease characterized by damage to chickens' respiratory and reproductive organs [1–5]. Diseases of the respiratory system caused by the infectious bronchitis virus (IBV) [5] have common features with the pathology of the respiratory system caused by SARS-CoV-2 in humans [6]. IB and SARS-CoV-2 viruses have many similarities: both are low copy coronavirus, both have a lipid envelope, and for both, the immune system reaction contributes significantly to the damage. These give the basis to believe that the mechanisms of action of these two viruses are similar.

Bacterial viruses (bacteriophages) are a very convenient model for studying various drug virucidal activities. Specifically, bacteriophage φ6 belonging to the Cystoviridae family is a promising model. These phage particles have a spherical or icosahedral shape and dsRNA type of nucleic acid. The presence of capsids in its composition makes it an effective model for testing the activity of virucidals and disinfectants against enveloped viruses, including coronaviruses. The virucidal activity of substances determined with the φ6 bacteriophage model allows the estimation of their activity against various viruses containing lipids in their capsid, such as herpesviruses 1 and 2 and coronavirus SARS-CoV-2, which infect eukaryotic cells [7–9]. The φ6 bacteriophage infects *P. phaseolicola* plant pathogen bacteria. Consequently, the advantage of working experimentally with a φ6 bacteriophage and plant pathogenic bacteria is that neither the phage nor the bacteria are pathogenic to humans, which minimizes the requirements for the laboratory biosafety level. Some research groups successfully applied the φ6 bacteriophage as a model for the SARS-CoV-2 virus [10,11].

Drugs possessing virucidal activity in the small intestine can be effective against viral infections in other tissues. Clinical observations showed the effectiveness of Ecocid, Triviron, Argovit, and lauric acid monoglyceride (C12) against viral mal-absorption in poultry farms (flavivirus, astroviral etiology). Ecocid shows high antibacterial and virucidal activity combined with low toxicity and stability in a living organism, making it a promising alternative as a topical virucidal agent [12] and a disinfectant that prevents the horizontal transfer of antibiotic resistance genes [13]. Triviron virucidal formulation began to be used in veterinary medicine relatively recently [14,15]. The mechanism of action has no analogs and belongs to a new pharmacological group of synthetic ribonucleases [2,16]. Silver nanoparticles [8,17–20] including Argovit AgNPs [21–24] have recently shown virucidal effects against some viruses, and are therefore also potentially effective against the SARS-CoV-2 virus. Lauric acid monoglyceride (Monolaurin) is known to inactivate lipid-coated viruses by binding to the lipid–protein envelope of the virus, thereby preventing it from attaching and entering host cells, making infection and replication impossible [25]. Other studies show that Monolaurin disintegrates the protective viral envelope, killing the virus [26,27]. Monolaurin has been studied to inactivate many pathogens, including the Herpes simplex virus [25–28].

The present study aimed to screen the virucidal activity of the four formulations mentioned above on in vitro (bacteriophage φ6) and in vivo (IB of chickens) models for SARS-CoV-2 infection.

2. Materials and Methods

2.1. Formulations

Argovit 1% is an aqueous suspension of highly dispersed silver nanoparticles (0.6 mg/mL of metallic silver) stabilized with polyvinylpyrrolidone, produced by the Center of Investigation and Production "Vector-Vita", Novosibirsk, Russia. Argovit™ is a stable suspension in water with an AgNP concentration of 200 mg/mL (20% *w*/*w*). The metallic Ag (content 1.2% *w*/*w*) is stabilized with polyvinylpyrrolidone (PVP 12.6 ± 2.7 kDa, content 18.8%) of AgNP total weight. The remaining 80% of the weight is distilled water. AgNPs have a spheroidal shape with a size varying from 1 to 90 nm and an average diameter of 35 ± 12 nm. The hydrodynamic diameter is 70 nm, the ζ potential is −15 mV, and a plasmonic resonance peak is registered at 420 nm [29]. Triviron (0.03% synthetic ribonuclease (1,5-bis-[N, N-1- (4-tetradecyl) diazoniabicyclo [2.2.2] octyl] pentane tetrobromide) was produced by «Trionisvet» Ltd., Korolev, Russia. Ecocid (0.05%), produced by Krka, d. d., Novo Mesto, Slovenia, contains a triple salt of potassium peroxomonosulfate (50%), as well as auxiliary substances: surfactants (sodium dodecylbenzene sulfonate), organic acids (malic, sulfamic), inorganic buffer systems (sodium chloride and sodium polyphosphate),

Azo Diestaff dye, and Citron as an aromatic additive with a lemon scent. Used in our work, Ecocid, Triviron, and Argovit are certified and are commercially produced. Lauric acid monoglyceride (0.005%) was synthesized and kindly provided by Dr. Fomenko Vladislav from institute N.N. Vorozhtsov, Novosibirsk Institute of Organic Chemistry of SB RAS.

2.2. Bacteriophage φ6 In Vitro Model

Bacteriophage φ6 (belonging to the *Cystoviridae* family) was used as an in vitro model to assess the formulations' virucidal activity. A *Pseudomonas phaseolicola* cell culture was used to determine phage particle activity. The tested preparations were incubated with a bacteriophage culture at room temperature at various concentrations. After 1, 5, 15, 30, and 60 min of incubation, inoculation of samples containing bacteriophages and formulations and only bacteriophage (control group) was carried out. The residual concentration of bacteriophage inoculations was determined in 10-fold dilutions. The survival rate of phage particles was determined by the method of two-layer agar (Grazia method). The studies were carried out at room temperature (26 ± 2 °C). All experiments were performed in duplicate.

2.3. Poultry

Cross Shaver male chickens aged 14 days and 198–210 g each were kept indoors and fed with standard granulated chicken feed. The experimental protocol involving animals was reviewed and approved by the Ethical Committee of Novosibirsk State Agrarian University of Siberian Federal Scientific Center of Agro-BioTechnologies of the Russian Academy of Sciences CM K PO 15-01-2019/No.3 of 10.03.2021.

2.4. Inoculum Administration

Chickens received a tenfold dose of a vaccine against IBV (from the H120 strain, live, dry) consisting of 5 lg Embryo Infectious Dose (40 per head). It was administered orally to each chicken. Experimental groups and a control group of 13 animals were formed.

2.5. Dosage Administration

The dosages of the preparations were as follows: Argovit and Triviron, 250 and 285 μL/animal, respectively. They were administered two times a day (in the morning and the evening) to each animal. The lauric acid monoglyceride dose was 0.1 mg/animal. Ecocide C (0.05%) dose chickens drank freely when they wanted (57 mL per day per head). For drinking water sanitation for animals, including birds, it is recommended to use 0.1% Ecocide C in water. Here, half of the recommended concentration was applied. All formulations were administered for 5 days. The animals were slaughtered on the 6th day. In Table 1, consumption per head per day for the studied formulations is summarized.

Table 1. Consumption per head per day for studied formulations.

No.	Formulation	Consumption of Active Component per Head per Day (μg)
1	Argovit C, 0.06% (metallic silver)	150
2	Triviron, 0.03%	85.5
3	Lauric acid monoglyceride (C12), 0.05%	100
4	Ecocid C, 0.05%	28,500

2.6. Histopathological Evaluation

The lung histopathological analysis was performed under an Imager D1 luminescence microscope (Zeiss) using AxioVision v 4.6.3.0 software (Zeiss, Jena, Germany). Briefly, the lung pieces (~1 mm thick) were placed in a 96-well plate filled with distilled water (150 μL per well), and 20 μL SYBRTM Green (1:1000) and 5 μL of SyproTM Ruby dye

(BioRad Laboratory) were added per well. Then, they were kept for 20 min to counterstain the preparations.

Luminescence microscopic analysis of lung pieces was carried out according to an original technique of a short period staining mode, whose principle consisted of staining a formalin-fixed biomaterial with two dyes. SYPRO Ruby intercalates proteins and glows, while SYBR Green I binds to DNA and fluoresces in orange-red and green ranges, respectively. The staining is carried out within a short period (20 min), during which dye diffusion occurs in a thin layer of cells, which avoids the intense background fluorescence of entire pieces of tissue. Thus, in contrast to histological examination, a short period staining mode is suitable for studying thin tissue layers. This mode makes it possible to visualize better epithelial and endothelial cells (for example, the inner surface of the bronchi and blood vessels and intestinal epithelium). Additionally, this mode provides a good visualization of blood capillaries, hemorrhages, and tissue saturation with hemoglobin, which is accompanied by a sharp decrease in fluorescence intensity. The difference of this luminescence microscopic mode from confocal microscopy lies in the lower price of the used equipment, faster staining process, more detailed observations of epithelium surfaces, and the influence of the diffusion process microscopic picture.

2.7. *qRT-PCR*

RNA was isolated from internal organs using silica columns, with preliminary lysis of cells with guanidine isothiocyanate. The copy number of the IBV was assessed by qRT-PCR [14].

3. Results

3.1. *Verification of the Virucidal Action with Bacteriophage φ6*

Experiments with a virus model (bacteriophage φ6) were conducted as a preliminary test to estimate active doses of the studied formulations. Triviron's effective concentration was ≥0.003% (30 μg/mL) for 1 min. At this point, the formulation completely suppressed the activity of phage particles (Table 2). Ecocid, at a concentration of 0.05% (500 μg/mL), completely suppressed viral activity in vitro within a minute, while Argovit, at a concentration of 0.00001% (0.10 μg/mL) of metallic silver, inactivated viral particles at ≥ 15 min of incubation. Lauric acid monoglyceride practically did not show a decrease in bacteriophage φ6 concentration for 60 min. So, experiments in vitro on the bacteriophage φ6 model demonstrated that among the four studied formulations at used concentrations, the virucidal activity decreases in the order: Ecocid ≈ Triviron > Argovit >> lauric acid monoglyceride.

The presence of capsids in the composition of the bacteriophage φ6 makes it an effective model for testing the activity of virucidal and disinfecting drugs against enveloped viruses, including coronaviruses. As follows from the obtained results (Table 1), Ecocid, Triviron, and Argovit in the used concentrations in in vitro experiments suppressed the activity of the enveloped virus, whereas lauric acid monoglyceride was practically inactive.

Table 2. Change in the concentration of bacteriophage φ6 after incubation with virucidal drugs.

Formulation Name and Concentration	**Bacteriophage φ6 Concentration, PFU */mL**					
	Initial	**After Exposure Time (min).**				
		1	**5**	**15**	**30**	**60**
Ecocid 0.05% (500 μg/mL)	$9.6 \pm 0.17 \times 10^5$	0	0	0	0	0
Triviron 0.003%, (30 μg/mL)	$1 \pm 0.00 \times 10^7$	0	0	0	0	0
Lauric acid monoglyceride (C12), 0.05%	$1 \pm 0.57 \times 10^7$	$7.4 + 1.3 \times 10^7$	$1.1 + 0.11 \times 10^7$	$2.0 + 3.3 \times 10^7$	$1.3 + 0.33 \times 10^7$	$8.9 + 6.3 \times 10^6$
Argovit, 0.00001% (10 μg/mL) of metallic silver	$4.2 \pm 0.15 \times 10^6$	$3.4 + 0.63 \times 10^4$	$6 + 0.115 \times 10^2$	0	0	0

* Plaque-forming unit (PFU).

3.2. Postmortem Examination

After the autopsy, the characteristic changes in the thymus (the IBV affects the medullar zone of the thymus) and lung were noted (Figure 1). The lungs were hyperemic, edematous, and sometimes triangular and diamond-shaped lesions were observed, indicating hematogenous drift of the infectious agent to these zones. The small intestine also showed hyperemia. The large intestine was unchanged. The kidneys were not inflamed. In most experimental groups, changes in the intestines, lungs, and thymus were completely absent or less pronounced than in the control group. The most severe lesions were observed in the control and then in the lauric acid monoglyceride-treated group organs.

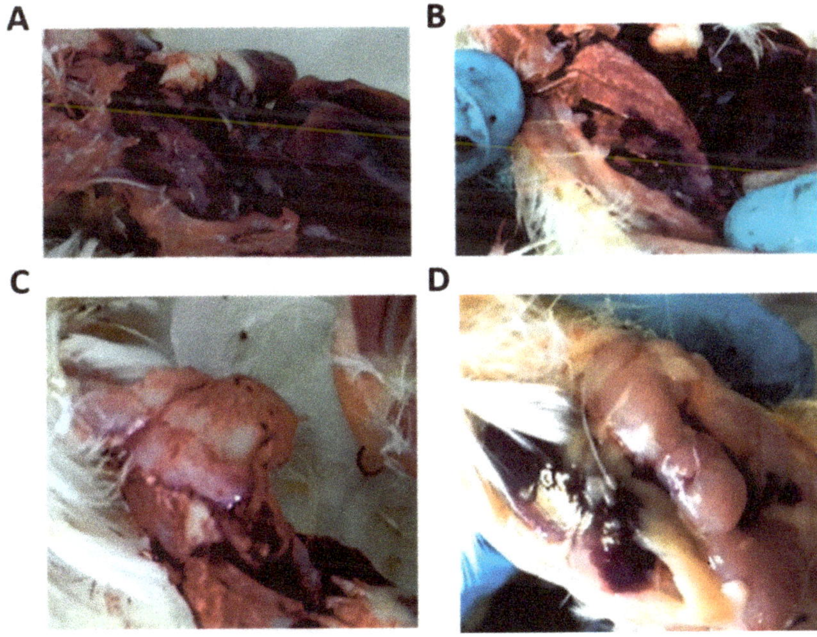

Figure 1. Macroscopic observations in lungs and thymus of control and experimental groups. Pneumonia signs were observed in the control group lungs (**A**) and experimental group lungs (**B**). Thymus inflammation in the control group (**C**) and thymus without noticeable changes in the experimental group (**D**).

Inflammatory changes in the intestines were observed in all groups, but these changes were lesser in the Ecocid and Triviron groups. Lung lesions relative to the control group were less pronounced in the Triviron-treated group. The smallest number of inflammatory reactions in the thymus was observed in poultry after Ecocid and Triviron application. So, the postmortem examination showed that the inflammatory responses in the studied groups increase in the following order: Triviron < Ecocid < Argovit < lauric acid monoglyceride < control.

3.3. Lung Luminescence Microscopy Histopathological Study

3.3.1. The Control Group

The control group is predominantly characterized by interstitial pneumonia in focal lesions involving the groups of segments of the parabronchus. In this case, some parabronchus does not present pathologic morphological changes, and others were excluded from the gas exchange, which leads to a decrease in overall gas exchange. The protein fluoresces in red; a reduction is due to infiltrative processes. An increase in luminosity can be associated with exudative processes and respiratory epithelium desquamation.

Hemorrhages were found (Figure 2A). A narrowing of the air capillaries further develops the inflammatory process due to thickening the interstitium and lamina propria associated with inflammatory infiltration (Figure 2B). More severe lesions are accompanied by the accumulation of serous exudate and total overlap of the lumens of the airway capillaries; exudate rich in cells accumulates in the parabronchus (Figure 2B), and diffuse hemorrhages appear. The fact that the lesion of the pulmonary parenchyma is topologically associated with hematogenous drift of the pathogen can be confirmed by identifying signs of vasculitis and perivascular hemorrhage infiltrates (Figure 2C). Another criterion for the primary involvement of blood vessels in the pathological process (from the pulmonary artery side) is the absence of a uniform radial lesion of the parabronchial segments (Figure 2D). As shown in Figure 2D, there is an overlap of the air capillary lumens on the pulmonary parenchyma areas adjacent to the site of interstitial inflammation. Opposing areas of the parabronchial segments present a well preserved network of air capillaries.

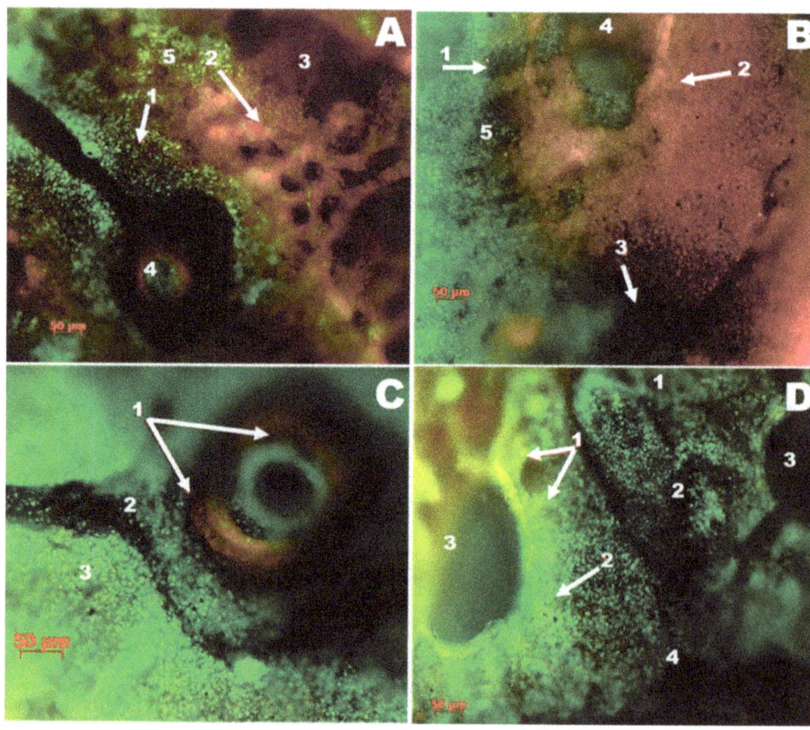

Figure 2. Luminescence microscopy photograph. Lungs and parabronchi of the control group stained with SYBR Green and Sypro Ruby (×50). (**A**) Interstitial inflammation in the lungs: 1—infiltrates, 2—air capillaries, 3—parabronchus, 4—blood vessel, 5—areas of the parabronchial segment with a decrease in the lumen of the air capillaries due to infiltration. (**B**) Focal changes in the lung: 1—inflammatory infiltration of the parabronchial segment, 2—saturation with serous-catarrhal exudate of the lung parenchyma and loss of the lumen of the airway capillaries, 3—hemorrhage, 4—the lumen of the parabronchus, 5—preserved air capillaries. (**C**) Vasculitis in the vessels: 1—fibrinoid inflammation of the vessel wall (artery), 2—suppression of fluorescence at the site of tissue imbibition by hemoglobin, 3—increased density of cells associated with inflammatory infiltration of the interstitium of the lung. (**D**) Inflammatory changes in the parabronchi: 1—air capillaries that have retained the lumen, 2—air capillaries that have lost their lumen due to inflammatory infiltration, 3—parabronchi, 4—hemorrhages around the affected blood vessel.

3.3.2. The Group Treated with Argovit

A comparative analysis of the results obtained in luminescence microscopy histopathological study revealed that the lungs were less affected in the group receiving the Argovit formulation. So, the Argovit-treated group was chosen for comparative analysis with the control group. A change interval from the complete absence of signs of inflammation to significant changes like the ones presented in Figure 2 was observed. Thus, Figure 3A,B show photographs of two lung samples characterized by a complete lack of inflammatory changes. The histoarchitecture of the respiratory surfaces of the parabronchus is not disturbed; there are no inflammatory infiltrates and hemorrhages. In contrast, lung changes compared to the control group in some individuals were observed (Figure 3C).

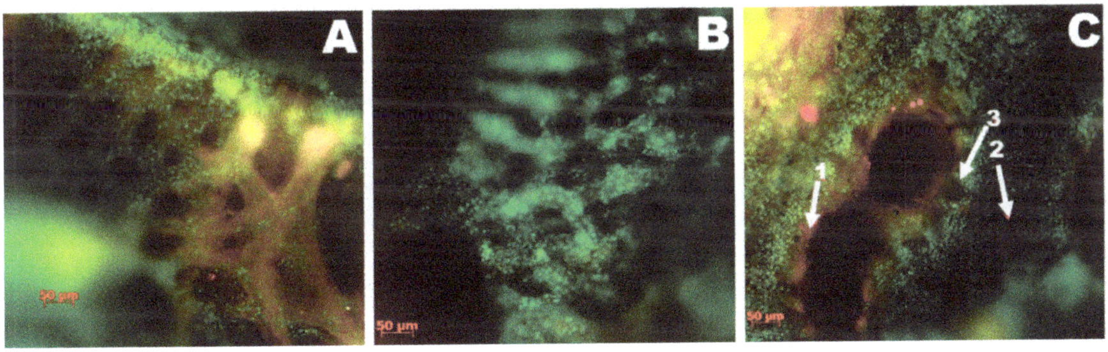

Figure 3. Luminescence microscopy photograph of lung preparations of the group treated with Argovit, stained with SYBR Green and Sypro Ruby (×50). (**A**,**B**) No inflammatory changes were observed. (**C**) The bronchus lumen: 1—the parabronchus wall, 2—the air capillary mouth, 3—the lung parenchyma in a state of vascular hyperemia.

3.3.3. IBV in Chickens with RT-PCR

Intestine

As shown in Figure 4, the virus concentration detected in the intestine in all studied groups varied in 8–10 orders of magnitude range, changing from 0.000038 to 2,482,450. At the same time, it is important to mention that the virus was detected in ≤69% of poultry intestines of each group. In the control group, viral concentration was detected in 69% of poultry intestines and varied 10 orders of magnitude from 0.000245 to 2,482,450 (Figures 4 and 5). After formulation treatments, virus concentration in the intestine changed in 8–9 orders (9 orders for Ecocid, Argovit, and C12 and 8 orders for Triviron), and poultry percentage with the detected virus was 38%, 54%, 54%, and 46%, respectively (Figures 4 and 5).

Lungs

Virus concentration in lungs varied 0–6 orders of magnitude: 4, 6, 0, 3, and 4 for Ecocid, Argovit, Triviron, C12, and the control group, respectively (Figure 4). For Ecocid, Argovit, Triviron, C12, and the control group, the number of chickens with the detected virus was four, two, one, five, and four among thirteen chickens in every group, which corresponds to 31, 15, 8, 39, and 31%, respectively (Figures 4 and 5).

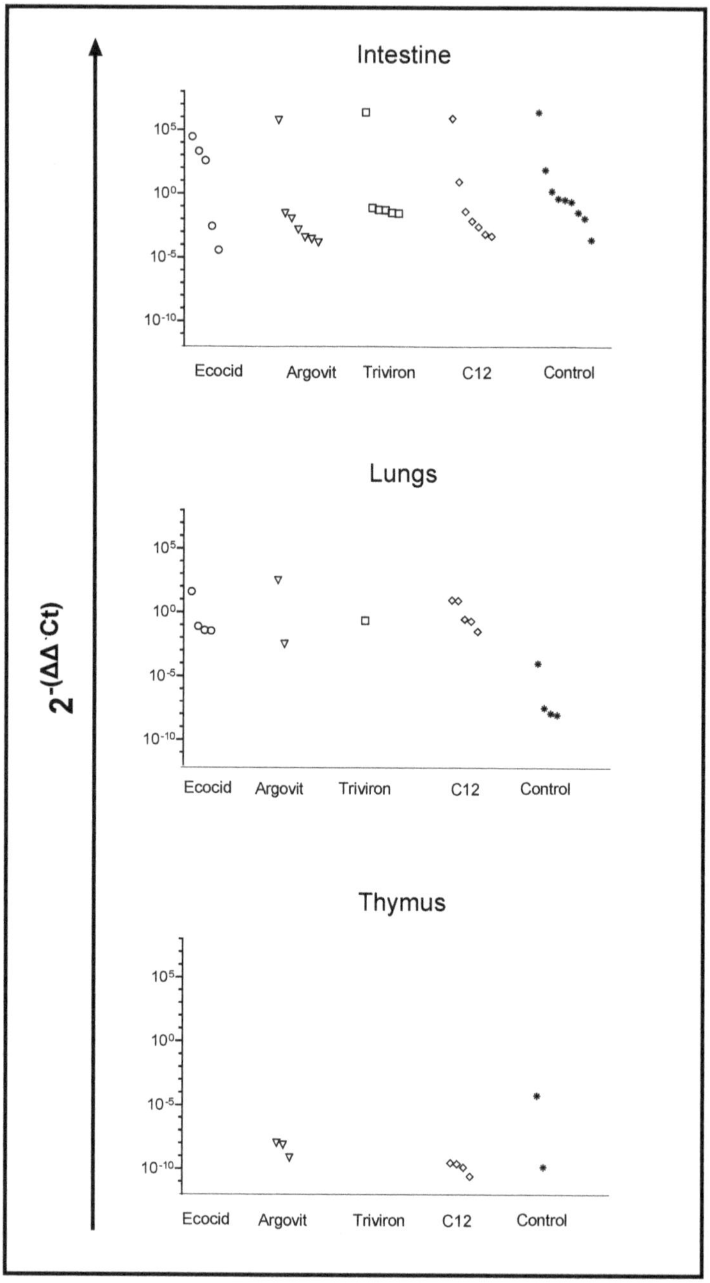

Figure 4. The concentration of IBV in the intestine, lungs, and thymus in chickens of the experimental and control groups. Data are presented in $2^{-(\Delta\Delta Ct)}$ units, which reflect the concentration of IBV from chickens treated with Ecocid (O), Argovit (∇), Trivirion (□), C12 (◊), Control (*). Every group contains data for 13 chickens.

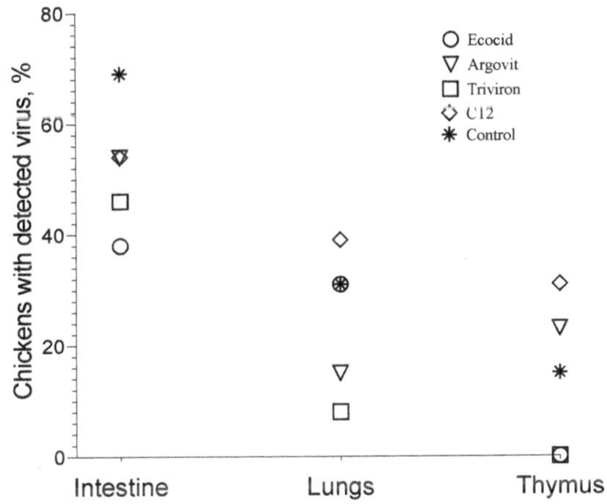

Figure 5. The percentage of chickens with virus detected in intestine, lungs, and thymus for all studied groups.

Thymus

After treatment with Ecocid and Triviron, the virus was not detected by PCR. Virus concentration in the thymus was detectable only for the Argovit, C12, and control groups (Figures 4 and 5). Virus concentration varied 1-5 orders of magnitude inside these groups, being 1, 1, and 5 in the range from 7.16×10^{-10} to 1.03×10^{-8}, from 2.64×10^{-11} to 2.95×10^{-10}, and from 1.43×10^{-10} to 6.01×10^{-5} for Argovit, C12, and the control group, respectively (Figure 4). The virus was detected on 23, 31, and 15% of poultry thymus, respectively (Figures 4 and 5).

4. Discussion

The results showed that treatment with all four studied formulations decreased the poultry percentage with virus detected in the intestine compared to the control group. The percentage of chickens in which the virus was detected on intestines was considerably higher in the control group (69%) than in experimental groups, being 38, 54, 46, and 54% for Ecocid, Argovit, Triviron, and C12, respectively (Figures 4 and 5). The variation in virus concentration in the control group (11 orders of magnitude) was slightly higher than in experimental groups (8–9 orders of magnitude).

The investigation of virus concentration in the lungs (Figures 4 and 5) showed a decrease in the fraction of chickens with detectable virus concentration, virus concentration, and an interval of variation in detected virus concentration in each group compared with the intestine. The fraction of chickens with detectable virus concentration in the lungs was lower than in the intestine, being 31% < 38% for Ecocid, 15% < 54% for Argovit, 8% < 46% for Triviron, 39% < 54% for C12, and 31% < 69% for the control group (Figure 5). These results showed that virus RNA in the lungs was detected on a minimum fraction of chickens for Triviron (8%) and Argovit (15%). These results are consistent with histopathological studies, where Argovit showed minimal lesions in epithelial and endothelial cells in the lungs (Figure 2). Maximum virus concentrations measured in the lungs were less than in the intestine for all studied groups. These concentrations decreased by 3, 7, 5, and 10 orders of magnitude for Ecocid, Triviron, C12, and the control group, respectively. However, the concentration only decreased by half for Argovit. The virus concentrations in lungs varied in all groups between 0 and 6 orders of magnitude: 4, 6, 0, 3, and 4 for Ecocid, Argovit, Triviron, C12, and the control, respectively.

Intestine virus concentrations varied in all groups between 8 and 11 orders of magnitude. This difference could indicate that intestinal tissue has conditions promoting viral replication, while lung tissue is more restrictive to viral replication. Unexpectedly, it was observed that virus concentrations in the lungs in the control group were much lower (between 9.19×10^{-9} and 1.07×10^{-4}) than in experimental groups (between 3.3×10^{-3} and 3.1×10^{2}). The explanation of this effect will need further studies. In contrast, thymus viral concentration was very low (Figure 5), and for Ecocid and Triviron, viral RNA was undetectable using qRT-PCR. For Argovit, C12, and the control group, most values were between 2.6×10^{-11} and 1.0×10^{-8}. Additionally, only one data point for the control group was higher (6.0×10^{-5}). This might be related to the fact that the thymus belongs to the immune system and, therefore, it highly restricts viral replication and/or eliminates viral particles.

All the above results demonstrate that all four studied formulations demonstrated virucidal activity against IB virus. Argovit, Triviron, and Ecocid showed some advantages compared with C12. However, in general, the difference in the virucidal activity of the four formulations was not significant. This work is only a first approximation of using these formulations against IBV, and further studies are needed to clarify the potential use of these formulations.

Obtained data showed that on the 6th day after IBV injection, the virus concentration was significantly higher in the intestine than in the lungs and thymus in all groups. The period of six days after inoculation was adequate time for the dissemination of the virus to all organs by blood circulation. Supposing that the intestine was not the organ with optimal conditions (among the three studied organs) for virus replication, after 6 days, the concentration of the virus would be maximal in other organs (if they have optimal conditions for replication), for example, in the lungs. However, this was not observed experimentally. Instead, the highest viral concentration was observed in the intestine. Hence, the obtained results suggest that the intestine is the organ with the highest cell secretory potential, receptors, and enzyme systems suitable for the reproduction of coronavirus and the maturation of viral particles, which creates optimal conditions for primary virus replication and accumulation. It implies that, specifically, the intestine serves as a source for virus dissemination to other organs. Low virus concentrations in the lungs and thymus could indicate that viral replication is occurring relatively slowly. However, lesions observed in our experiments in the lungs might show that the virus is not eliminated quickly from the lungs because even at low concentrations it manages to cause lesions. The hypothesis about the primary role of intestinal tissue in the replication/accumulation and source of dissemination of IBV was suggested in our previous publications [1,30].

The obtained results suggest the following mechanism of IB virus entering the organism. Firstly, the virus enters the nasopharynx and interacts with the mucus. The nasopharynx mucus layer (a porous gel net) acts as an impenetrable physical barrier to most pathogens. However, because the pore diameter of the gel net (approximately 500 nm) is significantly larger than the IB virus diameter (50–100 nm) [4], the IB virus easily penetrates mucus pores. Once swallowed, the mucus is digested by the gastrointestinal system, and the virus starts to replicate. It is important to mention that the size of the SARS-CoV-2 virus (approximately 60–140 nm) [31] is close to the IB virus diameter (50–100 nm).

Because IB and SARS-CoV-2 viruses are similar, the mechanism suggested above for IB virus could also be applied to the SARS-CoV-2 virus. The results of recent works confirm the validity of our hypothesis that the intestine is the principal organ for virus reproduction and accumulation, and it serves as a source of virus dissemination into other tissues. These results agree with the fact that that the highest ACE2 expression in human organisms occurs in the intestinal enterocyte brush border [32,33]. Although it was reported that ACE2 is expressed in the lung, liver, stomach, ileum, kidney, and colon, its expressing levels there are relatively low, especially in the lung. In lung AT2, this level is 4.7-fold lower than the average expression level value of ACE2-expressing cells of all 13 cell types studied in this work [32].

Some recently published clinical observations indicate the relevant role of the intestine in the infection caused by the SARS-CoV-2 virus, which agrees with our hypothesis. SARS-CoV-2 viral RNA can be detected in rectal swabs when nasopharyngeal testing was already negative, suggesting long-term gastrointestinal infection [34]. The recent observations showed that gastrointestinal symptoms were observed in 57% of the patients infected with SARS-CoV-2. In addition, these symptoms sometimes occurred without respiratory symptoms. Gastrointestinal manifestations are more common in patients with severe disease than in patients with non-severe disease. For approximately 48.1% of patients with severe acute respiratory syndrome COVID-19, a stool sample was positive for SARS-CoV-2 RNA virus [35]. This hypothesis [36] suggests that complications after COVID-19 could be caused by SARS-CoV-2-upregulated angiotensin II-caused disruption of mucosal barriers with following microbial and/or lipopolysaccharide movement from the gastrointestinal tract into various tissues.

Coronavirus modulates autophagy or its components for its benefit, primarily the autophagosomes used as replication and transcriptional niches. The IBV induces the formation of autophagosomes via the MAPK/ERK1/2 pathway dependent on Atg5 [37]. However, the induction of autophagy also facilitates the activation of apoptosis by generating a platform for activating caspase-8 or depleting endogenous inhibitors of this cell death pathway [38,39], as observed on tumor cells exposed to Bortezomib, which showed apoptosis via caspase-3 activation [40,41].

Although MAPK/ERK1/2 pathways generally promote cell survival, certain circumstances such as DNA or oxidative damage function as pro-apoptotic signaling [42]. Oxidative stress and DNA damage have been described on several tumor cells exposed to Argovit AgNPs, promoting cell death by an intrinsic apoptotic pathway [43,44]. Thus, the apoptotic pathway could be activated on cells harboring IBV due to the virus-promoted autophagosome increase and redox damage promoted by the presence of AgNPs, leading to cell death and the inhibition of virus proliferation. This fact could explain the viral titer decrease observed in the lungs after the administration of silver nanoparticles.

Furthermore, the absence of inflammatory infiltrates or hemorrhages observed in the lungs treated with AgNPs could be associated with the selective cytotoxic effect that should be kept only on those cells with increased autophagosomes, which are more susceptible to oxidative damage than non-infected cells. The above-mentioned cytotoxic selectivity was already observed on tumor and non-tumor cells exposed to AgNPs, where non-transformed cells showed no oxidative damage compared with tumoral cells [45].

It is also worth mentioning that recently published works describe the first experimental data on infection prevention with SARS-CoV-2 in humans at least for two formulations studied here (Argovit and C12). For example, recently, Argovit was studied in vitro and in vivo to prevent SARS-CoV-2 infection in health workers. The inhibitory effect of AgNPs in SARS-CoV-2 NL/2020 strain replication in cultured Vero E6 cells was confirmed [46]. A randomized study (with 114 and 117 participants in experimental and control groups, respectively) demonstrated that mouthwash and nose rinse with 1% Argovit-C reduced the SARS-CoV-2 infection rate 16 times in healthcare personnel attending on average 169 patients with COVID-19 per week per person [47]. It was demonstrated that the silver nanocluster/silica composite coating deposited on facial masks possessed a virucidal effect against SARS-CoV-2 [29]. The investigation of blood serum samples collected from 51 healthcare workers of an Italian COVID-19 hospital showed that a higher concentration of C12 was observed in protected workers compared with those infected with SARS-CoV-2 [46]. The authors suggested a potential defensive role of monolaurin against SARS-CoV-2 infection. They offered a randomized controlled trial of monolaurin supplements to confirm these observational findings before any therapeutic recommendations can be made.

Hence, the results of our study showed that it is vital to consider the prospects for combating coronavirus infections, including COVID-19, using virucidal drugs. Virucidal drugs are substances causing the inactivation of viral particles, thereby limiting the infection of new cells and preventing the damage of cells and body tissues. Silver nanoparticles should also be included in this class of drugs. Unlike classical antiviral drugs, for example, inhibitors of the activity of RNA-dependent RNA polymerase of coronaviruses (Areplivir, Remdesivir, etc.), virucidal drugs are not required to penetrate the cell to exert their virucidal effect. For this reason, the dosage of virucidal medicines may be less. Therefore, their toxic effect may be significantly less due to poor transport into cells (which reduces the risks of metabolic disorders).

One feature of the pharmacokinetics of many virucidal drugs is poor penetration through mucous barriers. However, the inactivation of viral particles on the surface of mucous membranes (intestines, nasopharynx) with virucidal medications, combined with an intensive self-cleaning of the surfaces of mucous membranes, creates good prospects for limiting the entry of the SARS-CoV-2 virus into the blood and lymphatic vessels, followed by hematogenous and lymphomatous transport to the pulmonary circulation (lungs). The intensity of biosynthetic processes (such as virus replication) in the epithelium of mucous membranes is extremely high. Limiting the reproduction of the virus in the nasopharynx and intestine can have a systemic effect on the development of COVID-19.

5. Conclusions

Four formulations (Ecocid, Triviron, Argovit, and lauric acid monoglyceride) possessing virucidal activity in the small intestine showed potential against chicken infectious bronchitis virus. Based on the obtained results, our hypothesis is that the transmission of IB virus in chickens occurs not through the respiratory system but through the intestine, where more RNA was determined. Then, from there, it is disseminated to other organs, including the lungs. To the best of our knowledge, our group is the first to propose such a route of infection for the IB virus.

Considering that IB is proposed as a model for SARS-CoV-2, because both viruses are low-copy coronaviruses with a lipid envelope and similar diameters, it was suggested that a similar mechanism based on primary virus replication and accumulation in the intestine could also be carried out for the SARS-CoV-2 virus. If confirmed, this paradigm may open up innovative treatments for COVID-19 and other respiratory diseases caused by a coronavirus, targeting intestinal viral load to minimize infection in other tissues.

Author Contributions: Conceptualization, V.A., N.B., A.P., Y.T.-M.; methodology, V.A., V.C., E.N., N.S. (Natalia Sigareva), N.D. (Natalia Davidova); validation, N.B., Y.T.-M., V.A., N.S. (Natalia Sigareva); formal analysis, N.S. (Nikolai Shkil), N.D. (Nikolai Donchenko), T.M., V.K., Y.K.; investigation, T.M., A.S.B., V.K., V.C., Y.K., E.N., A.P., N.S. (Natalia Sigareva); resources, N.D. (Nikolai Donchenko), A.P.; data curation, V.A.; writing—original draft preparation, T.M., E.N., A.S.B., V.C., Y.K., N.D. (Natalia Davidova), N.D. (Nikolai Donchenko), Y.T.-M., V.K.; writing—review and editing, N.B., Y.T.-M., A.P.; visualization, N.B., Y.T.-M.; supervision, N.D. (Natalia Davidova), N.S. (Nikolai Shkil); project administration, V.A., N.S. (Nikolai Shkil). All authors have read and agreed to the published version of the manuscript.

Funding: This work was supported by research project number 20-416-540006 of the Russian Foundation for Basic Research (RFBR).

Institutional Review Board Statement: The studies involving animals were reviewed and approved on March 10 of 2021 by the Ethical Committee of Novosibirsk State Agrarian University of the Siberian Federal Scientific Center of Agrobiotechnology of the Russian Academy of Sciences No. 3. All animal experiments were carried out in accordance with the ARRIVE guidelines and carried out in accordance with the principles of humanity formulated in the European Community Directive 86/609/EEC (Strasbourg, 1986).

Informed Consent Statement: Not applicable.

Data Availability Statement: The data presented in this study are available in article.

Acknowledgments: For the donation of the formulations used in the experiments, the authors thank the Center of Investigation and Production "Vector-Vita" Novosibirsk, Russia (veterinary drug Agovit); «Trionisvet» Ltd., Korolev, Russia (veterinary drug «Triviron»); Krka, d. d., Novo Mesto, Slovenia (veterinary drug Ecocid); and Vladislav Fomenko from institute NN. Vorozhtsov Novosibirsk Institute of Organic Chemistry of SB RAS (lauric acid monoglyceride), Tomsk Polytechnic University Development Program "Priority 2030".

Conflicts of Interest: The authors declare no conflict of interest.

References

1. Afonyushkin, V.N.; Mironova, V.S.C. Possible pathogenesis of coronavirus infections on the example of IBV as a model of infection associated with COVID-19 in humans. *BIO. J. Spec. Poult. Livest. Farms* **2020**, *4*, 4–6.
2. Afonyushkin, V.N.; Litvinov, A.V. An alternative way to prevent and treat PRRS. *Pig-Breeding* **2017**, 56–58. Available online: https://www.svinoprom.ru/ (accessed on 11 October 2021).
3. Cavanagh, D. Coronaviruses in poultry and other birds. *Avian Pathol.* **2005**, *34*, 439–448. [CrossRef] [PubMed]
4. Cowen, B.S.; Wideman, R.F.; Braune, M.O.; Owen, R.L. An infectious bronchitis virus isolated from chickens experiencing a urolithiasis outbreak. I. In vitro characterization studies. *Avian Dis.* **1987**, *31*, 878–883. [CrossRef]
5. Su, S.; Wong, G.; Shi, W.; Liu, J.; Lai, A.C.; Zhou, J.; Liu, W.; Bi, Y.; Gao, G.F. Epidemiology, genetic recombination, and pathogenesis of coronaviruses. *Trends Microbiol.* **2016**, *24*, 490–502. [CrossRef]
6. Huang, C.; Wang, Y.; Li, X.; Ren, L.; Zhao, J.; Hu, Y.; Zhang, L.; Fan, G.; Xu, J.; Gu, X.; et al. Clinical features of patients infected with 2019 novel coronavirus in Wuhan, China. *Lancet* **2020**, *395*, 497–506. [CrossRef]
7. Balagna, C.; Perero, S.; Percivalle, E.; Nepita, E.V.; Ferraris, M. Virucidal effect against coronavirus SARS-CoV-2 of a silver nanocluster/silica composite sputtered coating. *Open Ceram.* **2020**, *1*, 100006. [CrossRef]
8. Elechiguerra, J.L.; Burt, J.L.; Morones, J.R.; Camacho-Bragado, A.; Gao, X.; Lara, H.H.; Yacaman, M.J. Interaction of silver nanoparticles with HIV-1. *J. Nanobiotechnol.* **2005**, *3*, 1–10. [CrossRef] [PubMed]
9. Kawase, M.; Shirato, K.; van der Hoek, L.; Taguchi, F.; Matsuyama, S. Simultaneous Treatment of Human Bronchial Epithelial Cells with Serine and Cysteine Protease Inhibitors Prevents Severe Acute Respiratory Syndrome Coronavirus Entry. *J. Virol.* **2012**, *86*, 6537–6545. [CrossRef]
10. Fedorenko, A.; Grinberg, M.; Orevi, T.; Kashtan, N. Survival of the enveloped bacteriophage Phi6 (a surrogate for SARS-CoV-2) in evaporated saliva microdroplets deposited on glass surfaces. *Sci. Rep.* **2020**, *10*, 22419. [CrossRef] [PubMed]
11. Kakutani, K.; Matsuda, Y.; Nonomura, T.; Takikawa, Y.; Takami, T.; Toyoda, H. A Simple Electrostatic Precipitator for Trapping Virus Particles Spread via Droplet Transmission. *Int. J. Environ. Res. Public Health* **2021**, *18*, 4934. [CrossRef] [PubMed]
12. Afonyushkin, V.N.; Cherepushkina, V.S.; Tatarchuk, O.P.; Frolova, O.A. Study of anti-phage activity of disinfectants as a factor of suppressing horizontal gene transfer. *Bull. KSAU* **2020**, *4*, 88–96.
13. Afonyushkin, V.N.; Tabanyukhov, K.A.; Cherepushkina, V.S.; Khomenko, Y.S.; Tatarchuk, O.P. Effect of disinfectants based on potassium persulfate, hydrogen peroxide, glutaraldehyde and quaternary ammonium compounds on the genetic material of the pathogen bacteria specific to meat processing industry. *Theory Pract. Meat Process.* **2016**, *1*, 54–61. [CrossRef]
14. Afonyushkin, V.N.; Shirshova, A.N.; Shamovskaya, D.V.; Plomodyalov, D.N. A study of the antiviral effect of drug triviron on IBV. *Veterinary* **2018**, 24–28. Available online: https://agris.fao.org/agris-search/search.do?recordID=RU2019000107 (accessed on 11 October 2021).
15. Burakova, E.; Kovalev, N.; Zenkova, M.; Vlassov, V.; Silnikov, V. Structure–activity relationships in new polycationic molecules based on two 1,4-diazabicyclo[2.2.2]octanes as artificial ribonucleases. *Bioorg. Chem.* **2014**, *57*, 127–131. [CrossRef]
16. Afonyushkin, V.N. Influence of flaviviral infection on broiler productivity. *Veterinary* **2014**, *8*, 15–19.
17. Baram-Pinto, D.; Shukla, S.; Perkas, N.; Gedanken, A.; Sarid, R. Inhibition of Herpes Simplex Virus Type 1 Infection by Silver Nanoparticles Capped with Mercaptoethane Sulfonate. *Bioconjug. Chem.* **2009**, *20*, 1497–1502. [CrossRef]
18. Lara, H.H.; Ayala-Nuñez, N.V.; Ixtepan-Turrent, L.; Rodriguez-Padilla, C. Mode of antiviral action of silver nanoparticles against HIV-1. *J. Nanobiotechnol.* **2010**, *8*, 1. [CrossRef]
19. Lara, H.H.; Ixtepan-Turrent, L.; Garza-Treviño, E.N.; Rodriguez-Padilla, C. PVP-coated silver nanoparticles block the transmission of cell-free and cell-associated HIV-1 in human cervical culture. *J. Nanobiotechnol.* **2010**, *8*, 15. [CrossRef] [PubMed]
20. Sun, L.; Singh, A.K.; Vig, K.; Pillai, S.R.; Singh, S.R. Silver nanoparticles inhibit replication of respiratory syncytial virus. *J. Biomed. Nanotechnol.* **2008**, *4*, 149–158.
21. Romo-Quiñonez, C.R.; Álvarez-Sánchez, A.R.; Álvarez-Ruiz, P.; Chávez-Sánchez, M.C.; Bogdanchikova, N.; Pestryakov, A.; Mejia-Ruiz, C.H. Evaluation of a new Argovit as an antiviral agent included in feed to protect the shrimp Litopenaeus vannamei against White Spot Syndrome Virus infection. *PeerJ* **2020**, *8*, e8446. [CrossRef]
22. Borrego, B.; Lorenzo, G.; Mota-Morales, J.D.; Almanza-Reyes, H.; Mateos, F.; López-Gil, E.; de la Losa, N.; Burmistrov, V.A.; Pestryakov, A.N.; Brun, A.; et al. Potential application of silver nanoparticles to control the infectivity of Rift valley fever virus in vitro and in vivo. *Nanomed. Nanotechnol. Biol. Med.* **2016**, *12*, 1185–1192. [CrossRef] [PubMed]
23. Bogdanchikova, N.; Vázquez-Muñoz, R.; Huerta-Saquero, A.; Pena-Jasso, A.; Aguilar-Uzcanga, G.; Picos-Díaz, P.L.; Pestryakov, A.; Burmistrov, V.; Martynyuk, O.; Luna-Vazquez-Gomez, R.; et al. Silver nanoparticles composition for treatment of distemper in dogs. *Int. J. Nanotechnol.* **2016**, *13*, 225–235. Available online: https://www.researchgate.net/publication/315664816_Silver_nanoparticles_composition_for_treatment_of_distemper_in_dogs (accessed on 8 September 2021). [CrossRef]
24. Glotov, A.G.; Glotova, T.I.; Sergeev, A.A.; Sergeev, A.N. Study of Antiviral Activity of Different Drugs against Bovine Herpes Virus and Pestivirus. *Antibiot. Chemother.* **2004**, *49*, 6–9. Available online: https://www.researchgate.net/publication/8101788_Study_of_antiviral_activity_of_different_drugs_against_bovine_herpes_virus_and_pestivirus (accessed on 9 August 2021).
25. Isaacs, C.E.; Kims, K.; Thormar, H. Inactivation of Enveloped Viruses in Human Bodily Fluids by Purified Lipids. *Ann. N. Y. Acad. Sci.* **1994**, *724*, 457–464. [CrossRef]

26. Thormar, H.; Isaacs, C.E.; Brown, H.R.; Barshatzky, M.R.; Pessolano, T. Inactivation of enveloped viruses and killing of cells by fatty acids and monoglycerides. *Antimicrob. Agents Chemother.* **1987**, *31*, 27–31. [CrossRef] [PubMed]
27. Arora, R.; Chawla, R.; Marwah, R.; Arora, P.; Sharma, R.K.; Kaushik, V.; Goel, R.; Kaur, A.; Silambarasan, M.; Tripathi, R.P.; et al. Potential of Complementary and Alternative Medicine in Preventive Management of Novel H1N1 Flu (Swine Flu) Pandemic: Thwarting Potential Disasters in the Bud. *Evid.-Based Complement. Altern. Med.* **2011**, *2011*, 586506. [CrossRef]
28. Sands, J.; Auperin, D.; Snipes, W. Extreme Sensitivity of Enveloped Viruses, Including Herpes Simplex, to Long-Chain Unsaturated Monoglycerides and Alcohols. *Antimicrob. Agents Chemother.* **1979**, *15*, 67–73. [CrossRef]
29. Stephano-Hornedo, J.L.; Torres-Gutiérrez, O.; Toledano-Magaña, Y.; Gradilla-Martínez, I.; Pestryakov, A.; Sánchez-González, A.; García-Ramos, J.C.; Bogdanchikova, N. Argovit™ silver nanoparticles to fight Huanglongbing disease in Mexican limes (*Citrus aurantifolia* Swingle). *RSC Adv.* **2020**, *10*, 6146–6155. [CrossRef]
30. Mironova, T.E.; Afonyushkin, V.N.; Cherepushkina, V.S.; Kozlova, Y.N.; Bobikova, A.S.; Koptev, V.Y.; Sigareva, N.A.; Kolpakov, F.A. Study of the protective effects of virucidal drugs on the model of coronavirus pneumonia. *Vet. Korml.* **2020**, 35–38. [CrossRef]
31. Joseph, T. International Pulmonologist's Consensus on COVID-19, 2nd ed. 2020. Available online: https://teampata.org/portfolio/international-pulmonologists-consensus-on-covid-19/ (accessed on 11 October 2021).
32. Qi, F.; Qian, S.; Zhang, S.; Zhang, Z. Single cell RNA sequencing of 13 human tissues identify cell types and receptors of human coronaviruses. *Biochem. Biophys. Res. Commun.* **2020**, *526*, 135–140. [CrossRef]
33. The Human Protein Atlas Project Funded by the Knut & Alice Wallenberg Foundation. Available online: https://www.proteinatlas.org/ENSG00000130234-ACE2 (accessed on 1 October 2021).
34. Lamers, M.M.; Beumer, J.; van der Vaart, J.; Knoops, K.; Puschhof, J.; Breugem, T.I.; Ravelli, R.B.G.; Paul van Schayck, J.; Mykytyn, A.Z.; Duimel, H.Q.; et al. SARS-CoV-2 productively infects human gut enterocytes. *Science* **2020**, *369*, 50–54. [CrossRef]
35. Al-Beltagi, M.; Saeed, N.K.; Bediwy, A.S.; El-Sawaf, Y. Paediatric gastrointestinal disorders in SARS-CoV-2 infection: Epidemiological and clinical implications. *World J. Gastroenterol.* **2021**, *27*, 1716–1727. [CrossRef]
36. Sfera, A.; Osorio, C.; Zapata Martín del Campo, C.M.; Pereida, S.; Maurer, S.; Maldonado, J.C.; Kozlakidis, Z. Endothelial Senescence and Chronic Fatigue Syndrome, a COVID-19 Based Hypothesis. *Front. Cell. Neurosci.* **2021**, *15*, 673217. [CrossRef] [PubMed]
37. García-Pérez, B.E.; González-Rojas, J.A.; Salazar, M.I.; Torres-Torres, C.; Castrejón-Jiménez, N.S. Taming the Autophagy as a Strategy for Treating COVID-19. *Cells* **2020**, *9*, 2679. [CrossRef]
38. Marino, G.; Niso-Santano, M.; Baehrecke, E.H.; Kroemer, G. Self-consumption: The interplay of autophagy and apoptosis. *Nat. Rev. Mol. Cell Biol.* **2014**, *15*, 81–94. [CrossRef]
39. Young, M.M.; Takahashi, Y.; Khan, O.; Park, S.; Hori, T.; Yun, J.; Sharma, A.K.; Amin, S.; Hu, C.D.; Zhang, J.; et al. Autophagosomal Membrane Serves as Platform for Intracellular Death-inducing Signaling Complex (iDISC)-mediated Caspase-8 Activation and Apoptosis. *J. Biol. Chem.* **2012**, *287*, 12455–12468. [CrossRef] [PubMed]
40. Yerlikaya, A.; Okur, E.; Ulukaya, E. The p53-independent induction of apoptosis in breast cancer cells in response to proteasome inhibitor bortezomib. *Tumor Biol.* **2012**, *33*, 1385–1392. [CrossRef]
41. Bersani, F.; Taulli, R.; Accornero, P.; Morotti, A.; Miretti, S.; Crepaldi, T.; Ponzetto, C. Bortezomib-mediated proteasome inhibition as a potential strategy for the treatment of rhabdomyosarcoma. *Eur. J. Cancer* **2008**, *44*, 876–884. [CrossRef] [PubMed]
42. Zhimin, L.; Shuichan, X. ERK1/2 MAP Kinases in Cell Survival and Apoptosis. *IUBMB Life* **2006**, *58*, 621–631.
43. Valenzuela-Salas, L.M.; Girón-Vázquez, N.G.; García-Ramos, J.C.; Torres-Bugarín, O.; Gómez, C.; Pestryakov, A.; Villarreal-Gómez, L.J.; Toledano-Magaña, Y.; Bogdanchikova, N. Antiproliferative and Antitumour Effect of Nongenotoxic Silver Nanoparticles on Melanoma Models. *Oxid. Med. Cell. Longev.* **2019**, *2019*, 4528241. [CrossRef] [PubMed]
44. Cruz-Ramírez, O.U.; Valenzuela-Salas, L.M.; Blanco-Salazar, A.; Rodríguez-Arenas, J.A.; Mier-Maldonado, P.A.; García-Ramos, J.C.; Bogdanchikova, N.; Pestryakov, A.; Toledano-Magaña, Y. Antitumor Activity Against Human Colorectal Adenocarcinoma of Silver Nanoparticles: Influene of [Ag]/[PVP] Ratio. *Pharmaceutics* **2021**, *13*, 1000. [CrossRef] [PubMed]
45. Ruiz-Ruiz, B.; Arellano-García, M.E.; Radilla-Chávez, P.; Salas-Vargas, D.S.; Toledano-Magaña, Y.; Casillas-Figueroa, F.; Luna Vazquez-Gomez, R.; Pestryakov, A.; García-Ramos, J.C.; Bogdanchikova, N. Cytokinesis-Block Micronucleus Assay Using Human Lymphocytes as a Sensitive Tool for Cytotoxicity/Genotoxicity Evaluation of AgNPs. *ACS Omega* **2020**, *5*, 12005–12015. [CrossRef] [PubMed]
46. Almanza-Reyes, H.; Moreno, S.; Plascencia-López, I.; Alvarado-Vera, M.; Patrón-Romero, L.; Borrego, B.; Reyes-Escamilla, A.; Valencia-Manzo, D.; Brun, A.; Pestryakov, A.; et al. Evaluation of silver nanoparticles for the prevention of SARS-CoV-2 infection in health workers: In vitro and in vivo. *PLoS ONE* **2021**, *16*, e0256401. [CrossRef]
47. Barberis, E.; Amede, E.; Tavecchia, M.; Marengo, E.; Cittone, M.G.; Rizzi, E.; Pedrinelli, A.R.; Tonello, S.; Minisini, R.; Pirisi, M.; et al. Understanding protection from SARS-CoV-2 using metabolomics. *Sci. Rep.* **2021**, *11*, 13796. [CrossRef] [PubMed]

the total numbers obtained from 35 sections from 7 rats/group using an image analyzing system (Optimas 6.5) from CyberMetrics (Scottsdale, AZ, USA).

2.6. Immunohistochemistry

In this study, general immunohistochemistry was carried out to examine changes regarding the neurons, pro-inflammatory, and anti-inflammatory cytokines. For the immunohistochemistry, we used primary antibodies as follows: mouse anti-neuronal nuclei (NeuN; diluted 1:1100; Cat. No., MAB377; Chemicon International, Temecula, CA, USA), rabbit anti-TNF-α (diluted 1:1200) (Cat. No., ab66579; Abcam, Cambridge, UK), rabbit anti-IL-1β (diluted 1:250) (Cat. No., ab2105; Abcam, Cambridge, UK), goat anti-IL-4 (diluted 1:200) (Cat. No. sc-1260; Santa Cruz Biotechnology, Santa Cruz, CA, USA), and goat anti-IL-13 (diluted 1:200) (Cat. No., sc-393365; Santa Cruz Biotechnology, Santa Cruz, CA, USA). In short, as described previously [35], the sections were incubated with each diluted antibody for 12 h at 4 °C. After the sections were washed, they were reacted with biotinylated horse anti-mouse (diluted 1:200) (Cat. No., BA-2001;Vector Laboratories, Burlingame, CA, USA), goat rabbit (diluted 1:200) (Cat. No., BA-1000; Vector Laboratories, Burlingame, CA, USA), or rabbit anti-goat IgG (diluted 1:200) (Cat. No., BA-5000; Vector Laboratories, Burlingame, CA, USA) and, thereafter, developed by avidin-biotin complex (ABC) (diluted 1:300) (Cat. No. PK-4000; Vector Laboratories, Burlingame, CA, USA). Finally, they were visualized with 3,3′-diaminobenzidine solution (DAB; Cat. No., D8001; Sigma-Aldrich, St. Louis, MO, USA). The sections were identically reacted with DAB solution for 90 s at room temperature. In addition, negative control tests for NeuN, TNF-α, IL-1 β, IL-4, and IL-13 were performed for the specificity of each immunostaining, with pre-immune serum instead of each primary antibody. As a result, any immunostained structures were not shown in the tested sections.

For quantitative analysis of the number of NeuN immunoreactive motor neurons and their change, five sections/rat were chosen with a 120-μm interval. The numbers were counted as described in the Section 2.5.

For quantitative analysis of each immunoreactivity (TNF-α, IL-1 β, IL-4, and IL-13) in the ventral horn, the images were taken using the above-mentioned method and analyzed as described in our published paper [35]. Briefly, each image of the captured immunoreactivity was evaluated as optical density (OD): the OD was obtained after transforming each immunoreactive structure to mean gray level using the formula OD = log (256/mean gray level). Finally, each OD was compared as the relative optical density (ROD): a ratio of the ROD was evaluated as percent using Image J software (version 1.59) from NIH (Bethesda, MD, USA).

2.7. Statistical Analysis

In this study, SPSS software (version 15.0) from SPSS Inc (Chicago, IL, USA) was used to perform all statistical analysis. We used the Kolmogorov and Smirnov test for testing normal distributions and Bartlett test for testing the identical standard error of the means (SEMs), and all our data passed the normality test. The statistical significances of the mean among the experimental groups were determined by one-way analysis of variance followed by post hoc Tukey test for all pairwise multiple comparisons. Any differences lower than 0.05 of p value were considered significant.

3. Results

3.1. Changes in Physiological Function and Body Temperature

MAP and body temperature was recorded in each group before and after ACA operation as shown in Figure 2. Before ACA, MAP and body temperature were similar to the baselines observed in the Sham+vehicle group. Body temperature in the ACA/CPR+RIS group was not significantly different from that in the ACA/CPR+vehicle group (Figure 2A). Under 24 ± 1 °C of room temperature, a significant low body (rectal) temperature (33 ± 0.5 °C) in all RIS groups was detected from 1 to 2 h after ACA, which was due to RIS injection.

Thereafter, their body temperature was spontaneously and gradually increased with intermittently shivering to 37 ± 0.5 °C (Figure 2B).

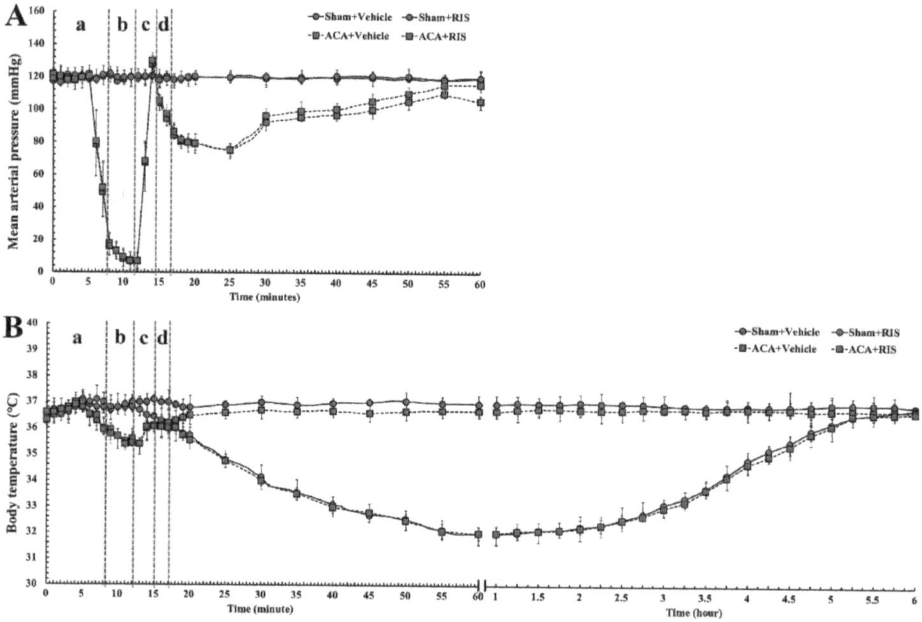

Figure 2. MAP (**A**) and body temperature (**B**) before, during, and after ACA in the Sham+vehicle, ACA/CPR+vehicle, and ACA/CPR+RIS groups. Note that body temperature in the ACA/CPR+RIS group was 33 ± 0.5 °C from 1 to 2 h after ACA. a, inducing ACA; b, maintaining ACA condition; c, conducting CPR; d, confirming ROSC. The bars indicate the means ± SEM (n = 7).

3.2. Survival Rate and Motor Deficit Score

The survival rate in the ACA/CPR+vehicle and ACA/CPR+vehicle groups was recorded by Kaplan-Meier analysis for 2 days after ACA/CPR (Figure 3A). In all sham groups, all rats survived (Figure 3A). In the ACA/CPR+vehicle group, the survival rate gradually reduced with time after ACA/CPR, showing 65.3% at 1 day and 4.3% at 2 days after ROSC (Figure 3A). In the ACA/CPR+RIS group, however, the survival rate was significantly high compared with that in the ACA/CPR+vehicle group, showing 92.4% at 1 day and 67.9% at 2 days after ACA/CPR (Figure 3A).

Hind limb motor deficit (paralysis) was evaluated with the Tarlov score at 1 day after ACA/CPR (Figure 3B). The rats of the Sham+vehicle group revealed normal function in their hind limbs. In the ACA/CPR+vehicle group, the score was significantly low (average 0.8 point) compared with that in the Sham+vehicle group (average 4.1 point) ($p < 0.01$). In the ACA/CPR+RIS group, however, motor function was significantly better (average 2.9 point) than that in the ACA/CPR+vehicle group ($p < 0.05$).

3.3. Neuroprotection by RIS

3.3.1. NeuN Immunoreactive Neurons

We examined neuronal damage/loss in the ventral horn of the lumbar part in the spinal cord after ACA/CPR using immunohistochemistry with NeuN: NeuN is well used to detect neuronal nucleus damage (Figure 4). In the Sham+vehicle and Sham+RIS groups, neurons in the anterior horn, which are called motor neurons, were well stained with NeuN in their nuclei (Figure 4A(a,b,e,f)). In the ACA/CPR+vehicle group, a few neurons stained with NeuN (NeuN$^+$ neurons) were shown in the anterior horn at 2 days after ACA/CPR

(Figure 4A(c,g)). The mean percentage of NeuN$^+$ neurons, in this group, was 24.6% of that in the Sham+vehicle group (Figure 4C). However, in the ACA/CPR+RIS group, many NeuN$^+$ neurons were found at 2 days after ACA/CPR (Figure 4A(d,h)), revealing that the mean percentage of the motor neurons was 91.7% of that in the Sham+vehicle group (Figure 4C).

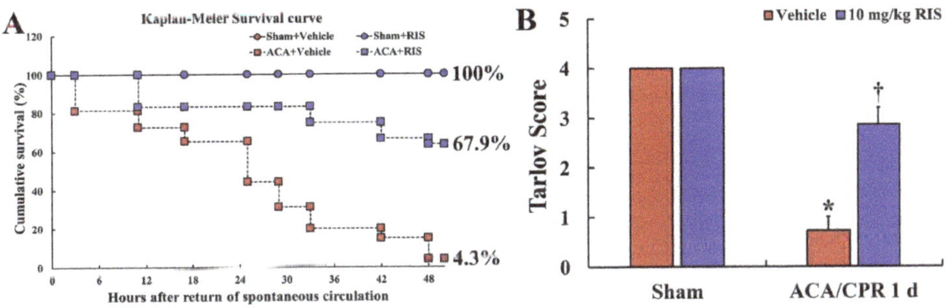

Figure 3. Kaplan-Meier survival curve and Tarlov score (**A**) Survival rate ($p < 0.05$) in the Sham+vehicle, Sham+IRS, ACA/CPR+vehicle, and ACA/CPR+RIS groups using Kaplan–Meier analysis for 2 days after ACA/CPR. The ACA/CPR+RIS group reveals a higher survival rate than the ACA/CPR+vehicle group. At 2 days after ACA/CPR, the cumulative survival rate in the ACA/CPR+vehicle group is 4.3% whereas the cumulate survival rate in the ACA/CPR+RIP group is 67.9%. (**B**) Motor function of both hind limbs in the Sham+vehicle, Sham+RIS, ACA/CPR+vehicle, and ACA/CPR+RIS groups using Tarlov Scoring System. At 1 day after ACA, a significant higher score in the ACA/CPR+RIP group is observed compared to that in the ACA/CPR+vehicle group. The bars indicate the means ± SEM ($n = 7$; * $p < 0.05$ vs. Sham+vehicle group; † $p < 0.05$ vs. ACA/CPR+vehicle group).

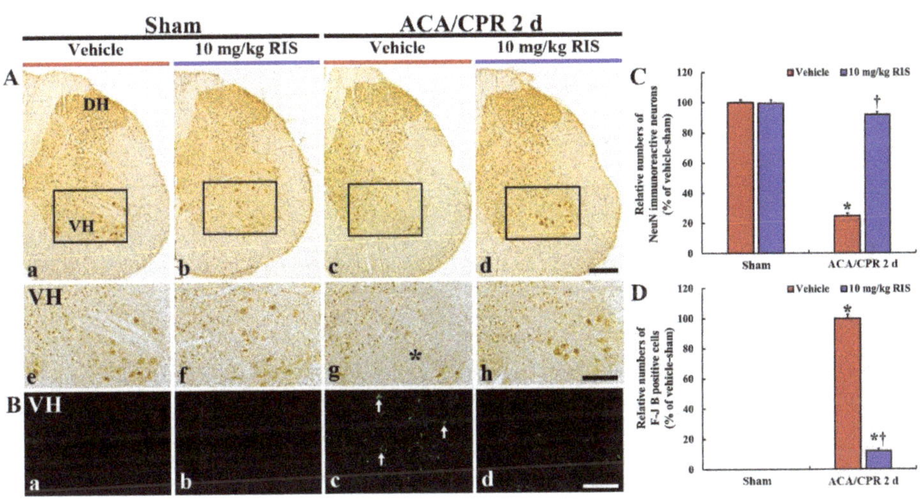

Figure 4. NeuN immunohistochemistry and F-J B histofluorescence (**A**) NeuN immunohistochemistry in the lumbar spinal cord of the Sham+vehicle (**a**,**e**), Sham+RIS (**b**,**f**), ACA/CPR+vehicle (**c**,**g**), and ACA/CPR+RIS (**d**,**h**) groups at 2 days after ACA/CPR. The middle panels are high magnified images for the squares in the upper panels. In the ACA/CPR+vehicle group, NeuN+ neurons are rarely shown (asterisk) in the ventral horn (VH). However, many NeuN+ cells are shown in the ACA/CPR+RIS group. DH, dorsal horn. Scale bar = 200 (**a–d**) and 100 (**e–h**) μm. (**B**) F-J B histofluorescence in the ventral horn of the Sham+vehicle (**a**), Sham+RIS (**b**), ACA/CPR+vehicle (**c**), and ACA/CPR+RIS (**d**) groups at 2 days after ACA/CPR. In the ACA/CPR+vehicle group, many F-J B+ cells (arrows) are shown, but the numbers of F-J B+ cells are decreased in the ACA/CPR+RIS group. Scale bar = 400 μm (**A**) and 100 μm. (**C**,**D**) Quantitative analyses of NeuN+ (**C**) and F-J B+ cells (**D**) in the VH. The bars indicate the means ± SEM ($n = 7$; * $p < 0.05$ vs. Sham+vehicle group; † $p < 0.05$ vs. ACA/CPR+vehicle group).

3.3.2. F-J B-Positive Cells

The neuroprotection by RIS from ACA/CPR in the ventral horn was analyzed by F-J B histofluorescence: F-J B is an excellent marker for detection of dead cells (neurons) (Figure 4B). No F-J B-positive (F-J B^+) cells were found in the Sham+vehicle and Sham+RIS groups (Figure 4B(a,b)). In the ACA/CPR+vehicle group, many F-J B^+ cells were found in the anterior horn at 2 days after ACA/CPR (Figure 4Bc,D). In the ACA/CPR+RIS group, the numbers of F-J B^+ cells were significantly decreased at 2 days after ACA/CPR (Figure 4Bd), showing that the mean percentage of the F-J B^+ cells was 12.1% of that in the ACA/CPR+vehicle group (Figure 4D).

3.4. Decreased Pro-Inflammatory Cytokines by RIS

3.4.1. TNF-α Immunoreactivity

TNF-α immunoreactivity shown in the Sham+vehicle group was shown in the motor neurons located in the anterior horn (Figure 5Aa). In the ACA/CPR+vehicle group, TNF-α immunoreactivity was gradually enhanced until 1 day after ACA, showing that the ROD of TNF-α immunoreactivity at 12 h and 1 day after ACA/CPR was 183.5% and 211.9%, respectively, compared with that in the Sham+vehicle group (Figure 5A(b,c),B). Thereafter, TNF-α immunoreactivity was decreased, but the ROD was 150.1% of that in the Sham+vehicle group (Figure 5Ad,B).

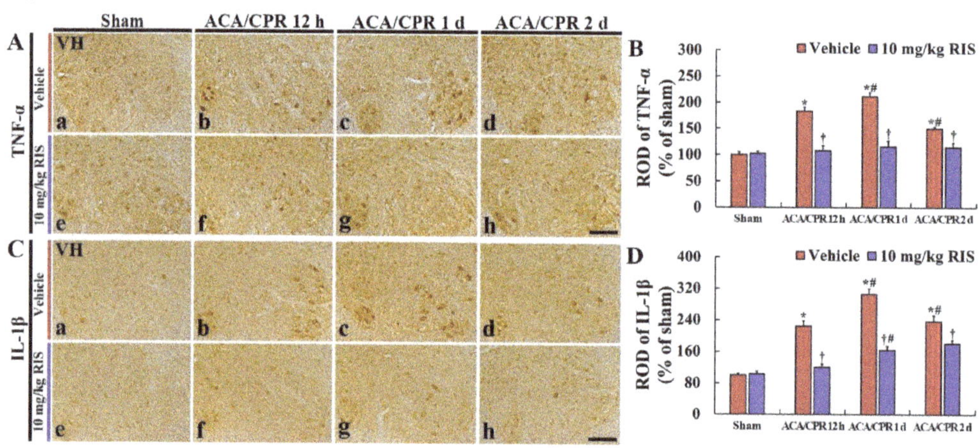

Figure 5. Immunohistochemical staining for TNF-α and IL-1β (**A**,**C**) Immunohistochemistry for TNF-α (**A**) and IL-1β (**C**) in the ventral horn of the Sham+vehicle (**a**), ACA/CPR+vehicle (**b**–**d**), Sham+RIS (**e**), and ACA/CPR+RIS (**f**–**h**) groups at 12 h, 1 day, and 2 days after ACA/CPR. In the ACA/CPR+vehicle group, TNF-α and IL-1β immunoreactivities are significantly increased from 12 h after ACA/CPR. However, in the ACA/CPR+RIS group, immunoreactivities of TNF-α and IL-1β are significantly low compared with that shown in the ACA/CPR+vehicle group. VH, ventral horn. Scale bar = 100 μm. (**A**,**C**) RODs of TNF-α (**B**) and IL-1β (**D**) immunoreactivity. The bars indicate the means ± SEM (n = 7; * $p < 0.05$ vs. Sham+vehicle group; † $p < 0.05$ vs. ACA/CPR+vehicle group; # $p < 0.05$ vs. Pre-time point of corresponding group).

In the Sham+RIS group, TNF-α immunoreactivity in the ventral horn was similar to that in the Sham+vehicle group (Figure 5Ae,B). In addition, in the ACA/CPR+RIS group, TNF-α immunoreactivity in the anterior horn showed no difference from that in the Sham+vehicle group (Figure 5A(f–h),B).

3.4.2. IL-1 β Immunoreactivity

In the Sham+vehicle group, IL-1β immunoreactivity was weakly shown in the motor neurons (Figure 5Ca). In the ACA/CPR+vehicle group, IL-1β immunoreactivity at 12 h, 1 day, and 2 days after ACA/CPR was intensely increased, showing that the ROD was

224.4%, 305.3, and 237.1%, respectively, compared with that in the Sham+vehicle group (Figure 5B(b–d),C).

In the Sham+RIS group, IL-1β immunoreactivity in the lumbar ventral horn was not significantly different from that found in the Sham+vehicle group (Figure 5Be,D). In the ACA/CPR+RIS group, IL-1β immunoreactivity was gradually enhanced after ACA, but the ROD at each point in time was significantly lower (41.2%, 45.4%, and 23.5%, respectively) than that in the ACA/CPR+vehicle group (Figure 5D).

3.5. Increased Anti-Inflammatory Cytokines by RIS

3.5.1. IL-4 Immunoreactivity

IL-4 immunoreactivity in the ventral horn of the Sham+vehicle group was shown in the motor neurons (Figure 6Aa). In the ACA/CPR+vehicle group, IL-4 immunoreactivity was dramatically and gradually decreased after ACA/CPR, showing that the ROD at 12 h, 1 day, and 2 days after ACA/CPR was 68.3%, 47.1%, and 31.3%, respectively, compared with that found in the Sham+vehicle group (Figure 6A(b–d),B).

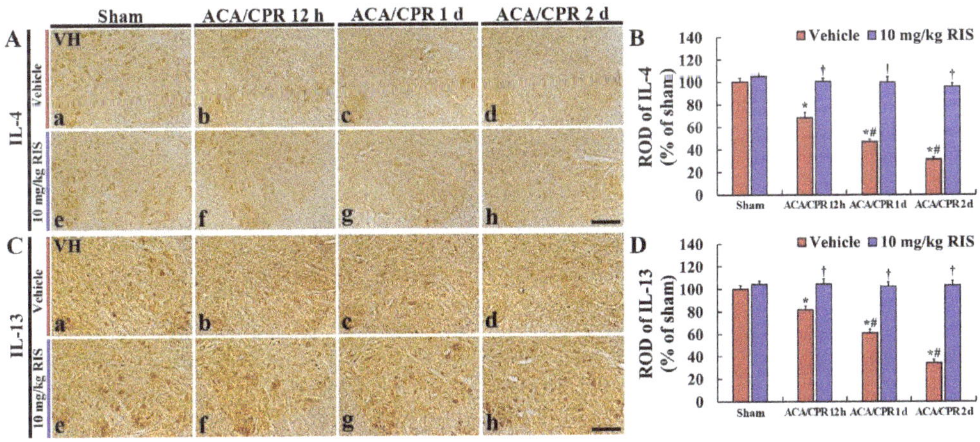

Figure 6. Immunohistochemical staining for IL-4 and IL-13 (**A**,**C**) Immunohistochemistry for IL-4 (**A**) and IL-13 (**C**) in the ventral horn of the Sham+vehicle (**a**), ACA/CPR+vehicle (**b**–**d**), Sham+RIS (**e**), and ACA/CPR+RIS (**f**–**h**) groups at 12 h, 1 day, and 2 days after ACA/CPR. In the ACA/CPR+vehicle group, immunoreactivities of IL-4 and IL-13 are significantly decreased from 12 h after ACA/CPR. However, in the ACA/CPR+RIS group, immunoreactivities of IL-4 and IL-13 are maintained after ACA/CPR. VH, ventral horn. Scale bar = 100 μm. (**A**,**C**) RODs of IL-4 (**B**) and IL-13 (**D**) immunoreactivity. The bars indicate the means ± SEM ($n = 7$; * $p < 0.05$ vs. Sham+vehicle group; † $p < 0.05$ vs. ACA/CPR+vehicle group; # $p < 0.05$ vs. Pre-time point of the corresponding group).

In the Sham+RIS group, IL-4 immunoreactivity in the lumbar ventral horn was similar to that shown in the Sham+vehicle group (Figure 6Ae,B). In the ACA/CPR+RIS group, IL-4 immunoreactivity in the anterior horn was maintained after ACA/CPR (Figure 6A(f–h),B).

3.5.2. IL-13 Immunoreactivity

In the ventral horn of the Sham+vehicle group, IL-13 immunoreactivity was also found in the motor neurons (Figure 6Ba). IL-13 immunoreactivity in the ACA/CPR+vehicle group was dramatically and gradually decreased after ACA/CPR (RODs: 81.7% at 12 h, 60.9% at 1 day, and 34.5% at 2 days after ACA/CPR) compared with that in the Sham+vehicle group (Figure 6B(b–d),D).

In the Sham+RIS group, IL-13 immunoreactivity in the ventral horn was not different from that shown in the Sham+vehicle group (Figure 6Ce,D). In the ACA/CPR+RIS group, IL-13 immunoreactivity in the anterior horn was also maintained after ACA/CPR (Figure 6C(f–h),D).

4. Discussion

Generally, the spinal cord is prominently susceptible to ischemic insults owing to diverse circulatory abnormalities [36–38]. In addition, the vulnerability and sensitivity of the spinal cord after ACA/CPR are completely different from those of the brain. In particular, the time course of cell death in the ischemic spinal cord is different from that in the ischemic brain after ACA/CPR [39]. This difference may be because much energy is required for the extensive activity of the motor neurons located in the anterior horn of the spinal cord [36]. It has been reported that thoracic aortic occlusion-induced spinal cord ischemia leads to neuronal damage in the ventral horn of the lumbar spinal cord from one day after spinal cord ischemia in rats [40]. In a rat model of ACA/CPR, neuronal death (loss) in the anterior horn in the lumbar part of the spinal cord occurs at one day after ACA/CPR [9]. In addition, Ahn et al. [10] recently reported that neuronal death in the central nervous system (CNS) autonomic control center (myelencephalon and thoracolumbar division of the spinal cord) occurred very early compared to the other CNS divisions after ACA/CPR in rats. In our current study, we found that ventral motor neurons at the level of the lumbar spinal cord were dead at two days after ACA/CPR in rats. These results indicated that the time course of neuronal damage/death in the CNS following global ischemia in the whole body (i.e., ACA) must be different according to the regions of the brain and spinal cord and that the spinal cord has a higher vulnerability to transient ischemia than the brain. In short, spinal cord damage following ACA/CPR occurred much faster than brain damage.

Hind-limb paralysis is one of the main disorders after ACA/CPR [8,9]. Duggal and Lach [39] reported that selective vulnerability of the lumbosacral part of the spinal cord was shown in patients with ACA/CPR and hypotension. Experimental studies on ischemic spinal cord injury have been conducted using animal models with aortic disease or local vascular change [40–42]. In a rabbit model of spinal cord ischemia, which is simply produced by occlusion of the spinal arteries that have no collateral circulation, paraplegia occurs when motor neurons in the lumbar spinal cord are damaged or dead after ischemic-reperfusion injury [43], and the death of motor neurons in the lumbar spinal cord is shown within one day after ischemia-reperfusion [44–46]. In our current study using a rat model of ACA/CPR, paralysis in the hind limbs was seen one day after ACA/CPR, and most motor neurons located in the anterior horn were not seen two days after ACA/CPR. Taken together, we suggest that paraplegia following ACA/CPR might occur with motor neuron damage or death because normal motor nerve fibers (general somatic efferent) via the spinal nerves cannot innervate muscles of the limbs [47].

For several decades, RIS, as a selective monoaminergic antagonist, has been widely used for the treatment of schizophrenia [17,18]. In addition, RIS has been reported to induce hypothermia [19,21,48]. It has been reported that hypothermia can display neuroprotection and improve damaged outcomes in experimental animal models of spinal cord and brain injury [49]. However, few data concerning the effects of hypothermia against spinal cord injury after ACA have been accumulated. In this regard, we examined the effect of RIS on motor deficits in the hind limbs and its related neuronal vulnerability in the spinal cord following ACA/CPR in rats. It was reported that RIS treatment after brain transient ischemia induced hypothermia within 30 min and lasted for four hours and that hypothermia displayed effective protection against the death of hippocampal neurons induced by transient brain ischemia by attenuating glial activation and maintaining antioxidant enzymes [21]. In our present study, the effects of RIS-induced hypothermia on spinal cord injury after ACA/CPR in rats were investigated, and, as expected, RIS-induced hypothermia significantly improved paraplegia and alleviated the damage/death (loss) of ventral motor neurons at two days after ACA/CPR. These results strongly suggest that RIS treatment after ACA improves neurological dysfunction by attenuating the damage of the ventral motor neurons in patients with spinal cord injury from ACA.

Over the past few years, a body of evidence has stressed the roles of inflammation in the pathophysiology of acute brain ischemia [50]. Cytokines include many groups of inflammatory mediators, and they act as signaling molecules to control inflammation and to induce positive or negative effects on neuronal survival [51]. It is well known that pro-inflammatory cytokines are involved in the amplification of inflammatory reactions and contribute to the pathogenesis of neurological disorders, whereas anti-inflammatory cytokines are decisively involved in resolving inflammation through downregulating the production of pro-inflammatory cytokines [52].

Some studies showed the anti-inflammatory properties of RIS in an in vivo and in vitro model. MacDowell K.S. et al. [53] demonstrated the anti-inflammatory effect of RIS. In detail, a single administration of RIS regulated various factors that triggered advanced inflammatory responses, such as the expression of inflammatory cytokines (interleukin (IL)-1β and tumor necrosis factor (TNF)-α) following lipopolysaccharide (LPS)-induced inflammation in the frontal cortex of rat brains. Additionally, a precedent study showed that RIS suppressed the production of pro-inflammatory cytokines and decreased the level of inducible NO synthase (iNOS), which are secreted by reactive microglia using a microglial cell line [54]. In this study, the immunoreactivity of pro-inflammatory cytokines (TNF-α and IL-1β) in the ventral horn of the ACA/CPR+vehicle group was increased with time after ACA/CPR, but, in the ACA/CPR+RIS group, the immunoreactivity of TNF-α and IL-1β was significantly lower than that in the ACA/CPR+vehicle group. It has been found that TNF-α and IL-1β are activated in the brains of animal models of transient brain ischemia as mediators in response to ischemic injury [55–57]. TNF-α and IL-1β play critical roles in post-ischemic inflammatory injury in the spinal cord [58–60]. Hasturk et al. [58] concluded that serum TNF-α and IL-1β levels significantly increased after spinal cord ischemia-reperfusion injury accompanied by tissue damage. In rat models of spinal cord ischemia-reperfusion injury, increased levels of cytokines induced by ischemic injury were observed to be associated with the deterioration of motor function and histological damage in the spinal cord [61] and TNF-α levels were significantly increased within 1.5 h, and peaked 3 h after ischemic injury [62]. In a swine model of spinal cord ischemia-reperfusion injury, TNF-α levels were significantly increased from 6 to 24 h after ischemic injury [63]. Additionally, IL-1 expression was significantly increased in the spinal cord 6 and 36 h following ischemic-reperfusion injury in mice [64]. Taken together, we suggest that pro-inflammatory cytokines might contribute to cell death in the spinal cord following ischemia-reperfusion injury. In addition, our current findings indicate that RIS treatment after ACA/CPT induces hypothermia and prevents the abnormal expressions of TNF-α and IL-1β in the ischemic spinal cord.

In our current study, significant decreases in the immunoreactivity of anti-inflammatory cytokines (IL-4 and IL-13) were observed in the anterior horn cord in the lumbar spinal cord after ACA/CPR. However, IL-4 and IL-13 immunoreactivity in the ACA/CPR+RIS group was not reduced compared to that in the sham+vehicle group. It was demonstrated that the sustained or increased expression of endogenous anti-inflammatory cytokines (IL-4 and IL-13) contributed to neuronal survival from ischemia-reperfusion injury in the gerbil hippocampus after transient forebrain ischemia [65,66]. Additionally, some studies showed that IL-4 and IL-13 suppressed the expression and production of pro-inflammatory cytokines (TNF-α and IL-1β) in the spinal cord of animal models of spinal cord ischemia [67,68]. Therefore, taken together, the findings suggested that the maintained expression of anti-inflammatory cytokines in the ACA/CPR+RIS group may contribute to the protection of motor neurons from ACA injury.

5. Conclusions

In brief, our present findings showed that RIS treatment after ACA/CPR induced hypothermic conditions, significantly reduced mortality, and attenuated hind-limb paralysis. In addition, neuronal damage/death (loss) in the ventral horn of the lumbar spinal cord was ameliorated. These might be associated with the significant decreases of pro-inflammatory cytokines and the maintenance of anti-inflammatory cytokines, which might be induced by the hypothermic condition induced by RIS treatment. Taken together, we suggest that immediate post-treatment with RIS after ACA can be utilized as a novel therapeutic approach of patients with ACA.

Author Contributions: Conceptualization, M.-H.W. and J.H.C.; Methodology, J.-C.L. and M.C.S.; Software, H.-I.K.; Validation, S.H. and J.H.A.; Investigation, T.-K.L., D.W.K. and H.-J.T.; Data Curation, T.-K.L. and J.H.P.; Writing—original draft preparation, T.-K.L. and J.-C.L.; Writing—review and editing M.-H.W.; Supervision, S.Y.C. and J.-C.L.; Project Administration, M.-H.W. and J.H.C.; Funding Acquisition, M.C.S., J.H.C. and S.Y.C. All authors have read and agreed to the published version of the manuscript.

Funding: This work was supported by Basic Science Research Program through the National Research Foundation of Korea (NRF) funded by the Ministry of Education (NRF-2020R1I1A3068251, NRF-2016R1D1A1B01011790 and NRF-2019R1A6A1A11036849).

Institutional Review Board Statement: The experimental protocol for this study was approved (approval no., KW-200113-1; approval date, 18 February 2020) by the Institutional Animal Care and Use Committee (IACUC). The content of the protocol adhered to the guidelines that are in compliance with the "Current International Laws and Policies" from the "Guide for the Care and Use of Laboratory Animals" (The National Academies Press, 8th Ed., 2011).

Informed Consent Statement: Not applicable.

Data Availability Statement: The data presented in this study are available on request from the corresponding author.

Acknowledgments: The authors would like to thank Seung Uk Lee and Hyun Sook Kim for their technical help in this study.

Conflicts of Interest: The authors declared that there are no conflict of interest to this work.

Abbreviations

ACA	asphyxial cardiac arrest
CNS	central nervous system
CPR	cardiopulmonary resuscitation
F-J B	fluoro-Jade B
IL	interleukin
NeuN	risperidone
RIS	major histocompatibility complex
ROSC	return of spontaneous circulation
ROD	relative optical density
TNF-α	tumor necrosis factor α
VH	ventral horn

References

1. Chalkias, A.; Xanthos, T. Post-cardiac arrest brain injury: Pathophysiology and treatment. *J. Neurol. Sci.* **2012**, *315*, 1–8. [CrossRef]
2. Schneider, A.; Bottiger, B.W.; Popp, E. Cerebral resuscitation after cardiocirculatory arrest. *Anesth Analg.* **2009**, *108*, 971–979. [CrossRef]
3. Lopez-Herce, J.; del Castillo, J.; Matamoros, M.; Canadas, S.; Rodriguez-Calvo, A.; Cecchetti, C.; Rodriguez-Nunez, A.; Carrillo, A. Post return of spontaneous circulation factors associated with mortality in pediatric in-hospital cardiac arrest: A prospective multicenter multinational observational study. *Crit. Care* **2014**, *18*, 607. [CrossRef]
4. Mongardon, N.; Dumas, F.; Ricome, S.; Grimaldi, D.; Hissem, T.; Pene, F.; Cariou, A. Postcardiac arrest syndrome: From immediate resuscitation to long-term outcome. *Ann. Intensive Care* **2011**, *1*, 1–11. [CrossRef]
5. Watts, J. Paraplegia and spinal cord ischaemia. *Anaesthesia* **1995**, *50*, 745–746. [CrossRef] [PubMed]
6. Grassner, L.; Klausner, F.; Wagner, M.; McCoy, M.; Golaszewski, S.; Leis, S.; Aigner, L.; Couillard-Despres, S.; Trinka, E. Acute and chronic evolution of mri findings in a case of posterior spinal cord ischemia. *Spinal Cord* **2014**, *52* (Suppl. S1), S23–S24. [CrossRef] [PubMed]
7. Nedeltchev, K.; Loher, T.J.; Stepper, F.; Arnold, M.; Schroth, G.; Mattle, H.P.; Sturzenegger, M. Long-term outcome of acute spinal cord ischemia syndrome. *Stroke* **2004**, *35*, 560–565. [CrossRef]

8. Ahn, J.H.; Lee, T.K.; Kim, B.; Lee, J.C.; Tae, H.J.; Cho, J.H.; Park, Y.; Shin, M.C.; Ohk, T.G.; Park, C.W.; et al. Therapeutic hypothermia improves hind limb motor outcome and attenuates oxidative stress and neuronal damage in the lumbar spinal cord following cardiac arrest. *Antioxidants* **2020**, *9*, 38. [CrossRef] [PubMed]
9. Lee, J.C.; Tae, H.J.; Cho, J.H.; Kim, I.S.; Lee, T.K.; Park, C.W.; Park, Y.E.; Ahn, J.H.; Park, J.H.; Yan, B.C.; et al. Therapeutic hypothermia attenuates paraplegia and neuronal damage in the lumbar spinal cord in a rat model of asphyxial cardiac arrest. *J. Therm. Biol.* **2019**, *83*, 1–7. [CrossRef]
10. Ahn, J.H.; Lee, T.K.; Tae, H.J.; Kim, B.; Sim, H.; Lee, J.C.; Kim, D.W.; Kim, Y.S.; Shin, M.C.; Park, Y.; et al. Neuronal death in the cns autonomic control center comes very early after cardiac arrest and is not significantly attenuated by prompt hypothermic treatment in rats. *Cells* **2021**, *10*, 60. [CrossRef] [PubMed]
11. Arrich, J.; Holzer, M.; Havel, C.; Mullner, M.; Herkner, H. Hypothermia for neuroprotection in adults after cardiopulmonary resuscitation. *Cochrane Database Syst. Rev.* **2016**, *2*, CD004128. [CrossRef] [PubMed]
12. Bernard, S.A.; Gray, T.W.; Buist, M.D.; Jones, B.M.; Silvester, W.; Gutteridge, G.; Smith, K. Treatment of comatose survivors of out-of-hospital cardiac arrest with induced hypothermia. *N. Engl. J. Med.* **2002**, *346*, 557–563. [CrossRef] [PubMed]
13. Gebhardt, K.; Guyette, F.X.; Doshi, A.A.; Callaway, C.W.; Rittenberger, J.C.; Post Cardiac Arrest, S. Prevalence and effect of fever on outcome following resuscitation from cardiac arrest. *Resuscitation* **2013**, *84*, 1062–1067. [CrossRef] [PubMed]
14. Cronberg, T.; Lilja, G.; Horn, J.; Kjaergaard, J.; Wise, M.P.; Pellis, T.; Hovdenes, J.; Gasche, Y.; Aneman, A.; Stammet, P.; et al. Neurologic function and health-related quality of life in patients following targeted temperature management at 33 degrees c vs 36 degrees c after out-of-hospital cardiac arrest: A randomized clinical trial. *JAMA Neurol.* **2015**, *72*, 634–641. [CrossRef] [PubMed]
15. Sterz, F.; Safar, P.; Tisherman, S.; Radovsky, A.; Kuboyama, K.; Oku, K. Mild hypothermic cardiopulmonary resuscitation improves outcome after prolonged cardiac arrest in dogs. *Crit. Care Med.* **1991**, *19*, 379–389. [CrossRef]
16. Kuboyama, K.; Safar, P.; Radovsky, A.; Tisherman, S.A.; Stezoski, S.W.; Alexander, H. Delay in cooling negates the beneficial effect of mild resuscitative cerebral hypothermia after cardiac arrest in dogs: A prospective, randomized study. *Crit. Care Med.* **1993**, *21*, 1348–1358. [CrossRef]
17. Schotte, A.; Janssen, P.F.; Gommeren, W.; Luyten, W.H.; Van Gompel, P.; Lesage, A.S.; De Loore, K.; Leysen, J.E. Risperidone compared with new and reference antipsychotic drugs: In vitro and in vivo receptor binding. *Psychopharmacology* **1996**, *124*, 57–73. [CrossRef] [PubMed]
18. Corena-McLeod, M. Comparative pharmacology of risperidone and paliperidone. *Drugs R D* **2015**, *15*, 163–174. [CrossRef] [PubMed]
19. Razaq, M.; Samma, M. A case of risperidone-induced hypothermia. *Am. J. Ther.* **2004**, *11*, 229–230. [CrossRef] [PubMed]
20. Brevik, A.; Farver, D. Atypical antipsychotic induced mild hypothermia. *S D J. Med.* **2003**, *56*, 67–70. [PubMed]
21. Yang, G.E.; Tae, H.J.; Lee, T.K.; Park, Y.E.; Cho, J.H.; Kim, D.W.; Park, J.H.; Ahn, J.H.; Ryoo, S.; Kim, Y.M.; et al. Risperidone treatment after transient ischemia induces hypothermia and provides neuroprotection in the gerbil hippocampus by decreasing oxidative stress. *Int. J. Mol. Sci.* **2019**, *20*, 4621. [CrossRef] [PubMed]
22. Jayaraj, R.L.; Azimullah, S.; Beiram, R.; Jalal, F.Y.; Rosenberg, G.A. Neuroinflammation: Friend and foe for ischemic stroke. *J. Neuroinflamm.* **2019**, *16*, 1–24. [CrossRef] [PubMed]
23. Lee, T.K.; Kang, I.J.; Kim, B.; Sim, H.J.; Kim, D.W.; Ahn, J.H.; Lee, J.C.; Ryoo, S.; Shin, M.C.; Cho, J.H.; et al. Experimental pretreatment with chlorogenic acid prevents transient ischemia-induced cognitive decline and neuronal damage in the hippocampus through anti-oxidative and anti-inflammatory effects. *Molecules* **2020**, *25*, 3578. [CrossRef] [PubMed]
24. Ceulemans, A.G.; Zgavc, T.; Kooijman, R.; Hachimi-Idrissi, S.; Sarre, S.; Michotte, Y. The dual role of the neuroinflammatory response after ischemic stroke: Modulatory effects of hypothermia. *J. Neuroinflamm.* **2010**, *7*, 1–18. [CrossRef] [PubMed]
25. Lakhan, S.E.; Kirchgessner, A.; Hofer, M. Inflammatory mechanisms in ischemic stroke: Therapeutic approaches. *J. Transl. Med.* **2009**, *7*, 1–11. [CrossRef] [PubMed]
26. Perini, F.; Morra, M.; Alecci, M.; Galloni, E.; Marchi, M.; Toso, V. Temporal profile of serum anti-inflammatory and pro-inflammatory interleukins in acute ischemic stroke patients. *Neurol. Sci.* **2001**, *22*, 289–296. [CrossRef]
27. Park, J.H.; Park, O.; Cho, J.H.; Chen, B.H.; Kim, I.H.; Ahn, J.H.; Lee, J.C.; Yan, B.C.; Yoo, K.Y.; Lee, C.H.; et al. Anti-inflammatory effect of tanshinone i in neuroprotection against cerebral ischemia-reperfusion injury in the gerbil hippocampus. *Neurochem. Res.* **2014**, *39*, 1300–1312. [CrossRef] [PubMed]
28. Yoo, K.Y.; Kim, I.H.; Cho, J.H.; Ahn, J.H.; Park, J.H.; Lee, J.C.; Tae, H.J.; Kim, D.W.; Kim, J.D.; Hong, S.; et al. Neuroprotection of chrysanthemum indicum linne against cerebral ischemia/reperfusion injury by anti-inflammatory effect in gerbils. *Neural Regen. Res.* **2016**, *11*, 270–277.
29. Martirosyan, N.L.; Patel, A.A.; Carotenuto, A.; Kalani, M.Y.; Bohl, M.A.; Preul, M.C.; Theodore, N. The role of therapeutic hypothermia in the management of acute spinal cord injury. *Clin. Neurol. Neurosurg.* **2017**, *154*, 79–88. [CrossRef]
30. Dietrich, W.D. Therapeutic hypothermia for spinal cord injury. *Crit. Care Med.* **2009**, *37*, S238–S242. [CrossRef]
31. Albus, U. *Guide for the Care and Use of Laboratory Animals*, 8th ed.; SAGE Publications Sage UK: London, UK, 2012.
32. Flecknell, P. *Laboratory Animal Anaesthesia*; Academic Press: Cambridge, MA, USA, 2015.
33. Lee, C.H.; Hwang, I.K.; Choi, J.H.; Yoo, K.Y.; Han, T.H.; Park, O.K.; Lee, S.Y.; Ryu, P.D.; Won, M.H. Calcium binding proteins immunoreactivity in the rat basolateral amygdala following myocardial infarction. *Cell Mol. Neurobiol.* **2010**, *30*, 333–338. [CrossRef]

34. Lee, J.C.; Park, J.H.; Kim, I.H.; Cho, G.S.; Ahn, J.H.; Tae, H.J.; Choi, S.Y.; Cho, J.H.; Kim, D.W.; Kwon, Y.G.; et al. Neuroprotection of ischemic preconditioning is mediated by thioredoxin 2 in the hippocampal ca1 region following a subsequent transient cerebral ischemia. *Brain Pathol.* **2017**, *27*, 276–291. [CrossRef] [PubMed]
35. Song, M.; Ahn, J.H.; Kim, H.; Kim, D.W.; Lee, T.K.; Lee, J.C.; Kim, Y.M.; Lee, C.H.; Hwang, I.K.; Yan, B.C.; et al. Chronic high-fat diet-induced obesity in gerbils increases pro-inflammatory cytokines and mtor activation, and elicits neuronal death in the striatum following brief transient ischemia. *Neurochem. Int.* **2018**, *121*, 75–85. [CrossRef] [PubMed]
36. Turkoz, A.; Gulcan, O.; Kizilkilic, O.; Kocum, A.; Turkoz, R. Spinal cord ischemia caused by cardiac arrest secondary to pericardial effusion. *J. Cardiothorac. Vasc. Anesth.* **2007**, *21*, 91–92. [CrossRef] [PubMed]
37. Cheng, M.Y.; Lyu, R.K.; Chang, Y.J.; Chen, C.M.; Chen, S.T.; Wai, Y.Y.; Ro, L.S. Concomitant spinal cord and vertebral body infarction is highly associated with aortic pathology: A clinical and magnetic resonance imaging study. *J. Neurol.* **2009**, *256*, 1418–1426. [CrossRef] [PubMed]
38. Dublin, A.B.; Latchaw, R.E.; Herrera, D.A.; Dahlin, B.C. Delayed complication after embolotherapy of a vertebral arteriovenous fistula: Spinal cord ischemia. *J. Vasc. Interv. Radiol.* **2010**, *21*, 392–393. [CrossRef] [PubMed]
39. Duggal, N.; Lach, B. Selective vulnerability of the lumbosacral spinal cord after cardiac arrest and hypotension. *Stroke* **2002**, *33*, 116–121. [CrossRef] [PubMed]
40. Lu, K.; Liang, C.L.; Chen, H.J.; Chen, S.D.; Hsu, H.C.; Liliang, P.C.; Lin, T.K.; Cho, C.L. Injury severity and cell death mechanisms: Effects of concomitant hypovolemic hypotension on spinal cord ischemia-reperfusion in rats. *Exp. Neurol.* **2004**, *185*, 120–132. [CrossRef] [PubMed]
41. Marsala, M.; Yaksh, T.L. Transient spinal ischemia in the rat: Characterization of behavioral and histopathological consequences as a function of the duration of aortic occlusion. *J. Cereb. Blood Flow Metab.* **1994**, *14*, 526–535. [CrossRef]
42. Lang-Lazdunski, L.; Matsushita, K.; Hirt, L.; Waeber, C.; Vonsattel, J.P.; Moskowitz, M.A.; Dietrich, W.D. Spinal cord ischemia. Development of a model in the mouse. *Stroke* **2000**, *31*, 208–213. [CrossRef]
43. Mazensky, D.; Flesarova, S.; Sulla, I. Arterial blood supply to the spinal cord in animal models of spinal cord injury. A review. *Anat. Rec.* **2017**, *300*, 2091–2106. [CrossRef]
44. Yoo, D.Y.; Cho, S.B.; Jung, H.Y.; Kim, W.; Choi, G.M.; Won, M.H.; Kim, D.W.; Hwang, I.K.; Choi, S.Y.; Moon, S.M. Tat-protein disulfide-isomerase a3: A possible candidate for preventing ischemic damage in the spinal cord. *Cell Death Dis.* **2017**, *8*, e3075. [CrossRef]
45. Lee, J.C.; Hwang, I.K.; Yoo, K.Y.; Jung, J.Y.; Cho, J.H.; Moon, S.M.; Kang, T.C.; Kim, W.K.; Kim, Y.S.; Won, M.H. Calbindin d-28k is expressed in the microvascular basal lamina in the ventral horn at early time after transient spinal cord ischemia in the rabbit. *Brain Res.* **2005**, *1047*, 123–128. [CrossRef]
46. Mechirova, E.; Danielisova, V.; Domorakova, I.; Dankova, M.; Stebnicky, M.; Mickova, H.; Burda, J. Bradykinin preconditioning affects the number of degenerated neurons and the level of antioxidant enzymes in spinal cord ischemia in rabbits. *Acta Histochem.* **2014**, *116*, 252–257. [CrossRef]
47. Sengupta, B.; Faisal, A.A.; Laughlin, S.B.; Niven, J.E. The effect of cell size and channel density on neuronal information encoding and energy efficiency. *J. Cereb. Blood Flow Metab.* **2013**, *33*, 1465–1473. [CrossRef] [PubMed]
48. van Marum, R.J.; Wegewijs, M.A.; Loonen, A.J.; Beers, E. Hypothermia following antipsychotic drug use. *Eur. J. Clin. Pharmacol.* **2007**, *63*, 627–631. [CrossRef]
49. Dietrich, W.D.; Atkins, C.M.; Bramlett, H.M. Protection in animal models of brain and spinal cord injury with mild to moderate hypothermia. *J. Neurotrauma* **2009**, *26*, 301–312. [CrossRef]
50. Shi, K.; Tian, D.C.; Li, Z.G.; Ducruet, A.F.; Lawton, M.T.; Shi, F.D. Global brain inflammation in stroke. *Lancet Neurol.* **2019**, *18*, 1058–1066. [CrossRef]
51. Becher, B.; Spath, S.; Goverman, J. Cytokine networks in neuroinflammation. *Nat. Rev. Immunol.* **2017**, *17*, 49–59. [CrossRef] [PubMed]
52. Dugue, R.; Nath, M.; Dugue, A.; Barone, F.C. Roles of pro-and anti-inflammatory cytokines in traumatic brain injury and acute ischemic stroke. *Mech. Neuroinflamm.* **2017**, *211*. [CrossRef]
53. Kato, T.; Monji, A.; Hashioka, S.; Kanba, S. Risperidone significantly inhibits interferon-gamma-induced microglial activation in vitro. *Schizophr. Res.* **2007**, *92*, 108–115. [CrossRef] [PubMed]
54. MacDowell, K.S.; Garcia-Bueno, B.; Madrigal, J.L.; Parellada, M.; Arango, C.; Mico, J.A.; Leza, J.C. Risperidone normalizes increased inflammatory parameters and restores anti-inflammatory pathways in a model of neuroinflammation. *Int. J. Neuropsychopharmacol.* **2013**, *16*, 121–135. [CrossRef] [PubMed]
55. Barone, F.C.; Arvin, B.; White, R.F.; Miller, A.; Webb, C.L.; Willette, R.N.; Lysko, P.G.; Feuerstein, G.Z. Tumor necrosis factor-alpha. A mediator of focal ischemic brain injury. *Stroke* **1997**, *28*, 1233–1244. [CrossRef] [PubMed]
56. Yamasaki, Y.; Matsuura, N.; Shozuhara, H.; Onodera, H.; Itoyama, Y.; Kogure, K. Interleukin-1 as a pathogenetic mediator of ischemic brain damage in rats. *Stroke* **1995**, *26*, 676–680, discussion 681. [CrossRef] [PubMed]
57. Murakami, Y.; Saito, K.; Hara, A.; Zhu, Y.; Sudo, K.; Niwa, M.; Fujii, H.; Wada, H.; Ishiguro, H.; Mori, H.; et al. Increases in tumor necrosis factor-alpha following transient global cerebral ischemia do not contribute to neuron death in mouse hippocampus. *J. Neurochem.* **2005**, *93*, 1616–1622. [CrossRef] [PubMed]

58. Hasturk, A.; Atalay, B.; Calisaneller, T.; Ozdemir, O.; Oruckaptan, H.; Altinors, N. Analysis of serum pro-inflammatory cytokine levels after rat spinal cord ischemia/reperfusion injury and correlation with tissue damage. *Turk. Neurosurg.* **2009**, *19*, 353–359. [PubMed]
59. Gokce, E.C.; Kahveci, R.; Gokce, A.; Sargon, M.F.; Kisa, U.; Aksoy, N.; Cemil, B.; Erdogan, B. Curcumin attenuates inflammation, oxidative stress, and ultrastructural damage induced by spinal cord ischemia-reperfusion injury in rats. *J. Stroke Cerebrovasc. Dis.* **2016**, *25*, 1196–1207. [CrossRef] [PubMed]
60. Zhu, P.; Li, J.X.; Fujino, M.; Zhuang, J.; Li, X.K. Development and treatments of inflammatory cells and cytokines in spinal cord ischemia-reperfusion injury. *Mediat. Inflamm.* **2013**, *2013*, 701970. [CrossRef]
61. Nakata, T.; Kawachi, K.; Nagashima, M.; Yasugi, T.; Izutani, H.; Ryugo, M.; Okamura, T.; Shikata, F.; Imagawa, H.; Yano, H.; et al. Transient ischemia-induced paresis and complete paraplegia displayed distinct reactions of microglia and macrophages. *Brain Res.* **2011**, *1420*, 114–124. [CrossRef] [PubMed]
62. Hirose, K.; Okajima, K.; Taoka, Y.; Uchiba, M.; Tagami, H.; Nakano, K.; Utoh, J.; Okabe, H.; Kitamura, N. Activated protein c reduces the ischemia/reperfusion-induced spinal cord injury in rats by inhibiting neutrophil activation. *Ann. Surg.* **2000**, *232*, 272–280. [CrossRef] [PubMed]
63. Reece, T.B.; Okonkwo, D.O.; Ellman, P.I.; Warren, P.S.; Smith, R.L.; Hawkins, A.S.; Linden, J.; Kron, I.L.; Tribble, C.G.; Kern, J.A. The evolution of ischemic spinal cord injury in function, cytoarchitecture, and inflammation and the effects of adenosine a2a receptor activation. *J. Thorac. Cardiovasc. Surg.* **2004**, *128*, 925–932. [CrossRef]
64. Smith, P.D.; Puskas, F.; Meng, X.; Lee, J.H.; Cleveland, J.C., Jr.; Weyant, M.J.; Fullerton, D.A.; Reece, T.B. The evolution of chemokine release supports a bimodal mechanism of spinal cord ischemia and reperfusion injury. *Circulation* **2012**, *126*, S110–S117. [CrossRef] [PubMed]
65. Yu, J.T.; Lee, C.H.; Yoo, K.Y.; Choi, J.H.; Li, H.; Park, O.K.; Yan, B.; Hwang, I.K.; Kwon, Y.G.; Kim, Y.M.; et al. Maintenance of anti-inflammatory cytokines and reduction of glial activation in the ischemic hippocampal ca1 region preconditioned with lipopolysaccharide. *J. Neurol. Sci.* **2010**, *296*, 69–78. [CrossRef] [PubMed]
66. Yan, B.C.; Kim, S.K.; Park, J.H.; Ahn, J.H.; Lee, C.H.; Yoo, K.Y.; Choi, J.H.; Lee, D.S.; Kim, M.J.; Kim, Y.M.; et al. Comparison of inflammatory cytokines changes in the hippocampal ca1 region between the young and adult gerbil after transient cerebral ischemia. *Brain Res.* **2012**, *1461*, 64–75. [CrossRef] [PubMed]
67. de Waal Malefyt, R.; Figdor, C.G.; Huijbens, R.; Mohan-Peterson, S.; Bennett, B.; Culpepper, J.; Dang, W.; Zurawski, G.; de Vries, J.E. Effects of il-13 on phenotype, cytokine production, and cytotoxic function of human monocytes. Comparison with il-4 and modulation by ifn-gamma or il-10. *J. Immunol.* **1993**, *151*, 6370–6381. [PubMed]
68. te Velde, A.A.; Huijbens, R.J.; Heije, K.; de Vries, J.E.; Figdor, C.G. Interleukin-4 (il-4) inhibits secretion of il-1 beta, tumor necrosis factor alpha, and il-6 by human monocytes. *Blood* **1990**, *76*, 1392–1397. [CrossRef] [PubMed]

veterinary sciences

MDPI

Article

Tumor Growth Progression in Ectopic and Orthotopic Xenografts from Inflammatory Breast Cancer Cell Lines

Sara Caceres [1,†], Angela Alonso-Diez [2,†], Belén Crespo [1], Laura Peña [2], Maria J. Illera [1], Gema Silvan [1], Paloma J. de Andres [2] and Juan C. Illera [1,*]

[1] Department of Physiology, Veterinary Faculty, Complutense University of Madrid, 28040 Madrid, Spain; sacacere@ucm.es (S.C.); belencre@ucm.es (B.C.); mjillera@ucm.es (M.J.I.); gsilvang@ucm.es (G.S.)

[2] Department of Medicine and Surgery, Veterinary Faculty, Complutense University of Madrid, 28040 Madrid, Spain; angalo02@ucm.es (A.A.-D.); laurape@ucm.es (L.P.); pjandres@ucm.es (P.J.d.A.)

* Correspondence: jcillera@ucm.es; Tel.: +34-19-1394-3865

† These authors contributed equally to this work.

Abstract: Xenografts can grow in immunosuppressed hosts, such as SCID mice, and tumor material can be injected into hosts either ectopically or orthotopically. Choosing the correct model to use is a crucial step in animal research. The aim of this study was to report the differences between ectopic and orthotopic xenografts in tumor progression, metastasis capacity, histological features, and steroid hormone profiles in xenografts from the cIMC (canine inflammatory mammary cancer) cell line IPC-366 and hIBC (human inflammatory breast cancer) cell line SUM149. To achieve this purpose, 40 female mice 6–8 weeks old were inoculated with IPC-366 and SUM149 cells subcutaneously (ectopic models) or into mammary fat pad (orthotopic models). Mice were monitored for tumor progression and appearance of metastases, and generated tumors were analyzed in terms of histological examination and steroid hormone production. The results revealed differences in tumor appearance and percentage of metastasis between ectopic and orthotopic models, which were higher in the ectopic xenografts from both cell lines. However, both models had similar characteristics of tumor progression, histological features, and steroid hormone secretion profiles. We show that the ectopic model can be validated as a good and useful model of tumor development in addition to, not contrary to, the orthotopic model in breast cancer research.

Keywords: xenograft; ectopic; orthotopic; hIBC; cIMC

Citation: Caceres, S.; Alonso-Diez, A.; Crespo, B.; Peña, L.; Illera, M.J.; Silvan, G.; de Andres, P.J.; Illera, J.C. Tumor Growth Progression in Ectopic and Orthotopic Xenografts from Inflammatory Breast Cancer Cell Lines. *Vet. Sci.* **2021**, *8*, 194. https://doi.org/10.3390/vetsci8090194

Academic Editors: Ana Faustino and Paula A. Oliveira

Received: 25 August 2021
Accepted: 10 September 2021
Published: 13 September 2021

1. Introduction

Human and canine inflammatory breast cancer are the most aggressive mammary neoplasms that affects women [1–3] and female dogs [4,5]. hIBC accounts for around 6% of human breast cancer diagnoses, presenting poor survival in women, and cIMC is more prevalent than hIBC (approximately 7.6%) [1,4,6]. These diseases are characterized by the invasion of dermal lymphatic vessels by neoplastic cells, which blocks lymph drainage and causes the characteristic edema [1,4,7]. In addition, this type of cancer is highly angiogenic and angioinvasive in both species [1,8–10]. Canine inflammatory mammary cancer has been suggested as the best spontaneous animal model for the study of human disease [1,5]. Several human inflammatory breast cancer cell lines have been established in order to study the in vitro mechanisms of this special type of breast cancer such as SUM149 [11–13]. Recently, the IPC-366 cell line, the first canine inflammatory mammary cancer triple-negative cell line, has been established and characterized [1,14].

Animal models developed for the study of human breast cancer have been useful tools for refining our understanding of breast cancer progression and metastasis [15–17]. Recently, xenografts for cIMC have been established [18]. In general, rodents, such mice, are being used for these studies because they are small in size, breed readily, and can be genetically modified [19]. The advantages of using xenografts are that many of these models are reproducible, are readily available, and a sufficient number can be used in studies to generate valid statistics. The disadvantages are that these models are costlier to run, the stromal component of the tumors is rodent, the hosts are immunodeficient, and most of the time the tumors are grown in a non-natural site [20].

Xenografts can grow in immunosuppressed hosts, such as athymic mice (nu/nu), severe combined immunodeficiency (NOD-SCID, NSG, or HuNSG) mice, or humanized mice, and tumor material can be injected into the host either ectopically (via subcutaneous injection, among others) or orthotopically (inoculation at the site of the primary tumor) [16,17,19]. In breast cancer research, ectopic xenografts are usually performed via subcutaneous and intravenous injections, and orthotopic ones by injecting cancer cells into the mice mammary fat pads [16,17].

Ectopic xenograft models are simple to perform and reproducible and result in a homogeneous tumor histology and growth rate. Accordingly, this type of xenograft is widely used in anticancer drug research [17,19]. In orthotopic xenograft models, the grafted tumor grows in the tissue of origin of the primary tumor. However, complex surgeries are often needed, leading to a limited number of mice used [17,19,20].

In breast cancer research, orthotopic models are the most widely used model. These xenografts better recapitulate the location of the disease and therefore better mimic human cancerous disease [21]. The mammary fat pad is considered the stromal microenvironment of the mammary gland [22], so the inoculation of cancer cells in this component is more similar to human mammary disease [21]. The disadvantages of using this model are that it requires surgery and the number of animals is limited. Another of the models used in breast cancer research is the ectopic model in which cancer cells are injected generally subcutaneously in the mammary chain. The two research models are valid, however, there is little literature on the differences between the two models in breast cancer.

Normal and neoplastic mammary glands are considered endocrine tissues due to the local biosynthesis of steroid hormones [23]. Several studies have shown a strong association between elevated levels of circulating estrogens and their metabolites with an increased risk of developing breast cancer [24]. Furthermore, data from in vitro studies suggest that androgens may also exert antiproliferative and apoptotic effects [25].

Recent studies have shown that the hormonal tumor environment is crucial for tumor development and progression [1]. In addition, male and female mice are capable of reproducing tumors, and their levels of intratumoral hormones will influence tumor progression [1,25]. Therefore, the form of inoculation of tumorigenic material could determine the hormonal microenvironment of the tumor.

This report describes two xenograft models from the hIBC cell line (SUM149) and cIMC cell line (IPC-366). Ectopic and orthotopic xenografts were performed to observe the differences between the two injections in terms of tumor growth, histology, and hormone secretion. Furthermore, this study intended to demonstrate that IMC could be a good animal model for the study of human disease by presenting similarities in tumor growth progression in vivo.

2. Materials and Methods

2.1. Cell Culture

Canine triple-negative inflammatory mammary carcinoma cell line IPC-366 was obtained from the Department of Physiology of the Veterinary Medicine School of the Complutense University of Madrid (established and characterized in our laboratory [14]). IPC-366 was cultured in Dulbecco's modified Eagle medium/nutrient mixture F-12 Ham (DMEM/F12) containing 5% fetal bovine serum, 1% L-glutamine, and 1% antibiotic-antimycotic. The human triple-negative inflammatory breast cancer cell line SUM149 was obtained from Asterand, Inc. (Detroit, MI, USA), (RRID: CVCL_3422). SUM149 was cultured in Ham's F12 (Thermo Fisher Scientific, Madrid, Spain) supplemented with 5% fetal bovine serum, 5 μg/mL insulin, 1 μg/mL hydrocortisone, and antibiotic-antimycotic (Sigma Aldrich, Madrid, Spain). All cell lines were maintained in a humidified atmosphere of 5% carbon dioxide at 37 °C. Cell culture was observed daily via phase-contrast microscopy.

2.2. Animals

Forty 6-to-8-week-old female NOD.CB-17-Prkdc scid-RJ mice were used in this study. The animals were housed in flexible-film isolators (Janvier Labs, Madid, Spain) in cages (1–2 animals per cage) in a room with controlled environmental conditions (20–22 °C, 50–55% relative humidity, 10–15 air changes per hour, 12 h/12 h light/dark cycle). Food and water, previously sterilized, were provided ad libitum. The required sample size needed to simultaneously compare the normal means of the groups was determined using the sample size determination module of the Statgraphics Centurion XVI statistical package (Statpoint Technologies Inc., Warrenton, VA, USA). Animals were anesthetized prior to all procedures with isoflurane at 4% for induction and 1.5% to maintain sedation, supplied in a fresh gas flow rate of 0.5 L oxygen/minute, and were observed until fully recovered. Animals were sacrificed using a lethal dose of isoflurane.

Clinical and experimental protocols of this study were approved by the Institutional Animal Care and Use Committee of Complutense University of Madrid, Spain (number: Proex 31/15). All procedures were completed in accordance with the Guide for the Care and Use of Laboratory Animals and conformed to the relevant EU Directive.

2.3. Cell Injections

A suspension of 10^6 IPC-366 and SUM149 cells was implanted subcutaneously (ectopically) into the fourth inguinal mammary gland or orthotopically into the fourth mammary fat pad (both $n = 20$, 10 IPC-366 and 10 SUM149). For orthotopical injections, animals were anaesthetized with isoflurane at 4% and an incision was made medial to the nipple. In order to expose the mammary gland, a cotton swab was used and cell suspension was injected directly into the mammary fat pad with a syringe and a 26-G needle. The wound was closed with a stapled suture.

Mice were inspected twice a week for the development of tumors. When tumors were detected, they were monitored weekly by palpation and measured using calipers. The greatest longitudinal diameter (length) and greatest transverse diameter (width) were measured. Tumor volume was estimated using the formula: volume = (length × width^2)/2 [26]. When tumors reached a volume of 1.5 cm^3 (endpoint), blood samples were obtained intracardially and collected in heparin-coated tubes. Prior to this procedure, animals were anaesthetized with isoflurane at 4% for induction and 1.5% to maintain sedation, supplied at a fresh gas flow rate of 0.5 L oxygen/minute. After blood collection, animals were euthanized using a lethal dose of isoflurane. Tumors were harvested at necropsy for subsequent analysis. The appearance of metastasis at the lung and liver were determined macroscopically at necropsy.

The collected tumors were divided into 2 fragments: one fragment for hitological examination was fixed in 4% paraformaldehyde and then embedded in paraffin, and the other fragment was frozen (−20 °C) for hormonal studies.

2.4. Histopathology and Immunohistochemistry

Tumors were histopathologically characterized on HE-stained sections following the routine method for histological description of neoplasms [5]. Immunohistochemical characterization of estrogen and progesterone receptors (ER, Ref. M7047, Dako; PR, Ref. 790-2223, Ventana, Oro Valley, AZ, USA) and human epidermal receptor-2 (HER-2, Ref. A0485, Dako, Santa Clara, CA, USA) was performed. Paraffin sections were placed in a PT module, heated for 20 min at 95 °C, and cooled down to 60 °C. Then, slides were rinsed in warm tap water and placed in an automatic immunostainer device (Lab Vision Corp., Fremont, CA, USA) for immunohistochemistry using a peroxidase detection system. After immunostaining, the slides were counterstained with hematoxylin and permanently mounted with Depex. Corresponding negative control slides were prepared by replacing the primary antibody with nonreactive antibody. Slides from human and canine mammary tumors with previously demonstrated reactivity to the primary antibody and tissue internal controls were used as positive controls [5].

For estrogen receptor, progesterone receptor, and HER-2 evaluation, 3+ positive scoring was considered, following the recommended guidelines of the American Society of Cancer Oncology (ASCO).

2.5. Steroid Determination in Serum, and Tumor Homogenates

For tumor homogenates, a total of 0.5 g of tumor collected at necropsy was homogenized in phosphate-buffered saline (PBS; pH 7.2) and centrifugated at $1200\times g$ for 20 min at 4 °C. Supernatants were collected, aliquoted individually, and frozen at −80 °C until hormones were assayed. Blood samples were centrifugated at $1200\times g$ for 20 min 4 °C, and the serum was separated and stored frozen at −20 °C before being assayed.

The hormones evaluated in this study were progesterone (P4), dehydroepiandrosterone (DHEA), androstenedione (A4), testosterone (T), dihydrotestosterone (DHT), estrone sulfate (E1SO4), and 17beta-estradiol (E2). The antibodies used were P4 (C914), A4 (C9111), T (R156), E1SO4 (R522-2), and E2 (C6E91). The antibodies were developed in the Department of Physiology (UCM, Madrid, Spain). DHEA and DHT determinations were performed using a commercially available EIA kit (Demeditech Diagnostic GmbH, Kiel, Germany) following the manufacturer's instructions.

Determined steroid hormones in tumor homogenates were assayed using previously validated competitive enzyme-linked immunosorbent assay (ELISA), and an amplified ELISA was used for blood samples [26]. Briefly, 96-well flat-bottom medium-binding polystyrene microplates (Greiner Bio-One, Madrid, Spain) were coated with the appropriate purified antibody dilution overnight at 4 °C. Afterward, for competitive ELISA, plates were washed and standards and tumor homogenate samples were diluted in working solution (CWS) and analyzed in duplicate. Plates were incubated at room temperature for 2 h. For amplified ELISA, standards and serum samples were added in duplicate and incubated overnight at 4 °C, then CWS was added to each well and incubated for 4 h at room temperature. For both ELISAs, after conjugate incubation plates were washed, to evaluate the amount of labelled steroid hormones, Enhanced K-Blue TMB substrate (Neogen, Lexington, KY, USA) was added to each well and incubated for an additional 15 min at room temperature. Finally, colorimetric reaction was stopped via the addition of 10% H_2SO_4 to each well. Absorbance was read at 450 nm using an automatic plate reader. Hormone concentrations were calculated by means of software developed for this technique (ELISA AID, Eurogenetics, Brussels, Belgium). A standard dose-response curve was constructed by plotting the binding percent ($B/B0 \times 100$) against each steroid hormone standard concentration. All hormone concentrations were expressed in ng/g for tumor homogenates and ng/mL for serum samples.

2.6. Statistics

The statistics software used for data analysis was SAS 9.4 (UCM, Madrid, Spain). The results were expressed as means ± SD. For tumor progression analysis (time of palpable tumor, % of tumor engraftment, time of 1.5 cm^3 volume (edpoint days), and % of animals with metastasis) and hormone determination to compare both cell lines (IPC-366 and SUM149) in each group, the one way ANOVA and Mann-Whitney rank-sum tests were performed. In all statistical comparisons, $p < 0.05$ was accepted as denoting significant difference.

3. Results

3.1. Differences in Tumor Appearance Time According to the Manner of Cell Injection

IPC-366 and SUM149 cells were injected ectopically and orthotopically in female SCID mice to observe if there were differences in tumor growth (Table 1, Figure 1). When IPC-366 cells were injected subcutaneously, all mice (100%) reproduced a tumor that was palpable approximately 2 weeks after injection (16.64 ± 1.72 days). When cells were injected into MFP, 70% of mice reproduced a tumor approximately 3 weeks after injection (21.40 ± 3.71 days). However, these differences were not statistically significant.

Table 1. Tumor growth parameters of IPC-366 and SUM149 cell lines in ectopic and orthotopic models.

Cell Line	Injection	% of Tumor Engraftment	Time of Palpable Tumor (Days)	Time of 1.5 cm^3 Volume (Edpoint, Days)	% of Animals with Metastasis
IPC-366 (n = 20)	Ectopic (n = 10)	100%	16.64 ± 1.72	42.02 ± 2.35	90%
	Orthotopic (n = 10)	70%	21.40 ± 3.71	49.81 ± 2.21 *	40% *
SUM149 (n = 20)	Ectopic (n = 10)	80%	26.82 ± 2.19 [a]	53.40 ± 4.86 [a]	80%
	Orthotopic (n = 10)	70%	30.35 ± 3.47	51.46 ± 3.67	60%

* $p < 0.05$; significant differences between ectopic and orthotopic models on each cell line. a; significant differences ($p < 0.05$) between cell lines.

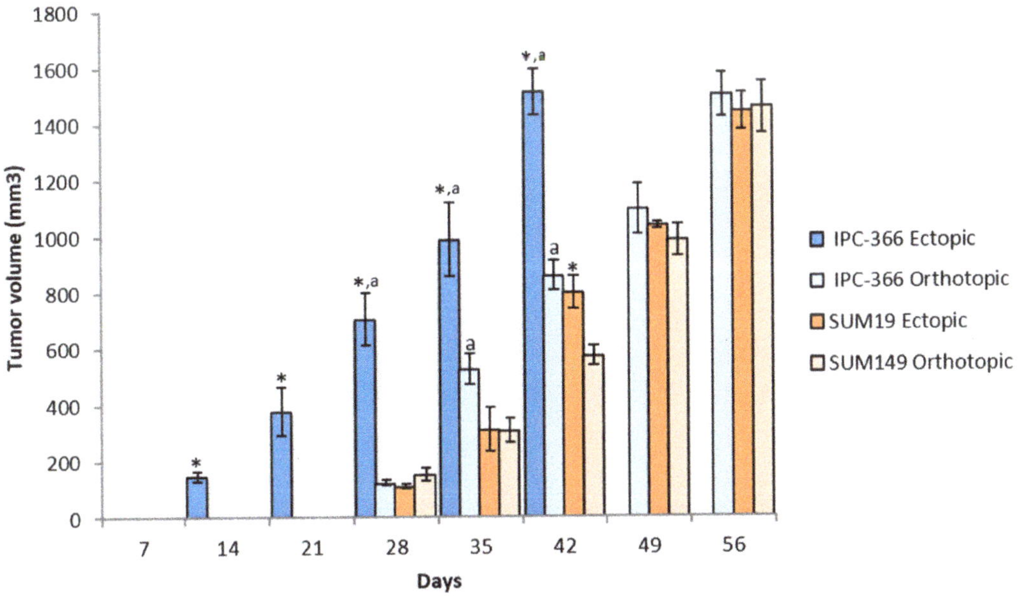

Figure 1. Tumor growth progression in ectopic and orthotopic models. The two injection models of both cell lines showed a fast pattern of tumor growth progression. Bar represents means ± SD * $p < 0.05$; significant differences between ectopic and orthotopic models on each cell line. a; significant differences ($p < 0.05$) between cell lines in each group (ectopic and orthotopic).

Approximately 4 weeks after injection, with SUM149 cells injected subcutaneously, 80% of mice reproduced a tumor (26.82 ± 2.19 days), and with cells injected into the MFP, 70% of mice reproduced a tumor (30.35 ± 3.47 days), and there was no significant difference in the time palpable tumors were found between the 2 groups.

Regarding differences between cell lines, ectopic injection of IPC-366 resulted in significantly earlier tumor appearance ($p < 0.05$) than SUM149. However, with orthotopic injections no significant differences were found between the two cell lines.

3.2. Tumor Progression in Ectopic and Orthotopic Models

After the emergence of tumors, their progression was monitored to observe if there were differences between ectopic and orthotopic models. Tumor progression with IPC-366 and SUM149 cell lines was similar (Figure 1). Both cell lines exhibited rapid growth in vivo, reaching a volume of 1500 mm^3 approximately 6–8 weeks after injection; it was significantly earlier in the IPC-366 ectopic model ($p < 0.05$), which reached final volume 6 weeks after injection (Table 1).

3.3. Occurrence of Metastasis According to the Manner of Cell Injection

These two models developed spontaneous distant metastases (Table 1). No significant differences were found between cell lines in the two models of injection. However, differences in the incidence of metastasis in IPC-366 were found between ectopic (90%) and orthotopic (40%) models, the appearance of metastases in the ectopic model being greater.

3.4. Histological Characteristics of Ectopic and Orthotopic Models

The histological examination of tumors from ectopic models revealed highly infiltrating, poorly demarcated, unencapsulated, densely cellular neoplastic growth extending into the adjacent dermis (Figure 1, inset) and striated muscle. Similarly, in the orthotopic models, both IPC-366 and SUM149 xenotransplanted mice had infiltrating, unencapsulated, and densely cellular masses infiltrating the adjacent adipose tissue and compressing the adjacent skin. In both orthotopic and ectopic xenografts, neoplastic cells were arranged in solid masses separated by a scant fibrovascular stroma. The neoplastic cells were medium size, round to oval, with indistinct cell borders and a moderate eosinophilic cytoplasm. The nucleus was medium to large, round to oval, with stippled chromatin, and one to two magenta nucleoli were evident. Anisocytosis and anisokaryosis were marked, the mitotic index was very high, and atypical mitoses were frequently observed (Figure 2).

In addition, some neoplastic cells presented morphological features of endothelial-like cells (ELCs): a rim of elongated, encircled cytoplasm that displaced an elongated nucleus to the periphery was a common finding, suggesting the presence of vasculogenic mimicry (Figure 2).

The presence of emboli in dermal capillaries and marked dermal edema, characterized by colorless spaces that separated dermal collagen fibers, confirmed the histological characteristics of inflammatory mammary carcinoma.

No morphological differences were found between IPC-366 and SUM149 in both orthotopic and ectopic xenografts, so IPC-366 can be considered a good model compared with its human counterpart SUM149 cell line.

Tumors generated by IPC-366 and SUM149 cell lines have common characteristics on the expression of ER, PR, and HER2 and were found to be negative in both ectopic and orthotopic xenografts (Table 2).

Table 2. Estrogen receptor (ER), Progesterone receptor (PR), and human epidermal growth factor receptor 2 (HER-2) expression on ectopic and orthotopic xenografts from IPC-366 and SUM149 cell lines.

Receptor	IPC-366 Ectopic	IPC-366 Orthotopic	SUM149 Ectopic	SUM149 Orthotopic
ER	Negative	Negative	Negative	Negative
PR	Negative	Negative	Negative	Negative
HER-2	Negative	Negative	Negative	Negative

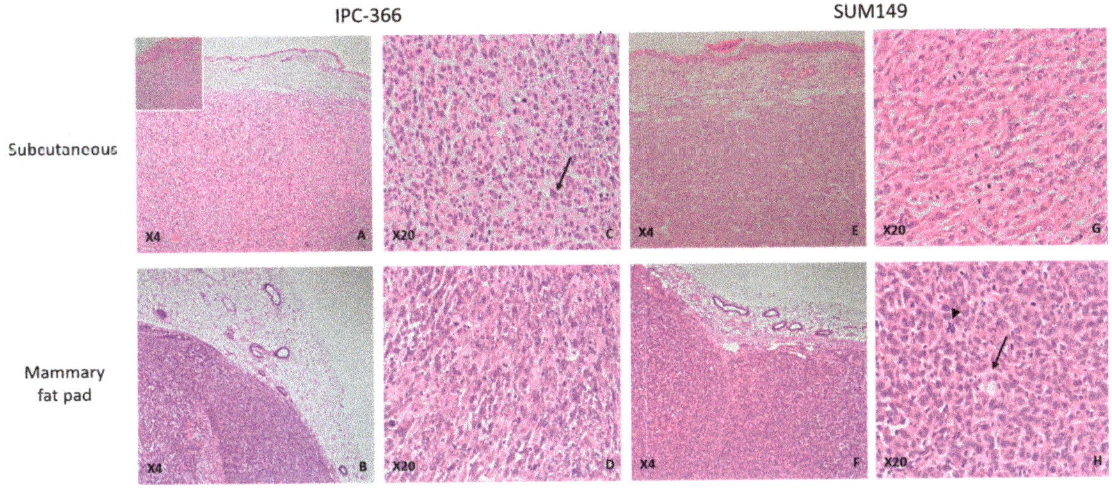

Figure 2. IPC-366 and SUM149 xenotransplanted mice, paraffin sections, H-E. (**A**) IPC-366 ectopic xenotransplanted mice. Neoplastic cells arranged in solid masses separated by a scant fibrovascular stroma infiltrating the adjacent dermis (inset: neoplastic cells infiltrating adjacent dermis). (**B**) IPC-366 orthotopic mice. Unencapsulated and densely cellular mass extending into the adjacent adipose tissue. (**C**,**D**) Ectopic and orthotopic IPC-366 xenotransplanted mice. Tumors are composed of highly pleomorphic cells with marked anisocytosis and anisokaryosis. Binucleated cells are commonly seen (arrow). (**E**,**F**) Ectopic and orthotopic SUM149 xenografted mice. Solid tumors infiltrate the dermis and adipose tissue. No histological differences were found between the types of SUM149 xenografts. (**G**) Orthotopic SUM149 xenograft. Medium to large round cells with a moderate eosinophilic cytoplasm and large nuclei with one or more evident nucleoli. (**H**) Orthotopic SUM149 xenograft. Presence of neoplastic cells with an elongated and empty cytoplasm that displaced the nuclei to the periphery, suggestive of endothelial-like cells (ELCs) (arrow). Atypical mitoses were frequently seen (arrowhead).

3.5. Similar Steroid Hormone Secretion in Ectopic and Orthotopic Models

Figure 3 shows the results of the concentrations of steroid hormones studied (P4, DHEA, A4, T, DHT, E1SO4, and E2), in both serum and tumor homogenate.

No significant differences in the plasma and intratumoral hormonal levels of the human and canine inflammatory carcinoma cell lines (SUM149 and IPC-366) were observed, nor are there differences depending on the type of inoculation (orthotopic or ectopic).

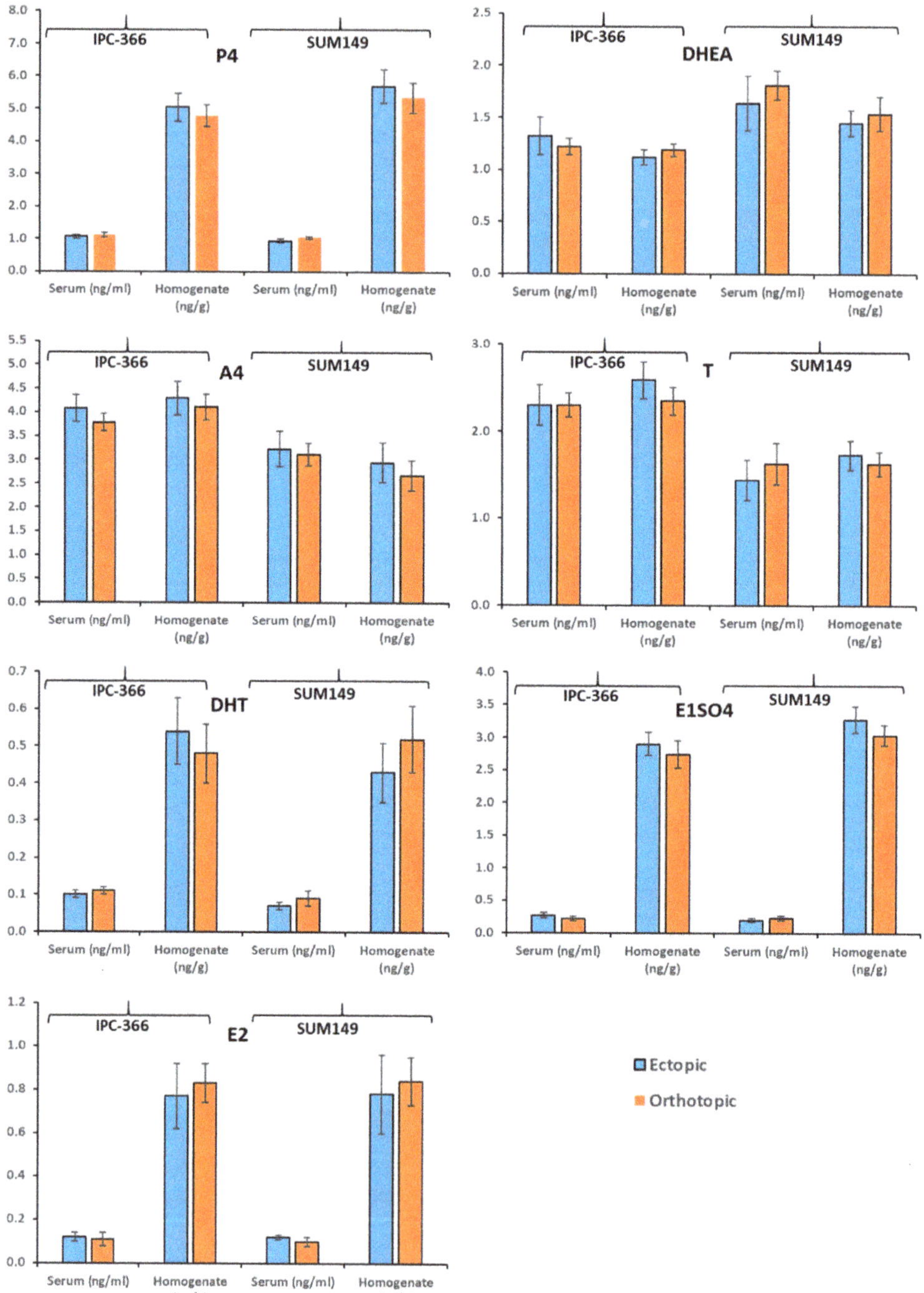

Figure 3. Steroid hormone secretion studied (P4, DHEA, A4, T, DHT, E1SO4, and E2), on ectopic (subcutaneous) and orthotopic (mammary fat pad) models of IPC-366 and SUM149 xenografts.

4. Discussion

Animal models of human breast cancer are valuable in cancer research for understanding the pathophysiology of cancer, including new target identification [16]. In general, tumor development in murine models is faster and more homogeneous. In the early 1970s, it was demonstrated that human tumor tissues could be successfully grown in athymic nu/nu mice, leading to ectopic tumor xenografts becoming a valuable approach to the study of cancer biology [16,17,19,20]. Nowadays, specific types of tumor models in rodents include ectopic xenografts of tumor-derived cell lines and orthotopic xenografts in which tumor cell lines are implanted into the primary tumor source [16,17,19].

The choice of the type of implantation of tumor cells is a critical step in cancer research, and a comparative study of these models is needed. In tumor models, the murine microenvironment affects the efficiency of engraftment, the rate of tumor growth, and their ability to metastasize [17,19,20,27]. Many breast cancer cell lines have the ability to grow subcutaneously, but ectopic xenograft models have limitations with regard tumor growth location, loss of tumor heterogeneity, and the absence of a specific murine microenvironment, which leads to tumor cells having paracrine interactions with noncancerous cells and tissues [17,18,28,29]. This is why several investigators have moved away from ectopic and use orthotopic xenografts in the MFP. It seems that orthotopic implantation has the advantage that the tumor growth is in the tissue of origin of the primary tumor and facilitating metastatic spread [20,27]. However, this implantation type also has disadvantages, such as the need for complex surgeries, the rodent microenvironment, variable tumor take-up rates, and the long time for primary tumor development [17–20].

This study intended to determine the differences in tumor characteristics in terms of progression, metastatic capacity, histological features, and hormonal secretion in ectopic and orthotopic models in order to evaluate the capacity of both models for their use in breast cancer research. To achieve this purpose, we decided to use two triple-negative cell lines of canine and human inflammatory carcinoma (IPC-366 and SUM149), which have been shown to have significant tumorigenic potential [30]. cIMC and hIBC are considered to be the most malignant and aggressive subtypes of breast cancer affecting female dogs and humans, respectively [4,6,25].

cIMC has been suggested as a model to study the human disease [4,5,25]. Recently, a triple-negative cIMC cell line (IPC-366) was established as a useful tool for TNBC research [1,14,25]. This study also intended to show that triple-negative cell lines from hIBC (SUM149) and cIMC (IPC-366) resemble each other, sharing in vivo characteristics. The results described above support the statement that cIMC is a good model for studying human disease.

On the other hand, the use of ectopic and orthotopic models in hIBC and cIMC research is limited. There are two murine models (patient derived xenografts) established for human inflammatory breast cancer, MARY-X and WIBC-9 [31,32], and recently a murine model for cIMC was established [18]. Therefore, the use of xenografts from hIBC and cIMC cell lines is crucial for research in this type of cancer.

In this study, the results reveal that with IPC-366 ectopic xenografts, 100% of mice reproduced a tumor 2 weeks post inoculation compared to 70% of mice with orthotopic xenografts, which reproduced a tumor 1 week later. However, SUM149 xenografts did not show significant differences in frequency of tumor appearance (80% ectopic and 70% orthotopic) or time of tumor emergence (4 weeks post inoculation in both). These results are in agreement with the literature [14,25,28]. Thus, ectopic models of cIMC may have an advantage over orthotopic models in that they have higher success rates. Possibly the difference in results between the human and canine models is due to the fact that the IPc-366 cell line presents a tumorogenic and malignancy potential greater than the human SUM149 cell line [30]. Therefore, the ectopic model can be validated as a good and useful model of tumor development in addition to, not contrary to, the orthotopic model. Furthermore, the results showed that IPC-366 and SUM149 grew rapidly in vivo and with tumor progression similar in both ectopic and orthotopic xenografts. Several studies found a correlation between tumor morphology and aggressiveness [4,6]. Agollah and colleagues (2014) showed that the orthotopic SUM149 model grew as multiple nodes/clusters and was capable of producing spontaneous metastasis, which is in accordance with our results in both cell lines [33].

Approximately 40% of hIBC patients have distant metastases to the brain, bones, and lymph nodes [33–35]. hIBC and cIMC are capable of spreading into the skin and distant sites through dermal lymphatic vessels [4,33,35]. According to the literature, the metastatic rates of ectopic xenografts are low compared with orthotopic models, in human studies [17]. In this study, the high rates found in both models differed from what is observed in most other tumor types based on literature [17]. A possible explanation could reside in the fact that ectopic subcutaneous injection of breast cancer cells can be performed very near the mouse mammary gland, while ectopic injections of other tumor types, such as brain tumors, are traditionally injected subcutaneously into the hind flank, far from the tissue of origin. In addition, we found a higher percentage of metastasis in the IPC-366 ectopic model than the orthotopic model. However, in SUM149 no differences were found between both models. However, dissemination patterns may vary not only between mice and humans but also among mouse strains [19,20]. Although we found these differences, both models reflect patterns of human and canine disease, such as metastasis, which help in the investigation of inflammatory breast carcinoma and validate the use of both models for a better understanding of breast cancer.

In order to further explore the differences between the ectopic and orthotopic models, we studied the histological characteristics of the tumors generated in both. No morphological differences were found between IPC-366 and SUM149 in orthotopic and ectopic xenografts, so IPC-366 is a good model compared with its human counterpart SUM149 cell line. In both models, neoplastic cells were distributed in solid masses and presented marked anisocytosis and anisokaryosis, characteristics that are similar to the appearance of tumors in the two species studied.

hIBC is predominantly ER-negative, PR-negative, and HER2-positive. It is known that triple-negative breast cancer (TNBC) is highly proliferative and sensitive to chemotherapy and has a poor prognosis [35–37]. In our study, the expression of ER, PR, and HER-2 receptors was negative in both the ectopic and orthotopic models, again validating the use of these models in breast cancer research.

It is important that xenograft models preserve inter- and intratumoral heterogeneity [38]. It has been shown that the hormonal tumor environment is crucial for tumor progression and dissemination [1,25]. In order to elucidate whether ectopic and orthotopic models share tumor microenvironment characteristics, an evaluation of their steroid hormone secretion and production profiles was performed.

In previous studies carried out by our group, possible local synthesis of some steroid hormones was indicated in normal and neoplastic mammary glands in canine mammary carcinoma, and more recently in hIBC and cIMC [1,26]. The formation of sex steroids in peripheral tissues in humans is well documented [39]. The action of progestogens, estrogens, and androgens (produced locally or not) is crucial in neoplastic growth and progression of breast cancer, due to their interactions with specific receptors [1,39]. From the results obtained, it can be seen that the plasma hormone levels (P4, DHEA, A4, T, DHT, E1SO4, and E2) were similar between the ectopic and orthotopic models, and similarly, the intratumoral hormone levels were similar between the 2 models. Thus, the ectopic and orthotopic models presented similar steroid hormone profiles, indicating that both models can be used.

The limitations to this study are that the results are only reproducible for these two human and canine inflammatory carcinoma cell lines and for their use in SCID mice; in other immunosuppressed mouse strains the results may vary. In addition, other factors such as the cell line pass number can influence the engraftment rate.

5. Conclusions

Ectopic and orthotopic models with hIBC and cIMC cell lines share characteristics in terms of tumor progression, metastatic rates, histological features, and hormonal secretion profiles, and both are useful for cancer research. Furthermore, the ectopic model can be validated as a good and useful model of tumor development in addition to and not contrary to the orthotopic model.

Author Contributions: Conceptualization, S.C., L.P. and J.C.I., methodology, S.C., B.C., G.S., A.A.-D., L.P., M.J.I. and J.C.I.; software, S.C., A.A.-D., M.J.I. and J.C.I.; validation, S.C., B.C., G.S. and J.C.I.; investigation, S.C., A.A.-D., B.C., G.S., M.J.I., P.J.d.A. and J.C.I.; writing—original draft preparation, S.C., A.A.-D. and J.C.I.; writing—review and supervision, S.C., A.A.-D., B.C., G.S., L.P., M.J.I., P.J.d.A. and J.C.I. All authors have read and agreed to the published version of the manuscript.

Funding: This research received no external funding.

Institutional Review Board Statement: Animals were housed and bred at the Animal Facility (Department of Animal Physiology, Veterinary Medicine School, Complutense University of Madrid (UCM), Spain). Experiments were approved by the Institutional Animal Care and Use Committee of the Complutense University of Madrid, Spain (number: Proex 31/15). All procedures were completed in accordance with the Guide for the Care and Use of Laboratory Animals and conformed to the relevant EU Directive.

Informed Consent Statement: Not applicable.

Data Availability Statement: The data that support the findings of this study are available from the corresponding author upon reasonable request.

Acknowledgments: The authors thank Carmen García-Velasco for her technical and animal care.

Conflicts of Interest: The authors declared no potential conflicts of interest with respect to the research, authorship, and/or publication of this article.

References

1. Caceres, S. Caracterización In Vitro e In Vivo de Una Nueva Línea Celular (IPC-366) de Cáncer Inflamatorio Mamario Canino (IMC): Comparación con el Cáncer de Mama Inflamatorio Humano. Ph.D. Thesis, University Complutense of Madrid, Madrid, Spain, 2016.
2. Hance, K.W.; Anderson, W.F.; Devesa, S.S.; Young, H.A.; Levine, P.H. Trends in inflammatory breast carcinoma incidence and survival: The surveillance, epidemiology, and end results program at the National Cancer Institute. *J. Natl. Cancer Inst.* **2005**, *97*, 966–975. [CrossRef]
3. Yamauchi, H.; Woodward, W.A.; Valero, V.; Alvarez, R.H.; Lucci, A.; Buchholz, T.A.; Iwamoto, T.; Krishnamurthy, S.; Yang, W.; Reuben, J.M.; et al. Inflammatory breast cancer: What we know and what we need to learn. *Oncologist* **2012**, *17*, 891–899. [CrossRef] [PubMed]
4. De Souza, C.H.M.; Toledo-Piza, E.; Amorin, R.; Barboza, A.; Tobias, K.M. Inflammatory mammary carcinoma in 12 dogs: Clinical features, cyclooxygenase-2 expression, and response to piroxicam treatment. *Can. Vet. J.* **2009**, *50*, 506–510. [PubMed]
5. Peña, L.; Pérez-Alenza, M.D.; Rodriguez-Bertos, A.; Nieto, A. Canine inflammatory mammary carcinoma: Histopathology, immunohistochemistry and clinical implications of 21 cases. *Breast Cancer Res. Treat.* **2003**, *78*, 141–148. [CrossRef]
6. Barkataki, S.; Javadekar, M.J.; Bradfield, P.; Murphy, T.; Witmer, D.D.; Van Golen, K.L. Inflamma-tory Breast Cancer: A Panoramic Overview. *J. Rare Dis. Res. Treat.* **2018**, *3*, 37–43.
7. Giordano, S.H.; Hortobagyi, G.N. Inflammatory breast cancer: Clinical progress and the main problems that must be addressed. *Breast Cancer Res.* **2003**, *5*, 284–288. [CrossRef]
8. Kleer, C.G.; van Golen, K.L.; Merajver, S.D. Molecular biology of breast cancer metastasis. Inflammatory breast cancer: Clinical syndrome and molecular determinants. *Breast Cancer Res.* **2000**, *2*, 423–429. [CrossRef]
9. Van der Auwera, I.; Van Laere, S.J.; Van den Eynden, G.; Benoy, I.; van Dam, P.; Colpaert, C.G.; Fox, S.B.; Turley, H.; Harris, A.L.; Van Marck, E.A.; et al. Increased angiogenesis and lymphangiogenesis in inflammatory versus noninflammatory breast cancer by real-time reverse transcriptase-PCR gene expression quantification. *Clin. Cancer Res.* **2004**, *10*, 7965–7971. [CrossRef]
10. Clemente, M.; Pérez-Alenza, M.D.; Peña, L. Metastasis of canine inflammatory versus non-inflammatory mammary tumors. *J. Comp. Path* **2010**, *143*, 157–163. [CrossRef]
11. Klopp, A.H.; Lacerda, L.; Gupta, A.; Debeb, B.G.; Solley, T.; Li, L.; Spaeth, E.; Xu, W.; Zhang, X.; Lewis, M.T.; et al. Mesenchymal Stem Cells Promote Mammosphere Formation and Decrease E-Cadherin in Normal and Malignant Breast Cells. *PLoS ONE* **2010**, *5*, e12180. [CrossRef]

12. Holliday, D.L.; Speirs, V. Choosing the right cell line for breast cancer research. *Breast Cancer Res.* **2011**, *13*, 215. [CrossRef] [PubMed]
13. Fernandez, S.V.; Robertson, F.M.; Pei, J.; Aburto-Chumpitaz, L.; Mu, Z.; Chu, K.; Alpaugh, R.K.; Huang, Y.; Cao, Y.; Ye, Z.; et al. Inflammatory breast cancer (IBC): Clues for targeted therapies. *Breast Cancer Res. Treat.* **2013**, *140*, 23–33. [CrossRef] [PubMed]
14. Caceres, S.; Peña, L.; de Andres, P.J.; Illera, M.J.; Lopez, M.S.; Woodward, W.A.; Reuben, J.M.; Illera, J.C. Establishment and characterization of a new cell line of canine inflammatory mammary cancer: IPC-366. *PLoS ONE* **2015**, *10*, e0122277. [CrossRef] [PubMed]
15. Lelekakis, M.; Moseley, J.M.; Martin, T.J.; Hards, D.; Williams, E.; Ho, P.; Lowen, D.; Javni, J.; Miller, F.R.; Slavin, J.; et al. A novel orthotopic model of breast cancer metastasis to bone. *Clin. Exp. Metastasis* **1999**, *17*, 163–170. [CrossRef] [PubMed]
16. Fantozzi, A.; Christofori, G. Mouse models of breast cancer metastasis. *Breast Cancer Res.* **2006**, *8*, 212. [CrossRef] [PubMed]
17. Ruggeri, B.A.; Camp, F.; Miknyoczki, S. Animal models of disease: Pre-clinical animal models of cancer and their applications and utility in drug discovery. *Biochem. Pharm.* **2014**, *87*, 150–161. [CrossRef]
18. Camacho, L.; Peña, L.; González Gil, A.; Caceres, S.; Diez, L.; Illera, J.C. Establishment and characterization of a canine xenograft model of inflammatory mammary carcinoma. *Res. Vet. Sci.* **2013**, *95*, 1068–1075. [CrossRef]
19. De Jong, M.; Maina, T. Of mice and humans: Are they the same?—Implications in cancer translational research. *J. Nucl. Med.* **2010**, *51*, 501–504. [CrossRef]
20. Teicher, B.A. Tumor models for efficacy determination. *Mol. Cancer Ther.* **2006**, *5*, 2435–2443. [CrossRef]
21. Kocatürk, B.; Versteeg, H.H. Orthotopic Injection of Breast Cancer Cells into the Mammary Fat Pad of Mice to Study Tumor Growth. *J. Vis. Exp.* **2015**, *967*, 51967. [CrossRef]
22. Hovey, R.C.; Aimo, L. Diverse and active roles for adipocytes during mammary gland growth and function. *J. Mammary Gland. Biol. Neoplasia* **2010**, *15*, 279–290. [CrossRef]
23. McNamara, K.M.; Sasano, H. The intracrinology of breast cancer. *J. Steroid Biochem. Mol. Biol.* **2015**, *145*, 172–178. [CrossRef]
24. Falk, R.T.; Brinton, L.A.; Dorgan, J.F.; Fuhrman, B.J.; Veenstra, T.D.; Xu, X.; Gierach, G.L. Relationship of serum estrogens and estrogen metabolites to postmenopausal breast cancer risk: A nested case-control study. *Breast Cancer Res.* **2013**, *15*, R34. [CrossRef]
25. Caceres, S.; Peña, L.; Silvan, G.; Illera, M.J.; Woodward, W.A.; Reuben, J.M.; Illera, J.C. Steroid Tumor Environment in Male and Female Mice Model of Canine and Human Inflammatory Breast Cancer. *BioMed Res. Int.* **2016**, *2016*, 8909878. [CrossRef]
26. Alonso-Diez, A.; Caceres, S.; Peña, P.; Crespo, B.; Illera, J.C. Anti-angiogenic treatments interact with steroid secretion in inflammatory breast cancer triple negative cell lines. *Cancers* **2021**, *13*, 3668. [CrossRef]
27. Lacroix, M.; Leclercq, G. Relevance of breast cancer cell lines as models for breast tumours: An update. *Breast Cancer Res. Treat.* **2004**, *83*, 249–289. [CrossRef]
28. Abate-Shen, C. A new generation of mouse models of cancer for translation research. *Clin. Cancer Res.* **2006**, *12*, 5274–5276. [CrossRef]
29. Szadvari, I.; Krizanova, O.; Babula, P. Athymic Nude Mice as an Experimental Model for Cancer Treatment. *Physiol. Res.* **2016**, *65*, 441–453. [CrossRef] [PubMed]
30. Caceres, S.; Peña, L.; Lacerda, L.; Illera, M.J.; de Andres, P.J.; Larson, R.A.; Gao, H.; Debeb, B.G.; Woodward, W.A.; Reuben, J.M.; et al. Canine cell line, IPC-366, as a good model for the study of inflammatory breast cancer. *Vet. Comp. Oncol.* **2016**, *15*, 980–995. [CrossRef] [PubMed]
31. Alpaugh, M.L.; Tomlinson, J.S.; Shao, Z.M.; Barsky, S.H. A novel human xenograft model of inflammatory breast cancer. *Cancer Res.* **1999**, *59*, 5079–5084. [PubMed]
32. Shirakawa, K.; Tsuda, H.; Heike, Y.; Kato, K.; Asada, R.; Inomata, M.; Sasaki, H.; Kasumi, F.; Yoshimoto, M.; Iwanaga, T.; et al. Absence of endothelial cells, central necrosis, and fibrosis are associated with aggressive inflammatory breast cancer. *Cancer Res.* **2001**, *61*, 445–451.
33. Agollah, G.D.; Wu, G.; Sevick-Muraca, E.M.; Kwon, S. In vivo lymphatic imaging of a hmodel. *J. Cancer* **2014**, *5*, 774–783. [CrossRef]
34. Jaiyesimi, I.A.; Buzdar, A.U.; Hortobagyi, G. Inflammatory breast cancer: A review. *J. Clin. Oncol.* **1992**, *10*, 1014–1024. [CrossRef] [PubMed]
35. Chavez, K.J.; Garimella, S.V.; Lipkowitz, S. Triple negative breast cancer cell lines: One tool in the search for better treatment of triple negative breast cancer. *Breast Dis.* **2010**, *32*, 35–48. [CrossRef] [PubMed]
36. Ueno, N.T.; Zhang, D. Targeting EGFR in Triple Negative Breast Cancer. *J. Cancer* **2011**, *2*, 324–328. [CrossRef] [PubMed]
37. Boyle, P. Triple-negative breast cancer: Epidemiological considerations and recommendations. *Ann. Oncol.* **2012**, *23*, 7–12. [CrossRef] [PubMed]
38. Choi, S.Y.; Lin, D.; Gout, P.W.; Collins, C.C.; Xu, Y.; Wang, Y. Lessons from patient-derived xenografts for better in vitro modeling of human cancer. *Adv. Drug Deliv. Rev.* **2014**, *15*, 222–237. [CrossRef]
39. Labrie, F.; Luu-The, V.; Labrie, C.; Simard, J. DHEA and itstransformation into androgens and estrogens in peripheraltarget tissues: Intracrinology. *Front. Neuroendocr.* **2001**, *22*, 185–212. [CrossRef]

veterinary sciences

Article

Imipramine Accelerates Nonalcoholic Fatty Liver Disease, Renal Impairment, Diabetic Retinopathy, Insulin Resistance, and Urinary Chromium Loss in Obese Mice

Geng-Ruei Chang [1], Po-Hsun Hou [2,3,4], Chao-Min Wang [1], Jen-Wei Lin [5], Wei-Li Lin [5,6], Tzu-Chun Lin [1], Huei-Jyuan Liao [1], Chee-Hong Chan [7,*] and Yu-Chen Wang [8,9,10,11,*]

1 Department of Veterinary Medicine, National Chiayi University, 580 Xinmin Road, Chiayi 60054, Taiwan; grchang@mail.ncyu.edu.tw (G.-R.C.); leowang@mail.ncyu.edu.tw (C.-M.W.); lin890090@gmail.com (T.-C.L.); pipi324615@gmail.com (H.-J.L.)
2 Department of Psychiatry, Taichung Veterans General Hospital, 4 Section, 1650 Taiwan Boulevard, Taichung 40705, Taiwan; peterhopo2@yahoo.com.tw
3 Faculty of Medicine, National Yang-Ming University, 2 Section, 155 Linong Street, Beitou District, Taipei 11221, Taiwan
4 College of Medicine, National Chung Hsing University, 145 Xingda Road, South District, Taichung 40227, Taiwan
5 Bachelor Degree Program in Animal Healthcare, Hungkuang University, 6 Section, 1018 Taiwan Boulevard, Shalu District, Taichung 433304, Taiwan; jenweilin@hk.edu.tw (J.-W.L.); ivorylily99@gmail.com (W.-L.L.)
6 General Education Center, Chaoyang University of Technology, 168 Jifeng Eastern Road, Taichung 413310, Taiwan
7 Division of Nephrology, Chang Bing Show Chwan Memorial Hospital, 6 Lugong Road, Lukang Township, Changhua 50544, Taiwan
8 Division of Cardiology, Asia University Hospital, 222 Fuxin Road, Wufeng District, Taichung 41354, Taiwan
9 Department of Medical Laboratory Science and Biotechnology, Asia University, 500 Lioufeng Road, Wufeng District, Taichung 41354, Taiwan
10 Division of Cardiovascular Medicine, China Medical University Hospital, 2 Yude Road, North District, Taichung 404332, Taiwan
11 College of Medicine, China Medical University, 91 Hsueh-Shih Road, North District, Taichung 404333, Taiwan
* Correspondence: cheehong.chan@gmail.com (C.-H.C.); richard925068@gmail.com (Y.-C.W.); Tel.: +886-975-617071 (C.-H.C.); +886-4-2332-3456 (Y.-C.W.)

Citation: Chang, G.-R.; Hou, P.-H.; Wang, C.-M.; Lin, J.-W.; Lin, W.-L.; Lin, T.-C.; Liao, H.-J.; Chan, C.-H.; Wang, Y.-C. Imipramine Accelerates Nonalcoholic Fatty Liver Disease, Renal Impairment, Diabetic Retinopathy, Insulin Resistance, and Urinary Chromium Loss in Obese Mice. *Vet. Sci.* **2021**, *8*, 189. https://doi.org/10.3390/vetsci8090189

Academic Editors: Ana Faustino, Paula A. Oliveira and W. Jean Dodds

Received: 6 July 2021
Accepted: 7 September 2021
Published: 9 September 2021

Abstract: Imipramine is a tricyclic antidepressant that has been approved for treating depression and anxiety in patients and animals and that has relatively mild side effects. However, the mechanisms of imipramine-associated disruption to metabolism and negative hepatic, renal, and retinal effects are not well defined. In this study, we evaluated C57BL6/J mice subjected to a high-fat diet (HFD) to study imipramine's influences on obesity, fatty liver scores, glucose homeostasis, hepatic damage, distribution of chromium, and retinal/renal impairments. Obese mice receiving imipramine treatment had higher body, epididymal fat pad, and liver weights; higher serum triglyceride, aspartate and alanine aminotransferase, creatinine, blood urea nitrogen, renal antioxidant enzyme, and hepatic triglyceride levels; higher daily food efficiency; and higher expression levels of a marker of fatty acid regulation in the liver compared with the controls also fed an HFD. Furthermore, the obese mice that received imipramine treatment exhibited insulin resistance, worse glucose intolerance, decreased glucose transporter 4 expression and Akt phosphorylation levels, and increased chromium loss through urine. In addition, the treatment group exhibited considerably greater liver damage and higher fatty liver scores, paralleling the increases in patatin-like phospholipid domain containing protein 3 and the mRNA levels of sterol regulatory element-binding protein 1 and fatty acid-binding protein 4. Retinal injury worsened in imipramine-treated mice; decreases in retinal cell layer organization and retinal thickness and increases in nuclear factor κB and inducible nitric oxide synthase levels were observed. We conclude that administration of imipramine may result in the exacerbation of nonalcoholic fatty liver disease, diabetes, diabetic retinopathy, and kidney injury.

Keywords: fatty liver disease; glucose; imipramine; insulin; obesity; renal impairment; retinal injury

1. Introduction

Imipramine, derived from dibenzazepine, is a prototypical tricyclic antidepressant (TCA). TCAs have a structure that is close to that of a phenothiazine, and they contain a tricyclic ring system. The central ring contains an alkylamine substituent [1]. On the basis of animal studies, imipramine is known as a candidate therapy for antipredator defensive behavior, sleep deprivation, and related anxiety- and depressive-like behaviors [2–4]. In individuals without depression, imipramine has no impact on arousal or mood; however, it may act as a sedative. In people with depression, their mood is positively affected by imipramine use [5]. Thus, it is among the most effective drugs for the treatment of severe long-term depression [6]. TCAs strongly inhibit the reuptake of norepinephrine and serotonin. 3° TCAs (i.e., tertiary TCAs), including amitriptyline and imipramine, have stronger inhibition of serotonin reabsorption than 2° TCAs, such as desipramine and nortriptyline [7]. The particular mechanism explaining the treatment-related benefits of imipramine is still not well understood. TCAs are able to blockade muscarinic, histamine H1, and α1-adrenergic receptors, which explains their anticholinergic, sedative, and hypotensive impacts, respectively [8]. The anticholinergic and sedative actions of imipramine are less pronounced than those of other 3° TCAs such as clomipramine and amitriptyline. In children, imipramine is used as an antidepressant and as a treatment for bedwetting. Off-label uses include the (1) treatment of panic disorders without or with agoraphobia, (2) second-line treatment of attention deficit hyperactivity disorder in young people, (3) management of bulimia, (4) treatment of postacute and posttraumatic stress, and (5) short-term treatment of acute depression in patients with schizophrenia and bipolar disorder [9–12]. Side effects such as drowsiness, dry mouth, excitement, and weight or appetite alterations may be induced by imipramine [13]. In-depth explorations of the side effects of imipramine have been inconclusive.

Obesity is a major risk factor for various conditions related to metabolic syndrome, such as hyperlipidemia, nonalcoholic fatty liver disease (NAFLD), insulin resistance (IR), and type 2 diabetes [14]. An imbalance between expenditure and consumption causes an increase in energy storage within the body, which could lead to weight gain and obesity [15]. Moreover, recent studies have suggested that atypical glucose–insulin homeostasis is linked to various depression severity indicators and could be a neurohormonal mediator of certain depression symptoms, such as neurocognitive impairments, and disorders that are comorbid with depression, such as cardiovascular disease [4,5,16]. Gupta et al. reported a significant association between imipramine administration and increased blood glucose levels of rabbits. They also indicated that hyperglycemic responses increased more when adrenaline and imipramine were administered simultaneously than when they were administered alone [17]. In addition, some scholars have observed the effects that the tricyclic antidepressants trimipramine and amitriptyline have on glucose homoeostasis in rats, indicating that their long-term administration could cause IR and diabetes [18]. Obesity is regularly linked to NAFLD, a group of conditions that can take the form of cirrhosis, nonalcoholic steatohepatitis, and fatty liver [19]. As well as drug-related factors including liver metabolism, daily dose, and chemical structure, different risk factors heighten the likelihood of liver injury induced by drugs [20]. How patients with fatty liver further develop nonalcoholic steatohepatitis is complicated; the mechanism could involve various factors such as the overproduction of reactive oxygen species, lowered reactive oxygen species (ROS) detoxification, and heightened profibrogenic and proinflammatory cytokine release [21]. In addition, tricyclic antidepressants may have adverse health effects associated with kidney damage, diabetes, and FLD, and their prolonged use may impede weight control and aggravate diabetes [22].

Additionally, obesity is a major renal disease risk factor. Similar to hypertension and diabetes, it heightens the risk of end-stage renal disease (ESRD) and chronic kidney disease (CKD) [23]. Moreover, in ESRD or CKD, associations exist between depression and unfavorable quality of life as well as poor health outcomes [24]. One study highlighted that symptoms of depression could be related to reduced kidney function and dialysis commencement; such symptoms should be monitored in all disease stages in CKD patients [25]. Long-term hyperglycemia and IR are among the key factors related to blood–retina barrier dysfunction and retina damage linked to diabetes [26]. The retina is crucial to human vision, and diabetes is, of course, a risk factor for diabetic retinopathy (DR). Heightened inflammation and oxidative stress are suggested to be the major mechanisms behind neural retina damage in diabetes [27,28]. Research has indicated that within 3 years of antipsychotic medication initiation in patients with schizophrenia who have visual disturbances, visual cortex function and retinal thickness gradually deteriorate [29]. Other studies have indicated that processes including vasorelaxation through alpha adrenergic blocking or direct impacts on the retinal vascular endothelium could be the reasons for cystoid macular edema induced by antidepressant use [30]. Thus, we sought to understand whether imipramine affects the retina and to elucidate the possible mechanisms.

The adverse effects of imipramine likely arise from differences between the conditions under metabolic syndrome that affect body weight and hyperglycemia associated with imipramine. Thus, in our investigation, mice were fed a high-fat diet (HFD) and administered imipramine to mimic obesity; this was conducted to investigate imipramine's effects on patients with obesity concomitant with psychotic disorders. The majority of research on the metabolism-related side effects of imipramine has centered on the central nervous system [4,31,32]. We examined glucose levels, lipid metabolism, and oxidative stress, and then we explored the effects of these on the kidneys caused by abnormal metabolism, particularly related to chromium; this topic is little understood. Chromium is involved in normal lipid, protein, and carbohydrate metabolism and can benefit individuals with diabetes, glucose intolerance, obesity, or nephropathy [22]. These results provide a deeper insight into metabolic impacts and the mechanisms of fatty liver as well as kidney damage stemming from chronic imipramine use as an antipsychotic in humans or animals. We also conducted investigations on whether imipramine exacerbated metabolic abnormalities, chromium changes, oxidative stress, liver and kidney damage, and DR.

2. Materials and Methods

2.1. Animals with Feed-Induced Obesity Administered Imipramine

Five-week-old male C57BL/6J mice were procured from Taiwan's National Laboratory Animal Center. Their treatment accorded with the Taiwanese government's Guidelines for the Care and Use of Laboratory Animals. The protocol for using the experimental mice was approved after a review by the institutional animal care and use committee of the first author's university (approval no. 109019). For 12 weeks, the mice were continuously administered an HFD from PMI Nutrition International (St. Louis, MO, USA; diet 58Y1, comprising 23.6% protein, 60.0% of energy from fat, 2.6 μg/g Cr, metabolizable energy: 5.16 kcal/g). To induce obesity, the HFD was administered for 12 weeks (in the literature, 4 weeks was the associated diet duration) [33]. We then categorized the group into two subgroups. During the 8-week treatment, while remaining on the HFD, one group was administered 10 mg/kg of oral imipramine (Sigma-Aldrich, St. Louis, MO, USA) through daily gavage, and the other was administered oral saline. At the start of the treatment, the control mice weighed 28.9 ± 1.48 g, and those receiving imipramine weighed 28.35 ± 1.09 g ($p > 0.05$). The imipramine dose was decided after consultation of the literature on imipramine administration in mouse model studies related to diabetes-like status, neurological behavior observations, antidepressant effects, stress, and memory deficits [34–37]. In addition, obese mice were administered oral imipramine (5 mg/kg/day) in our preliminary investigation; however, differences in weight gain and body weight were not significant between the imipramine-treated and control mice (Figure S1). Accordingly, we adopted 10 mg/kg as the imipramine dose in this study. Mice were individually housed in micro-isolation cages on HEPA-filtered and ventilated racks. The humidity (55% ± 5%) and temperature (22 ± 1 °C) were controlled, and the mice were maintained in a 12 h light/dark cycle. The mice freely accessed water and food. From the commencement of the experiment and on a weekly basis, food consumed and body weight were recorded. When the experiment was terminated, the mice were anesthetized for tissue and blood serum harvesting. Imipramine's impact on food intake, body weight, adipocyte concentration, levels of blood glucose, fatty liver, biochemical alterations, hepatic triglycerides, endocrine profiles, insulin signaling, and renal pathology was investigated. In addition, for urine sample collection, mice were kept for 12 h in individual metabolic cages (SN-783-0; AS ONE, Osaka, Japan) before they were killed.

2.2. Measurement of Food Intake, Body Weight, and Leptin and Insulin Levels

Every week of the study, we measured food consumed and body weight. To determine food intake, food that remained within each cage dispenser was weighed, as was the food that remained on the cage floor. Furthermore, after tissue and blood samples were harvested, serum leptin and insulin levels were measured following a 12 h fast using an ELISA mouse insulin kit (INSKR020; Crystal Chem, Downers Grove, IL, USA) and a leptin kit (INSKR020; Crystal Chem), respectively.

2.3. Measurement of Serum and Hepatic Triglycerides, Creatinine, Alanine Aminotransferase, Blood Urea Nitrogen, Aspartate Aminotransferase, Serotonin, Soluble Leptin Receptor, and Fibroblast Growth Factor-21

Serum triglyceride, alanine aminotransferase (ALT), aspartate aminotransferase (AST), blood urea nitrogen (BUN), and creatinine levels were determined from the collected blood samples by means of an automated analyzer (Catalyst One Chemistry Analyzer, IDEXX Laboratories, Westbrook, ME, USA) and commercial kits under the guidance of manufacturer-approved protocols; the variation coefficient between and within analyses was under 2%. Following Folch et al., we extracted hepatic triglycerides in a 2:1 (vol/vol) mixture of chloroform and methanol [38]. Subsequently, extract solubilization was performed; the extracts were twice heated gradually to 90 °C over 5 min and subsequently cooled to room temperature. Insoluble materials were removed through centrifugation. Finally, colorimetric assay-based triglyceride analysis was conducted using the supernatant and a BioVision triglyceride quantification kit (Milpitas, CA, USA). Mouse enzyme-linked immunosorbent assay kits were employed to measure serotonin, serum soluble leptin receptor, and fibroblast growth factor-21 (FGF21) levels (EL-M0543, EL-M0545, and EL-M0435, respectively; Zgenebio Biotech, Taipei, Taiwan).

2.4. Histological and Morphometric Analysis

We weighed the retroperitoneal and epididymal fat pads and kidneys, spleen, liver, and heart; the weights are presented as a proportion (%) of body weight. Hematoxylin and eosin staining (BioTnA, Kaohsiung, Taiwan) was used to reveal hepatic fat infiltration, with scores of 0, 1, 2, 3, and 4 indicating no visible fat, <5% fat infiltration on the liver surface, 5% to 25% infiltration of fat, 25% to 50% fat infiltration, and >50% infiltration of fat, respectively [14,39]. Moreover, many epididymal white adipose tissue (EWAT) and retroperitoneal WAT (RWAT) sections were collected and analyzed in terms of adipocyte number and size. Hematoxylin and eosin staining of sections was conducted. We analyzed ≥10 fields (approximately 100 adipocytes) per slide for every sample [40,41]. Image acquisition was conducted by means of a high-resolution digital microscope (Moticam 2300, Motic Instruments, Canada), and adipocyte size analysis was conducted using Motic Images Plus 2.0. Correlations between the sizes of adipocytes and their distributions (%) were conducted for HFD controls and HFD mice administered imipramine. In addition, the right eyeball was fixed in a solution containing 4% paraformaldehyde, followed by dehydration, which involved passing the eyeball over several graded concentrations of ethanol (70%, 80%, 95%, and 100%). To render them transparent, the eyeballs were dehydrated and placed in xylene. Subsequent to being embedded in paraffin, mouse eyeballs were sliced into sections for hematoxylin and eosin stains [42].

2.5. Intraperitoneal Glucose Tolerance Test (IPGTT)

After 7 weeks of the imipramine or saline protocol, we performed an IPGTT on mice with an obesity-like status that were starved overnight but had ad libitum water. The concentration used for the assay was 1 g of glucose per 1 kg of body weight; in animal obesity and diabetes models, this is appropriate for examining antidiabetes activity [22,43]. Blood glucose levels were determined 0, 30, 60, 90, and 120 min after intraperitoneal glucose was administered. We extracted tail vein blood using a One Touch glucose meter (LifeScan, Malpitas, CA, USA). Over the 0–120 min after administration of glucose, we conducted glucose tolerance tests based on area under the curve (AUC) values.

2.6. Insulin Sensitivity (IS) and IR Indexes

Fasting blood glucose is widely utilized to determine IR and IS indexes [15,33,39]. We therefore used these indexes for assessment of IR and β-cell secretion function following imipramine administration. The homeostasis model assessment estimated IR (HOMA-IR) value was calculated as HOMA-IR = [fasting insulin (mU/L) × fasting glucose (mmol/L)]/22.5 [23]. The IS index was determined as (1/[fasting insulin (mU/L) × fasting glucose (mmol/L)]) × 1000. The IR and IS models were constructed using fasting values of plasma insulin and glucose levels by using the HOMA approach, which has been validated against clamp measurements [33,43].

2.7. Western Blotting

The mice were killed at the end of the experiment. Their livers and gastrocnemius muscles were rapidly removed, minced roughly, and homogenized immediately. Western blotting was conducted using the approach detailed in another study [44]. We used Akt, phospho-Akt (Ser473), actin, GLUT4, and adiponectin antibodies procured from Cell Signaling Technology (Beverly, MA, USA). Sigma-Aldrich supplied antibodies against patatin-like phospholipid domain containing protein 3 (PNPLA3) and fatty acid synthase (FASN). Enhanced chemiluminescence reagents (Thermo Scientific, Rockford, MA, USA) were employed to obtain immunoreactive signals, which were detected using UVP ChemStudio (Analytik Jena, Upland, CA, USA). After that, Scion Image (Scion, Frederick, MD, USA) from the National Institute of Health was employed to measure protein expression and phosphorylation.

2.8. RNA Extraction and Real-Time Quantitative Polymerase Chain Reaction (PCR)

After the mice were sacrificed, total RNA was extracted from their eyeballs and liver tissues using TRIzol reagent (Sigma-Aldrich) per the manufacturer's protocol. We then examined RNA concentration based on absorbance levels of 260 to 280 nm and 230 to 260 nm with a Qubit fluorometer (Invitrogen, Carlsbad, CA, USA). Subsequently, the RNA (1 μg) was subjected to reverse transcription into cDNA by using an iScript cDNA synthesis kit (Bio-Rad, Hercules, CA, USA) following the producer's procedure. Subsequently, real-time polymerase chain reaction (PCR) was performed per the specifications of Bio-Rad's iTaq Universal SYBR Green Supermix kit and with the application of the Bio-Rad CFX Connect Real-Time PCR system. In brief, the cycling conditions were as follows: 95 °C for 5 min, 45 cycles at 95 °C for 15 s, and 60 °C for 25 s. The expression level of every target gene was determined relative to *β-actin* levels, with these levels expressed in the $2^{-\Delta\Delta Ct}$ manner. The primers used for RT-qPCR are listed in Table 1.

Table 1. Sequences of primers used for RT-qPCR.

Gene	Primer Sequences
Fatty acid-binding protein 4 (FABP4)	Forward: 5′-GATGAAATCACCGCAGACGACA-3′ Reverse: 5′-ATTGTGGTCGACTTTCCATCCC-3′
Sterol regulatory element–binding protein 1 (SREBP1)	Forward: 5′-CGGAAGCTGTCGGGGTAG-3′ Reverse: 5′-GTTGTTGATGAGCTGGAGCA-3′
Inducible nitric oxide synthase (iNOS)	Forward: 5′-CCTCCTCCACCCTACCAAGT-3′ Rreverse: 5′-CACCCAAAGTGCTTCAGTCA-3′
Cyclooxygenase-2 (COX-2)	Forward: 5′-TTCAAAAGAAGTGCTGGAAAAGGTTCT-3′ Rreverse: 5′-AGATCATCTCTACCTGAGTGTCCTT-3′
Nuclear factor κB (NF-κB)	Forward: 5′-GCAACTCTGTCCTGCACCTA-3′ Reverse: 5′-CTGCTCCTGAGCGTTGACTT-3′
Inhibitory κBα (IκBα)	Forward: 5′-AAGTGATCCGCCAGGTGAAG-3′ Reverse: 5′- CTGCTCACAGGCAAGGTGTA -3
Uncoupling protein 1 (UCP1)	Forward: 5′-GGCCTCTACGACTCAGTCCA-3′ Reverse: 5′-TAAGCCGGCTGAGATCTTGT-3′
β-actin	Forward: 5′-GGCTGTATTCCCCTCCATCG-3′ Reverse: 5′-CCAGTTGGTAACAATGCCATGT-3′

2.9. Analysis of Chromium Content

After all experiments were complete, blood, urine, and several types of tissue (kidney, blood, liver, bone, fat, and muscle) were collected. Concentrations of chromium in samples were measured according to a prior report [40]. In brief, each sample (0.1 g of tissue and 25 μL of blood and urine) was digested in 65% nitric acid, and then each sample was subjected to digestion with nitric acid overnight (temperature: 100 °C). Chromium concentrations were determined using an ICP Mass Spectrometer (NexION 350X, PerkinElmer, MA, USA). Next, distilled water was used to dilute the digested solution to a 5 mL solution prior to measurements. The relative chromium recovery rate was calculated at 10 ppb and 100 ppb of the quantification levels by 5% (n = 5) and 8% (n = 5), respectively. The absorption data were plotted onto a 1–500 ppb standard curve, and regression analysis was performed to identify the total chromium level in samples ($R^2 > 0.996$).

2.10. Measurement of Glutathione Peroxidase, Renal Catalase, and Superoxide Dismutase

We analyzed the antioxidant system's functional activity to determine the enzymatic activity of glutathione peroxidase (GPx), hepatic catalase (CAT), and superoxide dismutase (SOD). Ice-cold saline (0.9% sodium chloride) was used to perfuse kidneys, after which they underwent homogenization in chilled potassium chloride (1.17%) by means of a homogenizer and in accordance with a previous description [22]. Next, the homogenates were gathered for analysis after 5 min of centrifugation at 800× g (4 °C). Finally, the supernatant was subjected to 20 min centrifugation at 10,500× g (also 4 °C) to obtain the postmitochondrial supernatant for the kidney samples to measure SOD, catalase, and GPx activity. Commercially available colorimetrical kits were used for these procedures (catalase: #K773-100, #GPx: K762-100, and SOD: #K335-100; BioVision) following manufacturer-recommended protocols.

2.11. Statistical Analysis

Data are shown as the mean ± standard deviation. The t-test was employed for determining intergroup differences ($p < 0.05$ denoted a statistically significant difference). Additionally, we determined contingency data significance by means of Fisher's exact test.

3. Results

3.1. Imipramine Changed Food Efficiency, Morphometric Parameters, and Food Intake in Animals

Our preliminary investigation did not identify significantly different antiobesity effects in standard diet (SD) fed animals that were and were not administered imipramine (Figure S2). Mice on an HFD were administered imipramine for 8 weeks to elevate their body weight, metabolic parameters, serum leptin levels, and leptin receptor levels (Figure 1). The body weight of the imipramine-treated mice at the end of the experiment was 21% higher ($p < 0.01$) than that of the HFD-fed controls (Figure 1a). Furthermore, weekly food consumption was 21% higher in mice after they were administered imipramine, and their increase was parallel with the food consumption increase in controls (Figure 1b). These changes took place alongside significantly increased daily food efficiency in mice administered imipramine; this efficiency was 20% higher than the control animals (Figure 1c). We also revealed that imipramine had a significant effect on serum leptin (Figure 1d) and leptin receptor levels (Figure 1e), with the treatment group exhibiting a 15% higher serum leptin level and an 11% lower leptin receptor level than the control group. Leptin also plays a role in food intake regulation [22].

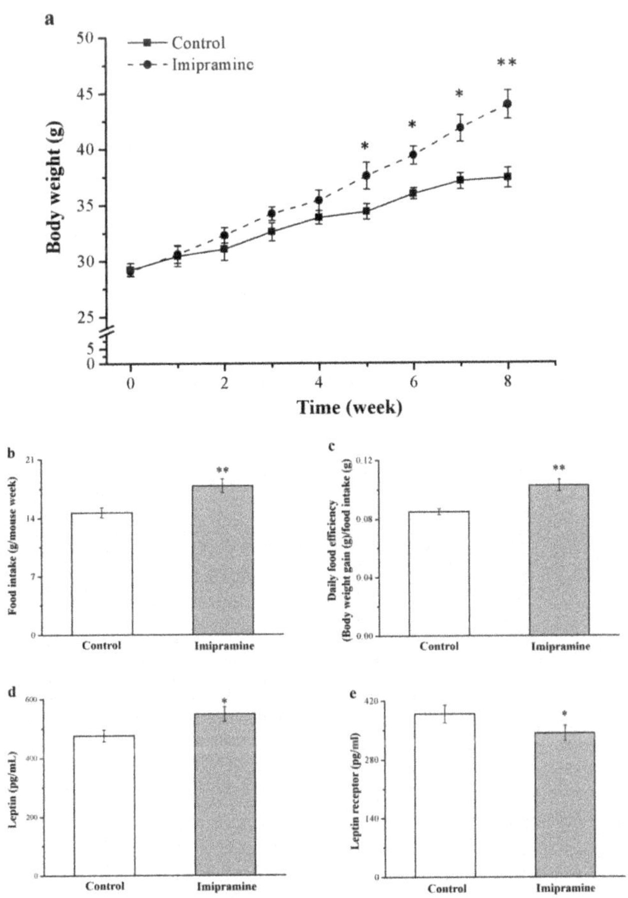

Figure 1. Effects of imipramine on (**a**) weekly body weight changes and (**b**) food intake in each group (measured weekly). (**c**) Daily food efficiency, (**d**) serum leptin levels, and (**e**) leptin receptor levels were measured in HFD-fed controls and imipramine-treated obese mice. For both groups, data are in the form of mean ± standard deviation (n = 8). * $p < 0.05$; ** $p < 0.01$.

3.2. Imipramine Increased Liver, Kidney, and Fat Pad Weights

Next, assessments were conducted to discover whether the indicated weight disparities were linked to body composition or adiposity alterations. Eight weeks after imipramine administration commenced, the body compositions of treated mice differed significantly from those of control mice in terms of their hearts, livers, kidneys, EWAT, and RWAT, but not in terms of their spleens (Figure 2). After 8 weeks of imipramine administration, the treated mice had significantly different body compositions from the control mice: their hearts, livers, kidneys, RWAT, and EWAT were 1.3, 1.3, 1.2, 1.2, and 1.3 times heavier than those of the control mice, respectively; moreover, when expressed as a proportion of body weight, these were 1.1, 1.2, 1.1, 1.2, and 1.1 times heavier than those of control mice, respectively. Spleen weights, as a proportion of the overall body weight, were not significantly different between the groups.

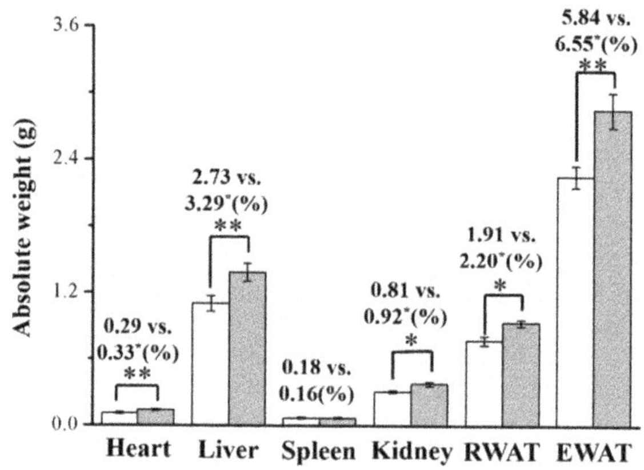

Figure 2. Effects of imipramine on total heart, retroperitoneal white adipose tissue (RWAT), spleen, liver, kidney, and epididymal white adipose tissue (EWAT) weights (shown as a percentage of total body weight). Both groups' data are in the form of mean ± standard deviation (n = 8). * $p < 0.05$; ** $p < 0.01$.

3.3. Imipramine Increased Hepatic Fat Accumulation and the Ratio of Large to Small Adipocytes but Reduced UCP1 mRNA Expression

A morphometric evaluation involving hematoxylin and eosin tissue staining for both groups revealed that the treated mice had considerably larger adipocytes in EWAT and RWAT and more liver fat compared with the controls (Figure 3a), indicating that imipramine promotes hepatic fat accumulation and may thus increase fat pad hypertrophy. Our analyses of alterations in fatty liver scores (Figure 3b) and RWAT (Figure 3c) and EWAT (Figure 3d) adipocyte size revealed significant differences between groups; the treatment mice had fatty liver scores almost 1.3 times higher than the control group's scores. Additionally, the mean EWAT and RWAT adipocyte size was significantly larger (1.4 and 1.2 times, respectively) in the imipramine group, which agrees with the observation of higher fat pad weights in the mice administered imipramine. Moreover, we determined that imipramine had a significant influence on the mRNA expression of *UCP1*, with the treatment group exhibiting a 12% lower level than the controls (Figure 3e). In other words, the mice administered imipramine had a lower proportion of RWAT and EWAT adipocytes with a diameter of 0–40 and 40–80 μm (Table 2) but a higher proportion of those with a diameter of 80–120 and >120 μm. Therefore, although an increasing trend in RWAT and EWAT adipocyte size was discovered in response to mice being fed an HFD, imipramine expedited this size increase.

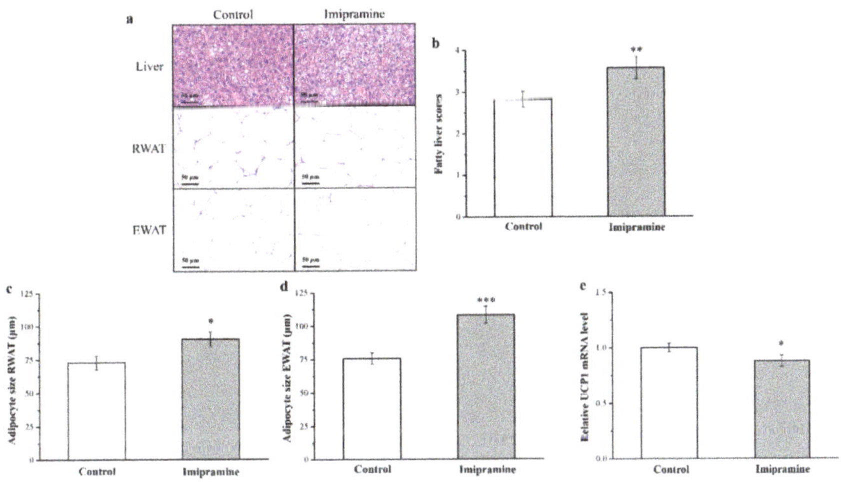

Figure 3. Effects of imipramine on (**a**) the mouse livers, retroperitoneal white adipose tissue (RWAT), and epididymal white adipose tissue (EWAT) of control and treated mice, as determined through hematoxylin and eosin staining (magnification, 200×). (**b**) Fatty liver score findings. (**c**) RWAT and (**d**) EWAT adipocyte sizes and (**e**) the mRNA expression of *UCP1* in brown adipose tissue were measured in HFD-fed control and imipramine-treated obese mice. Data are in the form of mean ± standard deviation (n = 8). * $p < 0.05$; ** $p < 0.01$; *** $p < 0.001$.

Table 2. Influence of imipramine on fat cell sizes in HFD-fed controls and imipramine-treated mice.

Variable	Control	Imipramine
Retroperitoneal white adipose tissue (RWAT)		
Adipocyte diameter		
0–40 μm (%)	17.59 ± 0.43	5.78 ± 0.04 ***
40–80 μm (%)	58.97 ± 4.31	30.98 ± 1.93 ***
80–120 μm (%)	21.44 ± 0.62	57.33 ± 3.86 ***
>120 μm (%)	0 ± 0	5.91 ± 0.77 ***
Epididymal white adipose tissue (EWAT)		
Adipocyte diameter		
0–40 μm (%)	5.62 ± 0.29	0 ± 0 ***
40–80 μm (%)	71.37 ± 5.82	20.47 ± 1.06 ***
80–120 μm (%)	23.01 ± 1.3	67.09 ± 4.84 ***
>120 μm (%)	0 ± 0	12.44 ± 0.92 ***

Data are in the form of mean ± standard deviation (n = 8). *** $p < 0.001$.

3.4. Imipramine Increased the Serum Levls of Serotonin, FGF21, ALT, and AST and the mRNA Levels of FABP4 and SREBP1

A report indicated that increased serotonin activity resulted in severe obesity and hepatic steatosis [45]. Our results indicate that the serum serotonin levels of imipramine-treated mice were 1.3 times higher than those of the controls (Figure 4a). In addition, the imipramine-treated group had 20% reduced FGF21 activity compared with that of the control group (Figure 4b). Serum ALT and AST were increased by 1.3 and 1.3 times, respectively, in the mice administered imipramine, indicating hepatic function (Figure 4c,d, respectively). FABP4 and SREBP1 are implicated in regulating many metabolic pathways, such as those related to type 2 diabetes, atherosclerosis, and hepatic lipid accumulation [38,43]. These genes reportedly have dominant effects in promoting FLD in rodents and humans [40]. In the current investigation, compared with control mice, mice administered imipramine had, respectively, 1.8 (Figure 4e) and 1.5 (Figure 4f) times higher liver mRNA levels of *FABP4* and *SREBP1*. The gene-associated activation of hepatic steatosis resulted in significantly increased AST and ALT levels; this was probably related to imipramine and its contribution to FLD, mediating the heightened mRNA expression of molecular mechanisms of hepatic lipid accumulation.

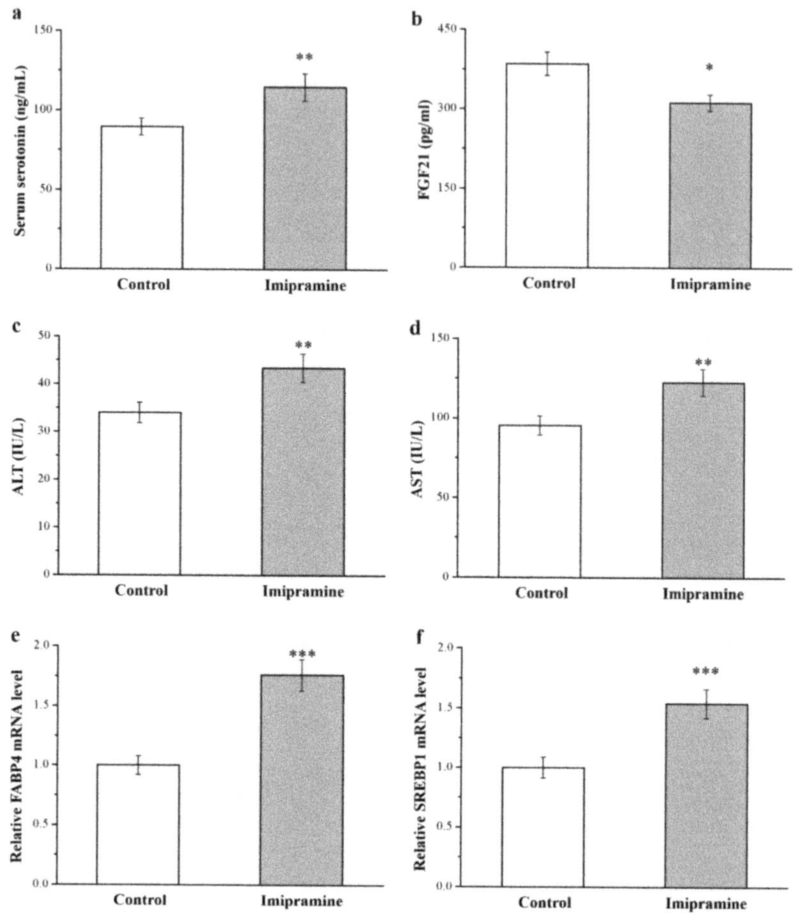

Figure 4. Effects of imipramine on (**a**) serum serotonin levels, (**b**) serum fibroblast growth factor-21 (FGF21) levels, (**c**) serum alanine aminotransferase (ALT) levels, (**d**) serum aspartate aminotransferase (AST) levels, and hepatic mRNA levels of (**e**) FABP4 and (**f**) SREBP1 measured in HFD-fed control and obese imipramine-treated mice. Data are in the form of mean ± standard deviation (n = 8). * $p < 0.05$; ** $p < 0.01$; *** $p < 0.001$.

3.5. Imipramine Increased Serum Triglycerides and Hepatic Triglycerides, FASN Levels, and PNPLA3 Levels but Reduced Hepatic Adiponectin Levels

Levels of blood triglycerides (Figure 5a) and hepatic triglycerides (Figure 5b) were significantly higher in imipramine-treated mice than in control mice. Furthermore, we noted a 1.2-fold and 1.3-fold increase in blood triglyceride levels and hepatic triglycerides, respectively, compared to control mice. In addition, Western blot analysis performed on the liver of mice determined adiponectin, FASN, and PNPLA3 expression (Figure 5c). Adiponectin has mediated stimulation of energy expenditure and regulates the expression of hepatic genes critical for glucose and lipid metabolism [40]. Alternatively, FASN is a crucial liver enzyme for lipid homeostasis and triglyceride synthesis [46], and PNPLA3 is a marker that is pathologically characterized by the regulation of lipogenesis in obesity, nonalcoholic fatty liver disease, and cardiovascular disease [33]. Remarkably, FASN and PNPLA3 expressions in the liver of imipramine-treated mice were significantly higher (by 1.5-fold and 1.5-fold; Figure 5d,e, respectively) compared with those in the control mice. Moreover, adiponectin expressions in the imipramine-treated mice were significantly lower by 63% compared with those in the control mice (Figure 5f). These results indicate that imipramine increased the upregulation of lipogenesis in the liver and exacerbated HFD-induced fatty liver scores.

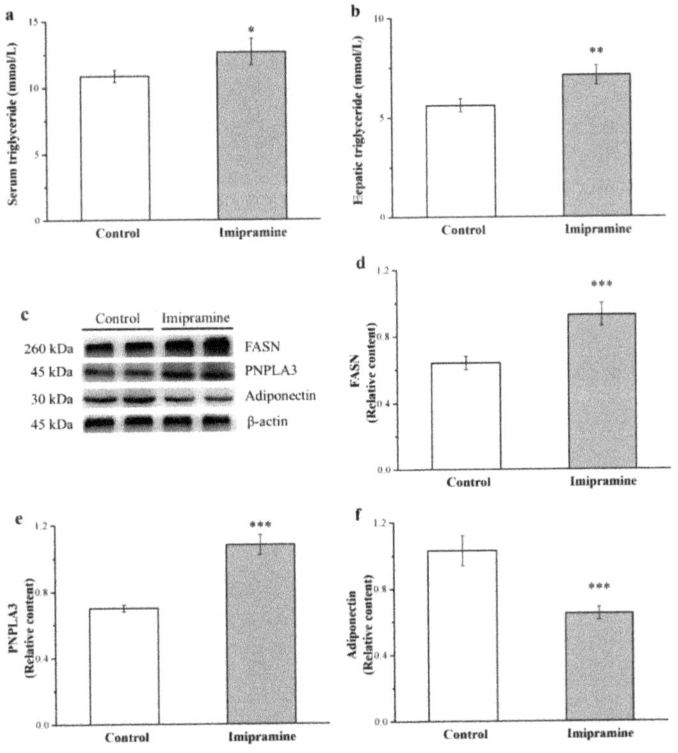

Figure 5. Effects of imipramine on (**a**) serum triglycerides and (**b**) hepatic triglycerides. (**c**) A representative image showing blots of liver extracts. The hepatic expression levels of (**d**) fatty acid synthase (FASN), (**e**) patatin-like phospholipid domain containing protein 3 (PNPLA3), and (**f**) adiponectin were measured. Data are in the form of mean ± standard deviation (n = 8). * $p < 0.05$; ** $p < 0.01$; *** $p < 0.001$.

3.6. Imipramine Reduced Glucose Tolerance and Lowered Insulin Levels

To examine the effect of imipramine on glucose homeostasis, we conducted glucose tolerance tests. Significant imipramine-induced impairment of glucose tolerance was discovered in the imipramine group compared with the control group, as demonstrated by the IPGTT data presented in Figure 6a. Notably, imipramine administration resulted in significantly heightened fasting blood glucose. Moreover, imipramine administration resulted in a significant increase in fasting blood glucose levels at all time points after glucose supplementation through injection; in addition, after 120 min, blood glucose levels in the imipramine group were 1.1 times higher after glucose injection than they were before it. Furthermore, when comparing the control and imipramine groups, the AUC for glucose levels at 120 min was 1.1 times higher in the mice administered imipramine; this difference reached statistical significance (Figure 6b). With glucose intolerance defined as a >10 mmol L^{-1} blood glucose level at 120 min after an intraperitoneal glucose injection, we revealed that a large proportion of mice administered imipramine had glucose intolerance (Figure 6c) [39]. Imipramine-treated mice had significantly lower levels of serum insulin (Figure 6d). Therefore, HFD-fed mice that received imipramine had exacerbated diabetes symptoms, with more severe hyperglycemia and impaired glucose tolerance associated with hypoinsulinemia.

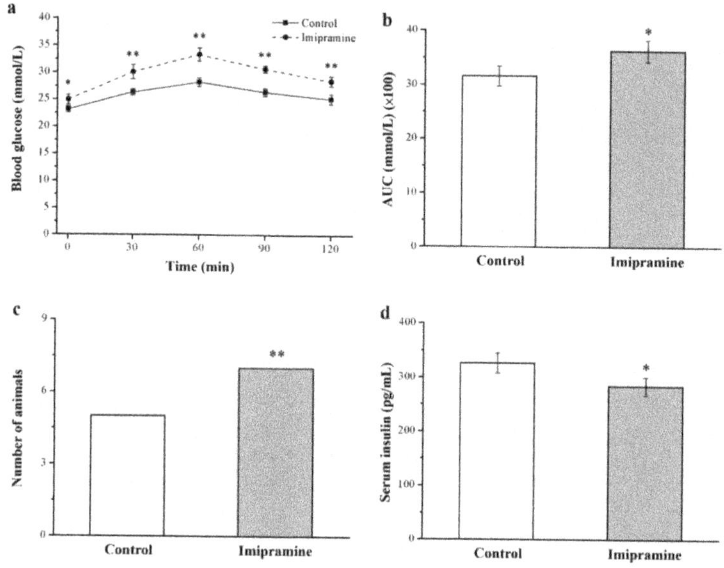

Figure 6. Impact of imipramine on (**a**) blood glucose tolerance score. (**b**) Area under the curve (AUC) at 120 min post-glucose injection of blood glucose levels. (**c**) Glucose intolerance (Fisher's exact test). (**d**) Serum insulin levels. Data are in the form of mean ± standard deviation (n = 8). * $p < 0.05$; ** $p < 0.01$.

3.7. Imipramine Reduced IS through Changing the Expression of Phosphorylated Akt and GLUT4

Long-term imipramine treatment increased IR and diminished insulin insensitivity in HFD-fed mice. This finding is backed up by a significant difference in the respective HOMA-IR value and IS index (Figure 7a,b, respectively). Moreover, compared with the control group, the imipramine group had a significantly (1.2 times) higher HOMA-IR (Figure 7a). The IS index of the imipramine group was significantly lower (by 17%) (Figure 7b). Furthermore, we made an attempt to understand the mechanisms underlying glucose homeostasis in the mice fed an HFD through examining Akt phosphorylation and GLUT4 expression within muscles after imipramine treatment (Figure 7c). Significantly lower (11% and 25%, respectively) muscle Akt activation and GLUT4 expression were observed in the imipramine group (Figure 7d,e).

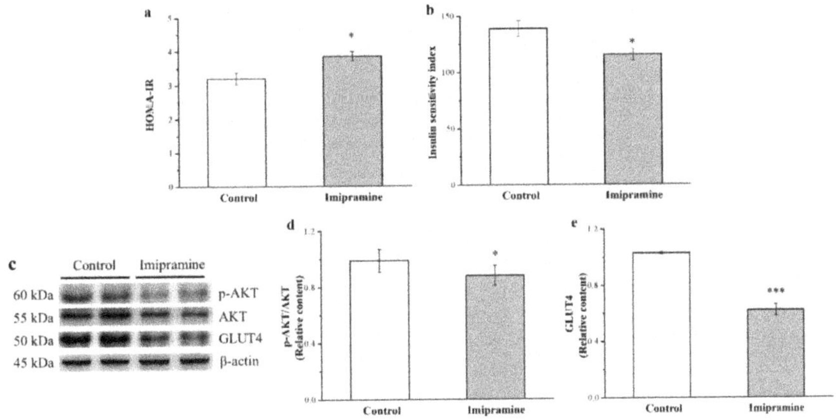

Figure 7. Effects of imipramine on the (**a**) homeostasis model assessment-estimated IR (HOMA-IR) index and (**b**) insulin sensitivity (IS) index. (**c**) Representative blot of gastrocnemius muscle extracts. Expression levels of (**d**) phosphorylated Akt and (**e**) GLUT4. Data are in the form of mean ± standard deviation ($n = 8$). * $p < 0.05$; *** $p < 0.001$.

3.8. Imipramine Affected Chromium Levels in Tissues and Organs and Increased Chromium Urinary Loss

Chromium (III) has important functions in glucose homeostasis, makes a major contribution to insulin action, and reduces IR [15,38]. Therefore, to evaluate whether imipramine-treated mice with HFD-induced obesity and glucose intolerance had altered chromium levels, we measured the chromium concentration in various harvested tissues and organs; the results are presented in Table 3. Chromium intake in the imipramine-treated mice was significantly higher (by 1.2 times) than that in the control mice. This finding was linked to the imipramine-treated mice being more hyperphagic than the control mice. Blood, bone, liver, muscle, and fat pad chromium levels were (significantly) 60%, 53%, 20%, 22%, and 24% lower, respectively, in the imipramine-treated mice than in the controls. By contrast, these chromium levels had an opposite trend in kidney and urine samples; average increases of 1.5 times and 2.2 times, respectively, were noted in the imipramine-treated mice compared with the controls, reaching the level of statistical significance.

Table 3. Chromium concentrations in the organs and tissues of HFD-fed controls and imipramine-treated mice.

Variable	Control	Imipramine
Chromium intake/mouse/week (μg)	16.42 ± 0.66	19.96 ± 0.91 *
Blood (ng/mL)	182.28 ± 8.51	72.36 ± 5.78 ***
Bone (ng/g)	326.58 ± 9.26	152.27 ± 6.31 ***
Liver (ng/g)	76.27 ± 6.83	60.61 ± 4.25 **
Muscle (ng/g)	52.36 ± 4.62	40.84 ± 3.73 **
Epididymal fat pads (ng/g)	50.59 ± 4.61	38.52 ± 2.54 **
Kidney (ng/g)	100.51 ± 3.75	151.73 ± 6.29 ***
Urine (ng/mL)	51.18± 3.65	112.41 ± 4.23 ***

Data for the two groups are in the form of mean ± standard deviation ($n = 8$). * $p < 0.05$; ** $p < 0.01$; *** $p < 0.001$.

3.9. Imipramine Induced Renal Injury, Increased Serum BUN and Creatinine, and Reduced Antioxidant Enzymes in the Kidneys

We next evaluated whether imipramine induced renal damage because evidence previously indicated that obesity, hyperlipidemia, and diabetes resulted in kidney injury [39]. Hematoxylin and eosin staining revealed that imipramine administration induced glomerulonephritis and inflammatory cell infiltration (Figure 8a). In addition, we observed significantly higher serum BUN and creatinine levels (1.2 times higher each) in the imipramine group (Figure 8b,c). Renal injury is linked to a reduction in antioxidant enzymes in the kidney; increased oxidant enzyme activity may result in alleviation of renal nephropathy and improve renal function [47,48]. Mice administered imipramine had significantly lower activities of antioxidant biomarkers CAT (Figure 8d), GPx (Figure 8e), and SOD (Figure 8f; 27%, 10%, and 13% lower, respectively).

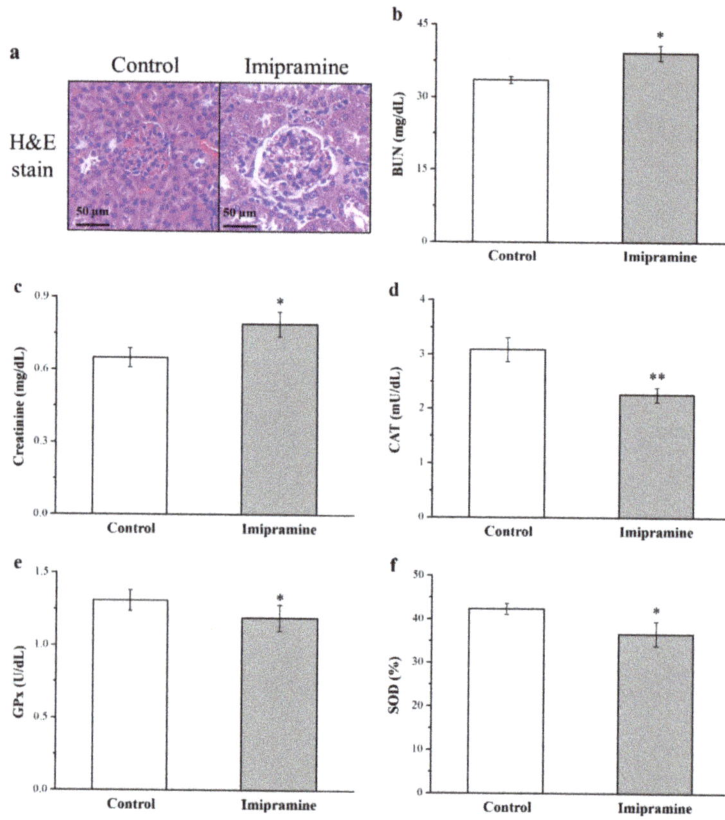

Figure 8. Effect of imipramine on (**a**) glomerular morphologies according to hematoxylin and eosin staining (magnification, 200×). (**b**) Serum BUN, (**c**) serum creatinine, (**d**) renal catalase (CAT) activity, (**e**) renal glutathione peroxidase (GPx) activity, and (**f**) renal superoxide dismutase (SOD) activity. Data are in the form of mean ± standard deviation ($n = 8$). * $p < 0.05$; ** $p < 0.01$.

3.10. Imipramine Aggravates Retina Damage and Increases the Gene Expression of iNOS, NF-κB, and COX-2 but Reduces IκBα Expression in the Eyeballs

Diabetes heightens oxidative stress, and such a heightened stress state may play a major role in diabetes complications, including retinal injury [47,48]. In hematoxylin and eosin-stained sections of retinas, the inner plexiform and nuclear layers (IPL and INL, respectively) of imipramine-treated mice were thinner than those of the control mice, and the ganglion cell layer was almost absent in the treated mice (Figure 9a). Ocular inflammation and associated complications are key causes of vision loss. In recent years, evidence has suggested that inflammation has a prominent role in the pathogenesis of several retinal conditions, such as DR [48]. Our results show that the gene expression of COX-2, iNOS, and NF-κB was higher in imipramine-treated mice compared with controls

(1.5, 1.8, and 1.2 times higher, respectively; Figure 9b,c, respectively). By contrast, the gene expression of IκBα in imipramine-treated mice was 18% lower than that in controls (Figure 9d). These results indicate that imipramine increased the risk of DR by elevating inflammation and glucose uptake.

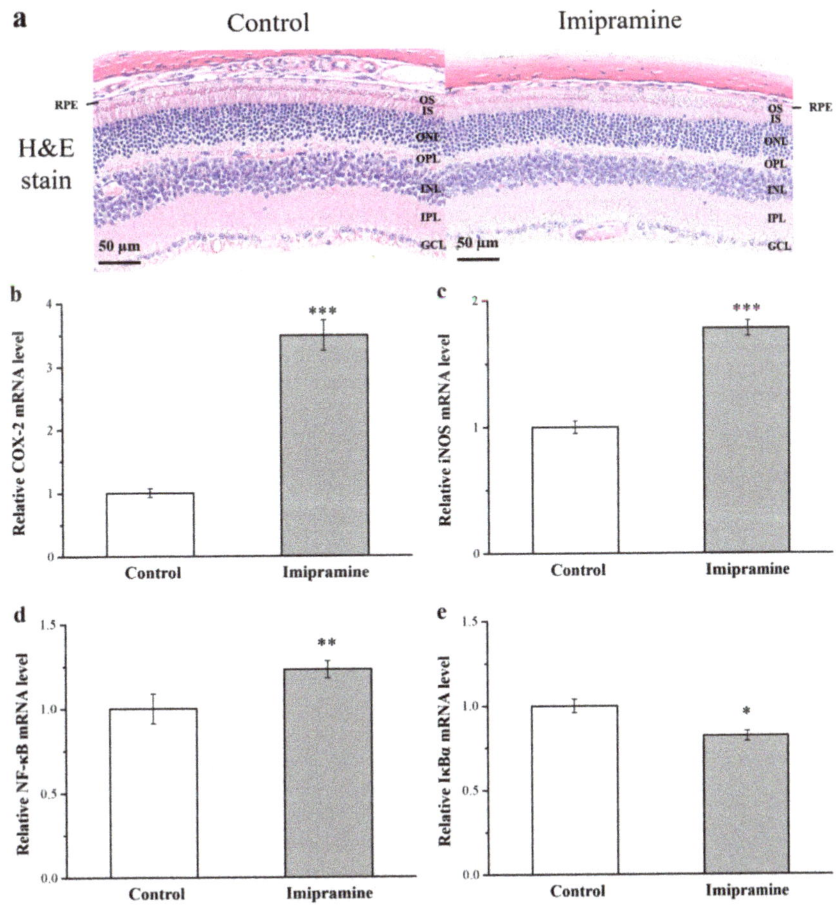

Figure 9. Effects of imipramine on (**a**) retinal morphology according to hematoxylin and eosin staining (magnification, 200×). (**b**) Cyclooxygenase-2 (COX-2), (**c**) inducible nitric oxide synthase (iNOS), (**d**) nuclear factor κB (NF-κB), and (**e**) inhibitory κBα (IκBα) mRNA levels. Data are in the form of mean ± standard deviation ($n = 8$). GCL, ganglion cell layer; IPL, inner plexiform layer; INL, inner nuclear layer; OPL, outer plexiform layer; ONL, outer nuclear layer; IS/OS, inner/outer segment junction; RPE, retinal pigment epithelium. * $p < 0.05$; ** $p < 0.01$; *** $p < 0.001$.

4. Discussion

We conducted this study to explore the impact of imipramine on obesity development in mice fed an HFD. Mice administered imipramine for 8 weeks had a higher obesity risk, more visceral fat, a higher fatty liver score, more severe glucose intolerance, increased IR, more severe kidney damage, and a larger AUC value 120 min after glucose was injected. Daily food efficiency, liver, kidney, and fat pad weight, fatty liver score, adipocyte size, serum and hepatic triglyceride levels, and serum BUN and creatinine also increased; however, reductions were noted in renal catalase, GPx, and SOD activities. Our results demonstrate that imipramine can increase weight gain and food intake and attenuate glucose homeostasis concomitant with hypoinsulinemia. Imipramine increases fatty liver scores and is linked to adipogenesis in the liver, including *FABP4* and *SREBP1* mRNA levels, in addition to adiponectin, FASN, and PNPLA3 activation. IR is typically associated with lower insulin signaling protein activity; our imipramine-treated hyperglycemic mice exhibited exacerbated glucose homeostasis induced by a reduction in phosphorylated Akt and GLUT4. Compared with the controls and after 56 days of imipramine treatment, the treated mice had significantly lower chromium levels in their bones, blood, livers, muscles, and fat pads; however, they had significantly higher levels in their urine and kidneys. Imipramine treatment also rendered mouse retinas abnormal and increased the expression of inflammatory-related genes, including COX-2, iNOS, and NF-κB, in the eyeballs. Finally, the treated mice also had increased obesity, hepatic fat accumulation, serum triglyceride levels, serum liver function indexes, and urinary chromium loss, and they exhibited insulin resistance, heightened glucose intolerance, renal injury, diabetic retinopathy, and thin IPLs and INLs corresponding to the development of hyperglycemia-associated fatty liver.

After mice were administered an HFD for 8 weeks, the diet made a contribution to obesity development. Weight gain increases occur when fat pad mass increases because of an increase in the number of new adipocytes from precursor cells or due to an enlarged adipocyte size because of fat storage [14,38]. Body weight gain is related to increased fat cell differentiation or fat pad weight caused by fat cell hypertrophy [14,39]. For this reason, the body weight of the imipramine-treated mice was greater than that of the control mice, and the weights of organs and tissue, including the kidneys, the liver, RWAT, and EWAT, were also greater. We suggest that the increases in body fat as well as RWAT, EWAT, and liver weights in the imipramine group may have been caused by increases in adipocyte size and fatty liver scores. Additionally, we obtained evidence that the increase in body fat may have been a result of lower energy expenditure, indicated by the mRNA expression in brown adipose tissue of UCP1. UCP1 expression occurs in the inner membrane of brown adipocyte mitochondria, creating heat through the uncoupling of oxidative phosphorylation, and thermogenesis induction is controlled by adrenaline. Hypothalamus feeding centers were linked to this thermogenic system for body temperature control, which enables the regulation of body weight and may well represent an efficient dual function related to the thermogenesis of brown fat [49]. Furthermore, leptin has roles in the hypothalamus regarding regulating food intake, and leptin synthesis occurs primarily within white adipose tissue [50]. The imipramine-treated mice ate more, and their serum leptin was higher than that of the controls. Leptin mediates food intake by means of sympathetic outflow to the hypothalamus, and leptin directly affects tissues through enhancement of lipid oxidation and a reduction in lipogenesis [51]. Imipramine therapy appeared to inhibit the leptin signaling pathway. Moreover, we found a decrease in leptin receptor levels in the imipramine-treated mice; the leptin receptor is a phenome of resistance to leptin. The finding that long-term use of imipramine may accelerate obesity through leptin signaling inhibition is valuable.

The accumulation of fat in the liver induces FLD, which is strongly associated with obesity [33,38]. The imipramine group had higher fatty liver scores (revealed through histopathology) and higher levels of the liver enzymes AST and ALT, which act as hepatic injury markers. Hepatocellular permeability is increased when liver injury is induced, and consequently, AST and ALT are released from intracellular spaces into plasma [52]. In addition, we analyzed *SREBP1* and *FABP4* mRNA expressions and discovered them to be associated with the expression of genes participating in lipid storage, hepatic steatosis, hepatic lipogenesis, and NAFLD pathogenesis [33,53]. Higher expression of SREBP1 and FABP4 mRNA in the liver was observed after long-term imipramine administration and caused liver injury owing to marked hepatic lipid infiltration. In addition, the activity of COX-2 in the obese imipramine-treated mice was significantly higher than in the control mice (Figure S3), which eventually led to inflammation and liver damage [54]. Serum AST and ALT levels were subsequently analyzed. Similar studies have found liver test abnormalities in patients administered tricyclic antidepressants, such as aminotransferase abnormalities and cholestatic hepatitis [55]. A report indicated that increased systemic serotonin activity through the knockout of the serotonin transporter resulted in IR, severe obesity, and hepatic steatosis [53]. Our findings corroborate those of other studies related to imipramine-treated HFD-fed mice; these mice exhibited substantially increased fat pad and liver weights, serum and hepatic triglyceride levels, and fatty liver scores in response to increased serum serotonin levels. For safety reasons, when imipramine use is linked to higher risks in individuals with depression and underlying liver disease, it is recommended that serum AST and ALT be regularly monitored during clinical trials despite the fact that no such recommendation has been formally made for related investigations.

Obesity is highly associated with elevated triglycerides [56]. In addition, the hallmark of NAFLD is accumulated triglycerides in hepatocytes' cytoplasm [57]. We explored serum and hepatic triglyceride levels. Our imipramine group exhibited elevated serum and hepatic triglyceride levels compared with the control group. This may have been related to imipramine decreasing fibroblast growth factor-21 activity, which has major effects on metabolic parameters, including on lipid and glucose homeostasis, and which promotes fast weight loss [58]. The finding may be linked to lower FGF21 levels preventing a reduction in excess lipid levels in hepatocytes; this may result in lipid overload-related stress and activate the release of several proinflammatory factors (e.g., tumor necrosis factor-α (TNF-α)) [59,60]. FASN may have a role in body weight regulation and the development of obesity [61]. Research has revealed a significant correlation of FASN expression with the degree of steatosis in primary human hepatocytes in vitro as well as in experimental murine models and in the livers of patients with NAFLD in vivo [62]. Furthermore, hepatic inflammation may be promoted by PNPLA3 during NAFLD progression through an increase in TNF-α expression and activation of endoplasmic reticulum stress-mediated and NF-kB-related inflammation in NAFLD [63]. In addition, adiponectin has antidiabetic characteristics owing to its insulin-mimetic and insulin-sensitizing effects, and furthermore, anti-inflammatory and anti-atherosclerotic effects have been consistently reported [64]. As expected, in this study, adiponectin was reduced in the imipramine-treated mice. Taken together, the results indicate that imipramine induced a marked intrahepatic accumulation of lipids, and this exacerbated liver damage in mice with HFD-induced NAFLD. The activation of multiple inflammatory signaling pathways by imipramine may accelerate the development of liver damage. Therefore, regulation of imipramine in these pathways may indicate novel treatment targets for these increasingly less common sequelae of hepatic side effects after imipramine administration.

Reports indicated that tricyclic antidepressants elevate and reduce glucose and insulin levels, respectively, in humans and animals, contributing to greater metabolic risks in non-diabetic patients with depression or the deterioration of glucose metabolism in individuals with depression and type 2 diabetes through the aggravation of glucose intolerance [65–67]. Moreover, evidence indicates that TCAs, such as doxepin and amitriptyline, induce hyperglycemia by inhibiting insulin release from the pancreas and inhibit glucose transport, resulting in decreased glucose uptake [68,69]. Our results indicate that imipramine exacerbates diabetes mellitus complications by reducing serum insulin levels, as evidenced by the fact that chronic imipramine use resulted in a reduced proportion of β-cells in mouse pancreas sections (Figure S4). Generally, a reduction in Akt phosphorylation and serum FGF21 level leads to the attenuation of insulin signaling, which impairs glucose homeostasis and enhances IR [33,39]. GLUT4 facilitates insulin-promoted glucose uptake into adipose tissue and muscle, and reduced GLUT4 expression levels lower insulin-mediated glucose uptake; exacerbated hyperglycemia may be the mechanism of action, as indicated by heightened glucose intolerance on the basis of an IPGTT [70]. Considerable basal glucose transport reductions suggest strong IR and glucose intolerance in mice that have diabetes and a selective deficiency of GLUT4 in their muscles [33]. Collectively, these findings indicate that HFD-fed mice administered imipramine have severe hyperglycemia and IS suppression.

Chromium administration may be a crucial adjunct therapy for type 2 diabetes because it may be involved in glucose metabolism through potentiating insulin's actions [38]. This action of chromium has a role in metabolism, and a consequence of this action is glucose metabolism alteration. Our results also indicate that chromium levels in the muscles, blood, bones, liver, and fat pads of HFD-fed mice administered imipramine were significantly lower than in control mice. These results show that imipramine negatively affects chromium accumulation in tissues [39]. Similarly, serum chromium levels were significantly decreased in patients with uncontrolled type 2 diabetes [71]. Therefore, prolonged imipramine intake could change chromium concentrations in various organs and tissues. Chromium mobilization or redistribution prompted the worsening of the hyperglycemic status of our mice administered imipramine.

Large amounts of trace elements are excreted in urine; however, the majority of them are absorbed again at the proximal renal tubule [34]. Although chromium levels were low in the metabolic tissue and blood of our imipramine group, they were high in urine and kidney samples. We surmised that the imipramine resulted in chromium release from metabolic tissues into the kidneys, thus increasing urinary chromium loss. Greater urine excretion and lower trace metal reabsorption are generally kidney damage responses [15]. This supposition is backed up by the higher renal function indexes (creatine and BUN) in our mice administered imipramine. Higher creatine and BUN concentrations are known to be kidney disease progression risk factors in those with moderate or severe CKD [72]. Additionally, imipramine caused lesions in mice and glomerulonephritis. Collectively, the results demonstrate that chronic intake of imipramine causes kidney injury; inflammatory cell infiltration in the renal interstitium was observed alongside increased biomarkers of nephropathy, namely, creatine and BUN. Detoxicant and antioxidant enzymes provide crucial protection to the kidney [73]. Due to their transport-related functions, the kidneys have an active oxidative metabolism, and this action causes ROS production [74]. We detected lowered GPx, catalase, and SOD activities in the treated mice compared with our controls. Furthermore, immunohistochemical (IHC) staining demonstrated that mice administered imipramine had heightened levels of renal inflammatory cytokines, including TNF-α (Figure S5). The results indicate that chronic and continuous imipramine administration can lead to hyperglycemia-induced kidney damage and long-term inflammation, exacerbating injuries caused by reduced antioxidant levels and oxidative stress. Researchers should investigate whether drugs other than imipramine can be used as replacement or adjunct therapies to alleviate kidney-related side effects in diabetes patients with depression.

Hyperglycemia is a common diabetes symptom, and it may induce oxidative stress and inflammatory injuries during DR development and later result in vascular dysfunctions [75]. DR can lead to blindness [76]. IR and chronic hyperglycemia are key factors related to blood–retina barrier dysfunction and retinal damage related to diabetes [77]. Various neuronal cells, including amacrine and ganglion cells, are affected by pathogenic changes that are manifestations of morphologic abnormalities and neuronal apoptosis [78]. Numerous physiological and molecular abnormalities develop in the retinas of individuals with diabetes, and the manifestations seem to be related to inflammation [79]. Upregulated COX-2 and iNOS have been noted in rodent and human retinas [80]. Moreover, NF-κB, which is inhibited by IκBα, is a widely expressed inducible transcription factor known to be a crucial regulator of various genes with roles in immune and inflammatory responses and cellular apoptosis and proliferation. NF-κB activation is induced by proinflammatory proteins such as iNOS [81]. Diabetes activates NF-κB in the retinas of rodents [82]. Through hematoxylin and eosin staining, we discovered that retina cells in the mice administered imipramine had more severe injury than such cells in control mice; RT-PCR revealed that the gene expression of inflammatory markers (e.g., iNOS, COX-2, and NF-κB) was heightened, and IκBα expression was reduced in mice administered imipramine.

Our findings suggest that an HFD can be used to create a mouse model of both hyperglycemia and obesity, and that prolonged imipramine intake in mice causes weight gain; increased food intake; increased kidney, liver, and fat pad weight; heightened serum and hepatic triglyceride, ALT, and AST levels; and increased fatty liver scores and adipocyte sizes. These findings could, indeed, be corroborated by lipid accumulation and lipogenesis activation for the regulation of nutrient metabolism. Furthermore, administering mice imipramine worsened their hyperglycemia, expedited glucose intolerance with glomerulonephritis, and caused retinopathy. The reduced expression of Akt and GLUT4 in skeletal muscles was possibly a factor in the indicated reduction in glucose metabolism. Imipramine altered the distribution of chromium in tissues and organs and increased the urinary loss of chromium. These findings point to the attenuation of glucose homeostasis and IS and the enhancement of IR. Moreover, the HFD slowed glucose homeostasis in these mice, but imipramine expedited the manifestation of diabetes symptoms. Thus, imipramine could have side effects related to kidney damage, FLD, and diabetes (Figure S6). Chronic imipramine intake could exacerbate conditions such as obesity, diabetes, CKD, and DR.

5. Conclusions

The present study reveals new information indicating that constant intake of imipramine in mice as part of an HFD had marked effects in terms of obesity and hyperglycemia inducement. Parallel alterations in weight gain (including fat pad and body weight), food efficiency, serum triglyceride levels, serum AST and ALT levels, fatty liver scores, adipocyte size, and glucose intolerance were indicated. The current investigation provides evidence of the mechanisms behind the imipramine-induced exacerbation of hyperglycemia through decreasing GLUT4 expression. Nephropathy in obese mice administered imipramine may have been due to the exacerbation of hyperglycemia, and reductions in antioxidant enzymes may have worsened nephropathy. DR as well as thinner IPLs and INLs among mice administered imipramine was noted, and the mechanisms behind these manifestations were the increased expression levels of inflammatory COX-2, iNOS, and NF-κB. Blood glucose alterations, kidney and liver function, and eyesight changes must be carefully assessed following initial imipramine intake to ensure early recognition of these rare side effects, especially in patients and animals receiving antidepressants. Such recognition would allow for early identification and treatment to prevent the development of DR, CKD, obesity, and hyperglycemia.

Supplementary Materials: The following are available online at https://www.mdpi.com/article/10.3390/vetsci8090189/s1, Figure S1. Effect of imipramine (5 mg/kg every day) on body weight and weekly body weight gain in HFD-fed controls and HFD-fed mice treated with imipramine during the 56-day experimental period. No significant differences in body weight and weekly body weight gain were observed between the two groups. Figure S2. Effect of imipramine (10 mg/kg every day) on body weight in SD-fed controls and SD-fed mice treated with imipramine during the 56-day experimental period. No significant differences in body weight were observed between the two groups. Figure S3. Effects of imipramine on (a) A representative image showing blots of liver extracts. The expression level of (b) COX-2 was measured. Data are in the form of mean ± standard deviation (n = 8). *** $p < 0.001$. Figure S4. Alterations in (a) islet morphologies (hematoxylin and eosin as well as immunohistochemical stain; 200× magnification) and (b) β-cell percentage in control and treatment groups. * $p < 0.05$. Figure S5. Alterations in (a) IHC stains (magnification: 200×) and (b) renal levels of TNF-α in treatment and control groups. ** $p < 0.01$. Western blotting with the antibody COX-2 (Sigma-Aldrich) and immunoreactive signals was employed along with enhanced chemiluminescence reagents (Thermo Scientific); UVP ChemStudio (Analytik Jena) was then used for signal detection. Primary antibodies against insulin (Bioss, Woburn, MA, USA) were used for IHC staining for insulin in pancreatic islets. Primary antibodies against TNF-α (Merck, Billerica, MA, USA) were utilized for further IHC staining for TNF-α in kidneys. IHC using a TAlink mouse/rabbit polymer detection system procured from BioTnA (#TACH04D, Kaohsiung, Taiwan) was used for determining the expression of proteins. Figure S6. Mechanistic presentation of side effects experienced by HFD-induced obese mice treated with imipramine for 56 days.

Author Contributions: Conceptualization, G.-R.C., C.-H.C. and Y.-C.W.; Methodology, P.-H.H., C.-M.W., J.-W.L., W.-L.L., T.-C.L. and H.-J.L.; Validation, G.-R.C., C.-H.C. and Y.-C.W.; Formal Analysis, P.-H.H., C.-M.W., J.-W.L., W.-L.L., T.-C.L., H.-J.L., C.-H.C. and Y.-C.W.; Investigation, G.-R.C., C.-H.C. and Y.-C.W.; Resources, P.-H.H. and C.-H.C.; Data Curation, G.-R.C. and C.-H.C. Writing—original draft preparation, G.-R.C., P.-H.H., C.-M.W., J.-W.L., W.-L.L., T.-C.L. and H.-J.L.; Writing—Review and Editing, C.-H.C. and Y.-C.W.; Visualization, C.-H.C. and Y.-C.W.; Supervision, C.-H.C., P.-H.H., and Y.-C.W.; Project Administration and Funding Acquisition: P.-H.H. and C.-H.C. All authors have read and agreed to the published version of the manuscript.

Funding: This study was supported, in part, by Grant BRD109031 from the Chang Bing Show Chwan Memorial Hospital (Taiwan), Ministry of Science and Technology (Taiwan; MOST 110-2313-B-415-001-MY2), the Taichung Veterans General Hospital (Taiwan), and National Chung-Hsing University (Taiwan; TCVGH-NCHU-1097612).

Institutional Review Board Statement: The review of our experimental protocol was conducted by National Chiayi University's Institutional Animal Care and Use Committee, who approved it under the approval No. 109019.

Informed Consent Statement: Not applicable.

Data Availability Statement: The data presented in this study are available on request from the corresponding author.

Acknowledgments: We thank LiTzung Biotechnology, Kaohsiung, Taiwan, for providing pathological assistance for this study.

Conflicts of Interest: The authors declare no conflict of interest. The funders had no role in the design of the study; in the collection, analyses, or interpretation of data; in the writing of the manuscript, or in the decision to publish the results.

References

1. Rose, J.B. Tricyclic antidepressant toxicity. *Clin. Toxicol.* **1977**, *11*, 391–402. [CrossRef]
2. Griebel, G.; Blanchard, D.C.; Agnes, R.S.; Blanchard, R.J. Differential modulation of antipredator defensive behavior in Swiss-Webster mice following acute or chronic administration of imipramine and fluoxetine. *Psychopharmacology* **1995**, *120*, 57–66. [CrossRef] [PubMed]
3. Serchov, T.; Clement, H.W.; Schwarz, M.K.; Iasevoli, F.; Tosh, D.K.; Idzko, M.; van Calker, D. Increased signaling via adenosine A1 receptors, sleep deprivation, imipramine, and ketamine inhibit depressive-like behavior via induction of Homer1a. *Neuron* **2015**, *87*, 549–562. [CrossRef]
4. Ramirez, K.; Sheridan, J.F. Antidepressant imipramine diminishes stress-induced inflammation in the periphery and central nervous system and related anxiety-and depressive-like behaviors. *Brain Behav. Immun.* **2016**, *57*, 293–303. [CrossRef] [PubMed]
5. Prado, C.E.; Watt, S.; Crowe, S.F. A meta-analysis of the effects of antidepressants on cognitive functioning in depressed and non-depressed samples. *Neuropsychol. Rev.* **2018**, *28*, 32–72. [CrossRef]
6. Tuomisto, J.; Tukiainen, E.; Voutilainen, R.; Tuomainen, P. Inhibition of 5-hydroxytryptamine and noradrenaline uptake in platelets and synaptosomes incubated in plasma from human subjects treated with amitriptyline or nortriptyline: Utilization of the principle for a bioassay method. *Psychopharmacology* **1980**, *69*, 137–142. [CrossRef] [PubMed]
7. Stoll, L.; Seguin, S.; Gentile, L. Tricyclic antidepressants, but not the selective serotonin reuptake inhibitor fluoxetine, bind to the S1S2 domain of AMPA receptors. *Arch. Biochem. Biophys.* **2007**, *458*, 213–219. [CrossRef]
8. Gillman, P.K. Tricyclic antidepressant pharmacology and therapeutic drug interactions updated. *Br. J. Pharmacol.* **2007**, *151*, 737–748. [CrossRef] [PubMed]
9. Pliszka, S. Practice parameter for the assessment and treatment of children and adolescents with attention-deficit/hyperactivity disorder. *J. Am. Acad. Child Adolesc. Psychiatry* **2007**, *46*, 894–921. [CrossRef]
10. Ben-Tovim, D.I.; Walker, K.; Gilchrist, P.; Freeman, R.; Kalucy, R.; Esterman, A. Outcome in patients with eating disorders: A 5-year study. *Lancet* **2001**, *357*, 1254–1257. [CrossRef]
11. Lehman, A.F.; Lieberman, J.A.; Dixon, L.B.; McGlashan, T.H.; Miller, A.L.; Perkins, D.O.; Regier, D. Practice guideline for the treatment of partients with schizophrenia. *Am. J. Psychiatry* **2004**, *161*, 1–56.
12. Rothbaum, B.O.; Astin, M.C. Integration of pharmacotherapy and psychotherapy for bipolar disorder. *J. Clin. Psychiatry* **2000**, *61*, 68–75.
13. Mavissakalian, M.R.; Perel, J.M. The side effects burden of extended imipramine treatment of panic disorder. *J. Clin. Psychopharmacol.* **2002**, *20*, 547–555. [CrossRef] [PubMed]
14. Chang, G.R.; Chiu, Y.S.; Wu, Y.Y.; Chen, W.Y.; Liao, J.W.; Chao, T.H.; Mao, F.C. Rapamycin protects against high fat diet–induced obesity in C57BL/6J mice. *J. Pharmacol. Sci.* **2009**, *109*, 496–503. [CrossRef]

15. Chang, G.R.; Chen, W.K.; Hou, P.H.; Mao, F.C. Isoproterenol exacerbates hyperglycemia and modulates chromium distribution in mice fed with a high fat diet. *J. Trace Elem. Med. Biol.* **2017**, *44*, 315–332. [CrossRef] [PubMed]
16. Horrobin, D.F.; Bennett, C.N. Depression and bipolar disorder: Relationships to impaired fatty acid and phospholipid metabolism and to diabetes, cardiovascular disease, immunological abnormalities, cancer, ageing and osteoporosis possible candidate genes. *Prostaglandins Leukot. Essent. Fatty Acids* **1999**, *6*, 217–234. [CrossRef] [PubMed]
17. Gupta, B.; Awasthi, A.; Jaju, B.P. Effect of acute & chronic imipramine treatment on glucose homeostasis. *Indian J. Med. Res.* **1992**, *96*, 65–71. [PubMed]
18. Chadwick, W.; Wilson, G.; Van De VENTER, M.; Oelofsen, W.; Roux, S. Shifts in metabolic parameters surrounding glucose homoeostasis resulting from tricyclic antidepressant therapy: Implications of insulin resistance? *J. Pharm. Pharmacol.* **2007**, *59*, 95–103. [CrossRef] [PubMed]
19. Begriche, K.; Massart, J.; Robin, M.A.; Bonnet, F.; Fromenty, B. Mitochondrial adaptations and dysfunctions in nonalcoholic fatty liver disease. *Hepatology* **2013**, *58*, 1497–1507. [CrossRef] [PubMed]
20. Weng, Z.; Wang, K.; Li, H.; Shi, Q. A comprehensive study of the association between drug hepatotoxicity and daily dose, liver metabolism, and lipophilicity using 975 oral medications. *Oncotarget* **2015**, *6*, 17031. [CrossRef] [PubMed]
21. Hardy, T.; Oakley, F.; Anstee, Q.M.; Day, C.P. Nonalcoholic fatty liver disease: Pathogenesis and disease spectrum. *Annu. Rev. Pathol.* **2016**, *11*, 451–496. [CrossRef]
22. Chang, G.R.; Hou, P.H.; Yang, W.C.; Wang, C.M.; Fan, P.S.; Liao, H.J.; Chen, T.P. Doxepin exacerbates renal damage, glucose intolerance, nonalcoholic fatty liver disease and urinary chromium loss in obese mice. *Pharmaceuticals* **2021**, *14*, 267. [CrossRef] [PubMed]
23. Kovesdy, C.P.; Furth, S.L.; Zoccali, C.; World Kidney Day Steering Committee. Obesity and Kidney Disease: Hidden Consequences of the Epidemic. *Can. J. Kidney Health Dis.* **2017**, *104*, 1–14. [CrossRef]
24. Cukor, D.; Fruchter, Y.; Ver Halen, N.; Naidoo, S.; Patel, A.; Saggi, S.J. A preliminary investigation of depression and kidney functioning in patients with chronic kidney disease. *Nephron Clin. Pract.* **2012**, *122*, 139–145. [CrossRef] [PubMed]
25. Tsai, Y.C.; Chiu, Y.W.; Hung, C.C.; Hwang, S.J.; Tsai, J.C.; Wang, S.L.; Chen, H.C. Association of symptoms of depression with progression of CKD. *Am. J. Kidney Dis.* **2012**, *60*, 54–61. [CrossRef]
26. Chang, G.R.; Liu, H.Y.; Yang, W.C.; Wang, C.M.; Wu, C.F.; Lin, J.W.; Lin, W.L.; Wang, Y.C.; Lin, T.C.; Liao, H.J.; et al. Clozapine worsens glucose intolerance, nonalcoholic fatty liver disease, kidney damage and retinal injury and increases renal reactive oxygen species production and chromium loss in obese mice. *Int. J. Mol. Sci.* **2021**, *22*, 6680. [CrossRef]
27. Kowluru, R.A.; Chakrabarti, S.; Chen, S. Re-institution of good metabolic control in diabetic rats and activation of caspase-3 and nuclear transcriptional factor (NF-kB) in the retina. *Acta Diabetol.* **2004**, *41*, 194–199. [CrossRef] [PubMed]
28. Zhang, T.; Ouyang, H.; Mei, X.; Lu, B.; Yu, Z.; Chen, K.; Ji, L. Erianin alleviates diabetic retinopathy by reducing retinal inflammation initiated by microglial cells via inhibiting hyperglycemia-mediated ERK1/2–NF-κB signaling pathway. *FASEB J.* **2019**, *33*, 11776–11790. [CrossRef]
29. Zhuo, C.; Ji, F.; Xiao, B.; Lin, X.; Chen, C.; Jiang, D.; Wang, W. Antipsychotic agent-induced deterioration of the visual system in first-episode untreated patients with schizophrenia maybe self-limited: Findings from a secondary small sample follow-up study based on a pilot follow-up study. *Psychiatry Res.* **2020**, *286*, 112906. [CrossRef]
30. Okamura, T.; Fujioka, H.; Ayajiki, K. Effects of nipradilol on alpha-adrenoceptor function in ocular arteries. *Pharmacology* **2002**, *65*, 110–118. [CrossRef] [PubMed]
31. Delle Chiaie, R.; Pancheri, P.; Scapicchio, P. Efficacy and tolerability of oral and intramuscular S-adenosyl-L-methionine 1, 4-butanedisulfonate (SAMe) in the treatment of major depression: Comparison with imipramine in 2 multicenter studies. *Am. J. Clin. Nutr.* **2002**, *76*, 1172S–1176S. [CrossRef]
32. Kumar, S.; Maheshwari, K.K.; Singh, V. Central nervous system activity of acute administration of ethanol extract of Punica granatum L. seeds in mice. *Indian J. Exp. Biol.* **2008**, *46*, 811–816.
33. Wu, C.F.; Hou, P.H.; Mao, F.C.; Su, Y.C.; Wu, C.Y.; Yang, W.C.; Lin, C.S.; Tsai, H.P.; Liao, H.Y.; Chang, G.R. Mirtazapine reduces adipocyte hypertrophy and increases glucose transporter expression in obese mice. *Animals* **2020**, *10*, 1423. [CrossRef]
34. Wróbel, A.; Serefko, A.; Wlaź, P.; Poleszak, E. The depressogenic-like effect of acute and chronic treatment with dexamethasone and its influence on the activity of antidepressant drugs in the forced swim test in adult mice. *Prog. Neuro-Psychopharmacol. Biol. Psychiatry.* **2014**, *54*, 243–248. [CrossRef] [PubMed]
35. Williams, B.; Correnti, J.; Oranu, A.; Lin, A.; Scott, V.; Annoh, M.; Beck, J.; Furth, E.; Mitchell, V.; Senkal, C.E.; et al. A novel role for ceramide synthase 6 in mouse and human alcoholic steatosis. *FASEB J.* **2018**, *32*, 130–142. [CrossRef] [PubMed]
36. Harquin Simplice, F.; David, E.T.; Herve Herve, N.A. Enhancing spatial memory: Anxiolytic and antidepressant effects of Tapinanthus dodoneifolius (DC) Danser in mice. *Neurol. Res. Int.* **2014**, *2014*, 9. [CrossRef] [PubMed]
37. Maciel, I.S.; Silva, R.B.; Morrone, F.B.; Calixto, J.B.; Campos, M.M. Synergistic effects of celecoxib and bupropion in a model of chronic inflammation-related depression in mice. *PLoS ONE* **2013**, *8*, e77227. [CrossRef]
38. Hou, P.H.; Chang, G.R.; Chen, C.P.; Lin, Y.L.; Chao, I.S.; Shen, T.T.; Mao, F.C. Long-term administration of olanzapine induces adiposity and increases hepatic fatty acid desaturation protein in female C57BL/6J mice. *Iran. J. Basic Med. Sci.* **2018**, *21*, 495–501.
39. Tsai, H.P.; Hou, P.H.; Mao, F.C.; Chang, C.C.; Yang, W.C.; Wu, C.F.; Liao, H.J.; Lin, T.C.; Chou, L.S.; Hsiao, L.W.; et al. Risperidone exacerbates glucose intolerance, nonalcoholic fatty liver disease, and renal impairment in obese mice. *Int. J. Mol. Sci.* **2021**, *22*, 409. [CrossRef] [PubMed]

40. Chang, G.R.; Chiu, Y.S.; Wu, Y.Y.; Lin, Y.C.; Hou, P.H.; Mao, F.C. Rapamycin impairs HPD-induced beneficial effects on glucose homeostasis. *Br. J. Pharmacol.* **2015**, *172*, 3793–3804. [CrossRef]
41. Tsai, M.Y.; Yang, W.C.; Lin, C.F.; Wang, C.M.; Liu, H.Y.; Lin, C.S.; Lin, J.W.; Lin, W.L.; Lin, T.C.; Fan, P.S.; et al. The ameliorative effects of fucoidan in thioacetaide-induced liver injury in mice. *Molecules* **2021**, *26*, 1937. [CrossRef]
42. Chen, B.; He, T.; Xing, Y.; Cao, T. Effects of quercetin on the expression of MCP-1, MMP-9 and VEGF in rats with diabetic retinopathy. *Exp. Ther. Med.* **2017**, *14*, 6022–6026. [CrossRef] [PubMed]
43. Chang, G.R.; Hou, P.H.; Chen, W.K.; Lin, C.T.; Tsai, H.P.; Mao, F.C. Exercise affects blood glucose levels and tissue chromium distribution in high-fat diet-fed C57BL6 mice. *Molecules* **2020**, *25*, 1658. [CrossRef] [PubMed]
44. Chang, G.R.; Wu, Y.Y.; Chiu, Y.S.; Chen, W.Y.; Liao, J.W.; Hsu, H.M.; Chao, T.H.; Hung, S.W.; Mao, F.C. Long-term administration of rapamycin reduces adiposity, but impairs glucose tolerance in high-fat diet-fed KK/HlJ mice. *Basic Clin. Pharmacol. Toxicol.* **2009**, *105*, 188–198. [CrossRef] [PubMed]
45. Chen, X.; Margolis, K.J.; Gershon, M.D.; Schwartz, G.J.; Sze, J.Y. Reduced serotonin reuptake transporter (SERT) function causes insulin resistance and hepatic steatosis independent of food intake. *PLoS ONE* **2012**, *7*, e32511. [CrossRef]
46. Ipsen, D.H.; Lykkesfeldt, J.; Tveden-Nyborg, P. Molecular mechanisms of hepatic lipid accumulation in non-alcoholic fatty liver disease. *Cell Mol. Life Sci.* **2018**, *75*, 3313–3327. [CrossRef] [PubMed]
47. Abd El-Twab, S.M.; Hozayen, W.G.; Hussein, O.E.; Mahmoud, A.M. 18β-Glycyrrhetinic acid protects against methotrexateinduced kidney injury by up-regulating the Nrf2/ARE/HO-1 pathway and endogenous antioxidants. *Ren. Fail.* **2016**, *38*, 1516–1527. [CrossRef] [PubMed]
48. Joussen, A.M.; Poulaki, V.; Le, M.L.; Koizumi, K.; Esser, C.; Janicki, H.; Adamis, A.P. A central role for inflammation in the pathogenesis of diabetic retinopathy. *FASEB J.* **2004**, *18*, 1450–1452. [CrossRef]
49. Kozak, L.P.; Anunciado-Koza, R. UCP1: Its involvement and utility in obesity. *Int. J. Obes.* **2008**, *32*, S32–S38. [CrossRef]
50. Khoza, S.; Barner, J.C. Glucose dysregulation associated with antidepressant agents: An analysis of 17 published case reports. *Int. J. Clin. Pharm.* **2011**, *33*, 484–492. [CrossRef]
51. Unger, R.H. Minireview: Weapons of lean body mass destruction: The role of ectopic lipids in the metabolic syndrome. *Endocrinology* **2003**, *144*, 5159–5165. [CrossRef] [PubMed]
52. Lescot, T.; Karvellas, C.; Beaussier, M.; Magder, S.; Riou, B. Acquired liver injury in the intensive care unit. *Anesthesiology* **2012**, *117*, 898–904. [CrossRef]
53. Yang, J.W.; Kim, H.S.; Im, J.H.; Kim, J.W.; Jun, D.W.; Lim, S.C.; Lee, K.; Choi, J.M.; Kim, S.K.; Kang, K.W. GPR119: A promising target for nonalcoholic fatty liver disease. *FASEB J.* **2016**, *30*, 324–335. [CrossRef] [PubMed]
54. Wu, Z.; Zhang, Y.; Gong, X.; Cheng, G.; Pu, S.; Cai, S. The preventive effect of phenolic-rich extracts from Chinese sumac fruits against nonalcoholic fatty liver disease in rats induced by a high-fat diet. *Food Funct.* **2020**, *11*, 799–812. [CrossRef] [PubMed]
55. Voican, C.S.; Corruble, E.; Naveau, S.; Perlemuter, G. Antidepressant-induced liver injury: A review for clinicians. *Am. J. Psychiatry* **2014**, *171*, 404–415. [CrossRef]
56. Lamon-Fava, S.; Wilson, P.W.; Schaefer, E.J. Impact of body mass index on coronary heart disease risk factors in men and women: The Framingham Offspring Study. *Arterioscler. Thromb. Vasc. Biol.* **1996**, *16*, 1509–1515. [CrossRef]
57. Kawano, Y.; Cohen, D.E. Mechanisms of hepatic triglyceride accumulation in non-alcoholic fatty liver disease. *J. Gastroenterol.* **2013**, *48*, 434–441. [CrossRef]
58. Emanuelli, B.; Vienberg, S.G.; Smyth, G.; Cheng, C.; Stanford, K.I.; Arumugam, M.; Kahn, C.R. Interplay between FGF21 and insulin action in the liver regulates metabolism. *J. Clin. Investig.* **2014**, *124*, 515–527. [CrossRef]
59. Ritchie, M.; Hanouneh, I.A.; Noureddin, M.; Rolph, T.; Alkhouri, N. Fibroblast growth factor (FGF)-21 based therapies: A magic bullet for nonalcoholic fatty liver disease (NAFLD)? *Expert Opin. Investig. Drugs* **2020**, *29*, 197–204. [CrossRef]
60. Asrih, M.; Jornayvaz, F.R. Inflammation as a potential link between nonalcoholic fatty liver disease and insulin resistance. *J. Endocrinol.* **2013**, *218*, R25–R36. [CrossRef] [PubMed]
61. Mobbs, C.V.; Makimura, H. Block the FAS, lose the fat. *Nat. Med.* **2002**, *8*, 335–336. [CrossRef] [PubMed]
62. Dorn, C.; Riener, M.O.; Kirovski, G.; Saugspier, M.; Steib, K.; Weiss, T.S.; Hellerbrand, C. Expression of fatty acid synthase in nonalcoholic fatty liver disease. *Int. J. Clin. Exp. Pathol.* **2010**, *3*, 505.
63. Yuan, S.; Liu, H.; Yuan, D.; Xu, J.; Chen, Y.; Xu, X.; Liang, H. PNPLA3 I148M mediates the regulatory effect of NF-kB on inflammation in PA-treated HepG2 cells. *J. Cell Mol. Med.* **2020**, *24*, 1541–1552. [CrossRef] [PubMed]
64. Trujillo, M.E.; Scherer, P.E. Adiponectin–journey from an adipocyte secretory protein to biomarker of the metabolic syndrome. *J. Intern. Med.* **2005**, *257*, 167–175. [CrossRef] [PubMed]
65. Gupta, B.; Shakarwal, M.K.; Kumar, A.; Jaju, B.P. Modulation of glucose homeostasis by doxepin. *Methods Find Exp. Clin. Pharmacol.* **1992**, *14*, 61–71.
66. Erenmemisoglu, A.; Ozdogan, U.K.; Saraymen, R.; Tutus, A. Effect of some antidepressants on glycaemia and insulin levels of normoglycaemic and alloxan-induced hyperglycaemic mice. *J. Pharm. Pharmacol.* **1999**, *51*, 741–743. [CrossRef]
67. Deuschle, M. Effects of antidepressants on glucose metabolism and diabetes mellitus type 2 in adults. *Curr. Opin. Psychiatry.* **2013**, *26*, 60–65. [CrossRef]
68. Joost, H.G.; Poser, W.; Panten, U. Inhibition of insulin release from the rat pancreas by cyproheptadine and tricyclic antidepressants. *Naunyn Schmiedebergs Arch. Pharmacol.* **1974**, *285*, 99–102. [CrossRef]

69. Ardizzone, T.D.; Bradley, R.J.; Freeman, A.M.; Dwyer, D.S. Inhibition of glucose transport in PC12 cells by the atypical antipsychotic drugs risperidone and clozapine, and structural analogs of clozapine. *Brain Res.* **2001**, *923*, 82–90. [CrossRef]
70. Chang, G.R.; Hou, P.H.; Wang, C.M.; Wu, C.F.; Su, H.K.; Liao, H.J.; Chen, T.P. Chronic everolimus treatment of high fat diet mice leads to a reduction in obesity but impaired glucose tolerance. *Pharmacol. Res. Perspect.* **2021**, *9*, e00732. [CrossRef]
71. Chen, G.; Liu, P.; Pattar, G.R.; Tackett, L.; Bhonagiri, P.; Strawbridge, A.B.; Elmendorf, J.S. Chromium activates glucose transporter 4 trafficking and enhances insulin-stimulated glucose transport in 3T3-L1 adipocytes via a cholesterol-dependent mechanism. *Mol. Endocrinol.* **2006**, *20*, 857–870. [CrossRef]
72. Rajendran, K.; Manikandan, S.; Nair, L.; Karuthodiyil, R.; Vijayarajan, N.; Gnanasekar, R.; Kapil, V.V.; Mohamed, A.S. Serum chromium levels in type 2 diabetic patients and its association with glycaemic control. *J. Clin. Diagn. Res.* **2015**, *9*, OC05–OC08. [CrossRef]
73. Seki, M.; Nakayama, M.; Sakoh, T.; Yoshitomi, R.; Fukui, A.; Katafuchi, E.; Kitazono, T. Blood urea nitrogen is independently associated with renal outcomes in Japanese patients with stage 3–5 chronic kidney disease: A prospective observational study. *BMC Nephrol.* **2019**, *20*, 115. [CrossRef] [PubMed]
74. Maser, R.L.; Vassmer, D.; Magenheimer, B.S.; Calvet, J.P. Oxidant stress and reduced antioxidant enzyme protection in polycystic kidney disease. *J. Am. Soc. Nephrol.* **2002**, *13*, 991–999. [CrossRef] [PubMed]
75. Cheung, N.; Mitchell, P.; Wong, T.Y. Diabetic retinopathy. *Lancet* **2010**, *376*, 124–136. [CrossRef]
76. Sánchez-Thorin, J.C. The epidemiology of diabetes mellitus and diabetic retinopathy. *Int. Ophthalmol. Clin.* **1998**, *38*, 11–18. [CrossRef]
77. Mysona, B.A.; Shanab, A.Y.; Elshaer, S.L.; El-Remessy, A.B. Nerve growth factor in diabetic retinopathy: Beyond neurons. *Expert Rev. Ophthalmol.* **2014**, *9*, 99–107. [CrossRef]
78. Ali, S.A.; Zaitone, S.A.; Dessouki, A.A.; Ali, A.A. Pregabalin affords retinal neuroprotection in diabetic rats: Suppression of retinal glutamate, microglia cell expression and apoptotic cell death. *Exp. Eye Res.* **2019**, *184*, 78–90. [CrossRef]
79. Tang, J.; Kern, T.S. Inflammation in diabetic retinopathy. *Retin. Eye Res.* **2011**, *30*, 343–358. [CrossRef]
80. Zheng, L.; Kern, T.S. Role of nitric oxide, superoxide, peroxynitrite and PARP in diabetic retinopathy. *Front. Biosci.* **2009**, *14*, 3974–3987. [CrossRef]
81. Zheng, L.; Szabó, C.; Kern, T.S. Poly (ADP-ribose) polymerase is involved in the development of diabetic retinopathy via regulation of nuclear factor-κB. *Diabetes* **2004**, *53*, 2960–2967. [CrossRef] [PubMed]
82. Kowluru, R.A.; Kowluru, V.; Xiong, Y.; Ho, Y.S. Overexpression of mitochondrial superoxide dismutase in mice protects the retina from diabetes-induced oxidative stress. *Free Radic. Biol. Med.* **2006**, *41*, 1191–1196. [CrossRef] [PubMed]

veterinary sciences

MDPI

Article

Establishment of a Newborn Lamb Gut-Loop Model to Evaluate New Methods of Enteric Disease Control and Reduce Experimental Animal Use

Ambre Baillou [1,2,†], Nathalie Kasal-Hoc [3,†], Céline Barc [3], Juliette Cognié [4], Anne Pinard [3], Jérémy Pezant [3], Julie Schulthess [2], Pauline Peltier-Pain [2], Sonia Lacroix-Lamandé [1,*,‡] and Fabrice Laurent [1,*,‡]

1 UMR1282 Infectiologie et Santé Publique, INRAE Centre Val de Loire, Université François Rabelais de Tours, 37380 Nouzilly, France; ambre.baillou@inrae.fr
2 Phileo by Lesaffre, 137 rue Gabriel Péri, 59700 Marcq-en-Barœul, France; j.schulthess@phileo.lesaffre.com (J.S.); p.peltierpain@phileo.lesaffre.com (P.P.-P.)
3 UE1277 Plateforme d'Infectiologie Expérimentale (PFIE), INRAE Centre Val de Loire, 37380 Nouzilly, France; nathalie.kasal-hoc@inrae.fr (N.K.-H.); celine.barc@inrae.fr (C.B.); anne.pinard@inrae.fr (A.P.); jeremy.pezant@inrae.fr (J.P.)
4 UMR85 Physiologie de la reproduction et des Comportements, INRAE Centre Val de Loire, CNRS, IFCE, Université François Rabelais de Tours, 37380 Nouzilly, France; juliette.cognie@inrae.fr
* Correspondence: sonia.lamande@inrae.fr (S.L.-L.); fabrice.laurent@inrae.fr (F.L.)
† Shared co-first authorship.
‡ Shared senior contribution.

Citation: Baillou, A.; Kasal-Hoc, N.; Barc, C.; Cognié, J.; Pinard, A.; Pezant, J.; Schulthess, J.; Peltier-Pain, P.; Lacroix-Lamandé, S.; Laurent, F. Establishment of a Newborn Lamb Gut-Loop Model to Evaluate New Methods of Enteric Disease Control and Reduce Experimental Animal Use. *Vet. Sci.* **2021**, *8*, 170. https://doi.org/10.3390/vetsci8090170

Academic Editors: Ana Faustino and Paula A. Oliveira

Received: 30 July 2021
Accepted: 20 August 2021
Published: 24 August 2021

Abstract: Enteric infectious diseases are not all well controlled, which leads to animal suffering and sometimes death in the most severe cases, in addition to economic losses for farmers. Typical symptoms of enteric infections include watery diarrhea, stomach cramps or pain, dehydration, nausea, vomiting, fever and weight loss. Evaluation of new control methods against enteric infections requires the use of many animals. We aimed to develop a new method for an initial in vivo screen of promising compounds against neonatal diseases such as cryptosporidiosis while limiting experimental animal use. We therefore adapted an in vivo method of multiple consecutive but independent intestinal loops to newborn lambs delivered by cesarean section, in which endotoxin responsiveness is retained. This new method allowed for the screening of natural yeast fractions for their ability to stimulate immune responses and to limit early *Cryptosporidium parvum* development. This model may also be used to investigate host–pathogen interactions and immune responses in a neonatal controlled environment.

Keywords: newborn lamb; multiple intestinal loop model; 3R; *Cryptosporidium parvum*; yeast cell wall fractions

1. Introduction

Infectious diseases are mainly controlled by the use of vaccines and antimicrobials. Despite major discoveries, some infectious diseases remained refractory to vaccination and treatments and therefore require specific attention together with the new infectious zoonotic threats that increasingly emerge these last decades. The evaluation of new antimicrobials is usually first performed by bioinformatics and/or by in vitro approaches sometimes resulting from High-Throughput Screening (HTS) on miniaturized 2D cell-based assays, evaluating 100,000 or more samples per day [1,2]. Many of these candidates failed when entering clinical trials and have required the unnecessary use of numerous experimental animals. Ex vivo cultures of multipotent or pluripotent stem cells in a three-dimensional (3D) matrix represent major improvements and contribute to reducing animal use [3,4]. However, despite their obvious advantages, these systems cannot yet fully reproduce all the complex characteristics and interactions encountered in vivo, such as the (I) multiplicity and spatial organization of cell types, (II) the immune cell recruitments, (III) the ability to experiment in presence of a full cultivable and non-cultivable microbiota or (IV) to reproduce the peculiarity of a specific age of development e.g., neonatal period, etc. In addition, these models are usually unsuitable for assessing the mechanism of physiopathology resulting from infectious processes. Until technical progress can resolve these limitations, it is therefore important to develop or improve in vivo methods to minimize the use of animals and take full advantage of each experiment in accordance with the 3R rules (Replacement, Reduction and Refinement) [5].

Neonatal enteric infections result essentially from the infection of bacterial, parasitic, or viral pathogens and they have profound effects on intestinal absorption, nutrition, and youth development as well as on global mortality. Cryptosporidiosis due to the zoonotic protozoan parasite *Cryptosporidium parvum* (*C. parvum*) is characterized by infection of the epithelial cells of the small intestine, mainly in the ileum, leading to acute diarrhea and dehydration that may lead to death in severe cases [6]. This disease represents a true one-health threat with severe consequences for human and animal health [7,8]. Extensive new epidemiology studies revealed cryptosporidiosis to be the second leading cause of death in children due to diarrheal disease worldwide [9], and is the first cause of diarrheal enteric disease in young ruminants in France [10]. There is no vaccine and a very limited chemotherapy available for animals and humans [11].

We aim to evaluate natural alternatives for controlling cryptosporidiosis using lambs as a target species and also as a model for larger animals such as calves. Indeed, lambs represent a cost-effective model and harbor similar development of gut lymphoid tissues at birth to other young ruminants. We intent to stimulate the immune responses of animals from birth with colostrum supplemented with natural products such as yeast cell wall (YCW) fractions that contain TLR2, TLR4 and Dectin-1 receptor ligands. However, intestinal immune tolerance initiates rapidly after birth in response to microbial colonization gained during vaginal delivery and subsequently via colostrum, milk and multiple contacts with the environment. This immune tolerance is characterized by a rapid hypo-responsiveness to microbial antigens as demonstrated in mouse models [12,13].

We therefore developed an in vivo model suitable to investigate both host–pathogen interactions in a controlled environment and to evaluate new natural antiparasitic compounds. Our model relies on two previously described gut-loop models: one performed with fetal lambs (120 to 130 days of gestation) but with just a single loop [14] and another one made with 4–6 month-old lambs [15]. The surgical procedure was therefore successfully adapted right after birth to cesarean-born lambs and allowed to produce an isolated intestinal segment in the ileum area free of microbiota and immune system stimulation. Multiple loops were created in the isolated intestinal segment which allowed for the evaluation endotoxin responsiveness, immune responses, parasite replication in presence or not of YCW fractions. This new model combines different specificities which were required for our investigations such as (1) a newborn model for neonatal enteric disease study, (2) sterility for evaluating the immunomodulatory properties of natural compounds provided in the first colostrum and (3) a large number of intestinal loops per animal to perform multiple comparisons of selected compounds while severely limiting the use of experimental animals.

2. Materials and Methods

2.1. Development of a New Animal Model

2.1.1. Ethic Statements

Animal needs were met in accordance with the European Community Council Directive 2010/63/EU (Decree: 2013-118 01/02/2013). The experimental facilities had received authorization to house experimental animals from the local bureau of veterinary services (Indre-et-Loire, France, authorization N°: D 37-175-3), and all the experimental procedures were approved by the Val de Loire Ethics Committee (authorization N° APAFIS#16870-201809261558973 v2). All animal experimentations have been performed in the Infectiology of Farm, Model and Wildlife Animals Facility (PFIE, Centre INRAE Val de Loire available online: https://doi.org/10.15454/1.5572352821559333E12 (accessed on 23 August 2021); member of the National Infrastructure EMERG'IN). All the personnel involved had special training in animal care, handling and experimentation, as required by the French Ministry of Agriculture.

2.1.2. Neonatal Care of Newborn Lambs after Cesarean Surgery

The pregnant Préalpes-du-Sud ewe selected to give birth to two lambs was subjected to classical cesarean surgery. After stimulation of the respiratory function, right after birth, they were positioned in lateral decubitus and received intranasally one to two drops of doxapram (Dopram®, Vetoquinol, Lure, France) to stimulate the respiratory rate. The umbilical cord was disinfected with Vetedine® (Vetoquinol, Lure, France) and newborns were then placed together in close contact in the housing box, in a dry and warm environment, under a heating lamp. After 30 min, one of the two lambs was operated.

2.1.3. Intestinal Loop Surgery

Preparation: The lamb was placed on a heated neonatal resuscitation table and received isoflurane via a face mask for general anesthesia and buprenorphine (Buprecare® Multidose 0.01 mg/kg, Axience, Pantin, France) for analgesia by intramuscular injection. The lamb was then intubated with an endotracheal tube (internal diameter 4 to 5 mm, RÜSCH®, Teleflex, Wayne, NJ, USA) and put on assisted ventilation (CPV mode: controlled pressure ventilation, Fabius® Tiro, Dräger, Lübeck, Germany), with a 50/50 oxygen and air mixture. The monitoring system installed (BLT, M8500) allowed to continuously monitor rectal temperature, heart and respiratory rates, oxygen saturation (SpO_2) and End Tidal CO_2 ($ETCO_2$). An intravenous catheter was placed in the cephalic vein and a warm infusion of sterile isotonic Ringer-Lactate saline solution (Osalia, Paris, France) was started at a rate of around 10 mL/kg/h via a flow regulator.

Surgical intervention: The surgical area was disinfected and a subcutaneous injection of 1 mL of lidocaine was performed at the level of the midline. After a few minutes, a 10-cm skin incision was made in the midline, caudal to the umbilicus. After dissection of the subcutaneous tissue and opening of the abdominal cavity, the urachus duct was reclined. Sterile compresses moistened with prewarmed (+37 °C) sterile saline were placed all around the surgical wound. The intestinal tract was exteriorized on the compresses, the area of interest (from the ileocecal fold) was identified and the rest was reintroduced into the abdomen. The externalized portion was regularly humidified with sterile warm saline throughout the procedure.

Creation of loops and inter-loops: A silk ligature (dec. 3, SILK®, SMI, St Vith, Belgium) was made approximately at 5 cm from the ileocecal fold (retrograde direction). This 5 cm area is the site of the distal enterotomy. Then, moving up the small intestine, ligatures were placed at approximately 2–3 cm intervals to generate three consecutive loops, and an interloop region which separates each set of triplicates. This can be repeated as many times as necessary before the continuous ileal Peyer's patch disappears (Figure 1b). Long ends were performed to allow the identification of each triplicate of loops plus the interloop region to be easily clamped for reliable tracking during the loop injection stage. The lumen of the gastrointestinal tract was not rinsed before the ligatures were made, the intestinal contents remained therefore intact.

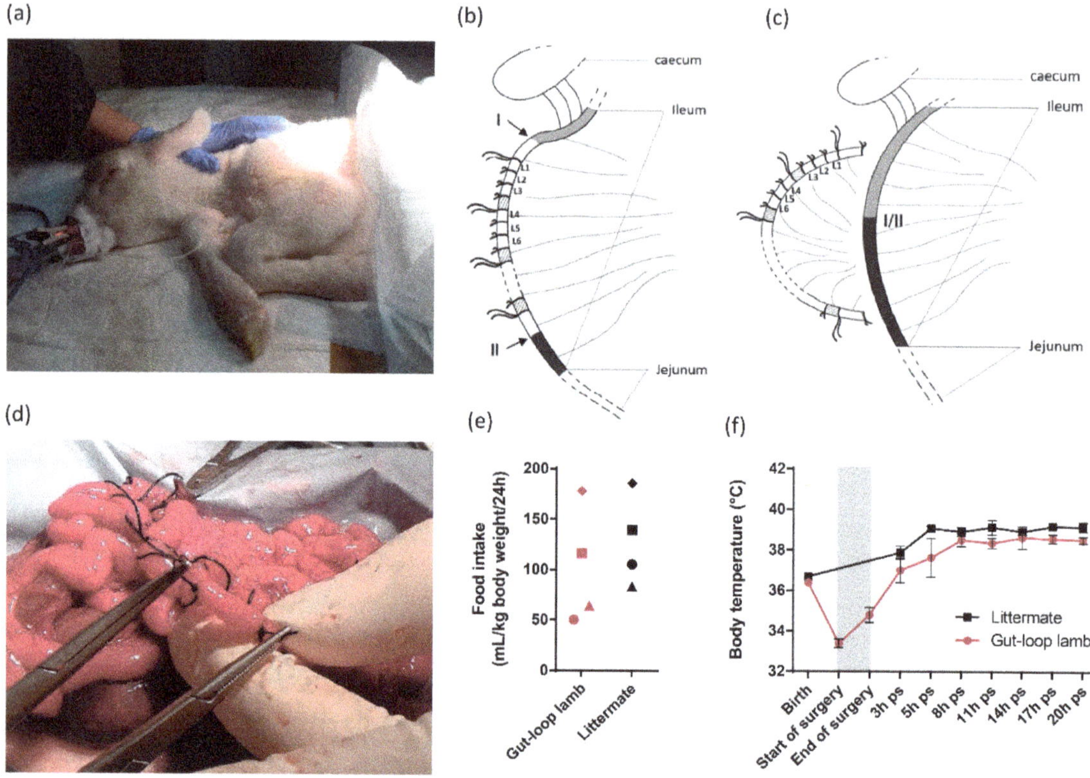

Figure 1. Intestinal loops surgery on the ileum of cesarean-section born lamb. (**a**) Picture of lamb that was anesthetized before surgery, intubated with endotracheal tube and put on assisted ventilation; (**b**) Schematic representation of the intestinal loop generation in the ileum of caesarean-born lamb. Series of triplicates of loops (L1, L2, L3) were generated between the two areas (I and II) indicated by arrows. Interloop region separates series of triplicates and are indicated in shaded areas. The first ligature of the triplicates harbors longer ends, so do the one of each triplicate to facilitate the identification; (**c**) Schematic representation of end-to-end anastomosis is visualized (I/II) allowing intestinal transit restauration and the separation of the intestinal segment containing the isolated loops; (**d**) Picture of a loop creation within intestinal segment during surgery; (**e**) Food intake represented by the volume of milk drunk per ml and per kg of lamb body weight per 24 h for the gut-loop lamb and its littermate. Each point shape corresponds to an independent experimentation (n = 4 for each group); (**f**) Body temperature (°C) (mean ± SEM) was monitored from birth to 20-h post-surgery (ps) for the gut-loop lamb and its littermate. The grey box corresponds to the surgery intervention (from start to end).

Double enterotomy, end-to-end anastomosis: A double enterotomy was performed on each extremity of the area containing loops (Figure 1c). The four ends created were immediately wiped with a sterile compress moistened with warm saline. To avoid leakage or adhesion, each end of the isolated intestinal segment was sutured with a single suture (dec. 1.5, MONOCRYL®, Johnson & Johnson, New Brunswick, NJ, USA) and a jejuno–ileal anastomosis was performed to allow normal intestinal transit before gently placing the entire externalized digestive tract back into the abdominal cavity. It should be noted that the vascularization of the intestinal loops after the separation of the ileal segment was maintained. The abdominal cavity was rinsed several times with sterile warm saline. The abdominal wall was stitched with a single suture (dec. 2, SAFIL®, B. BRAUN, Melsungen, Germany) followed by subcutaneous and cutaneous sutures and finally, the wound was disinfected. Isoflurane supply was cut off, the animal was put on assisted ventilation with a 100% oxygen mixture and then positioned in a lateral recumbence to accelerate awakening. The animal was regularly "switched" to spontaneous ventilation (Man/Spont) and received an intravenous injection of doxapram (0.87 mg/kg, Dopram-V Injectable®, Vetoquinol, Lure, France), repeated once if necessary, to help spontaneous breathing. The animal was then extubated, the catheter was removed and the animal was placed in the housing box with heat lamp, in direct contact with the control lamb from the litter.

Post-operative care: The lambs were fed as the first signs of awakening were observed (Colostromix®, Technovet Eurotonic, Briec, France). The volume and number of feedings were adapted upon lamb request. Analgesia was continued with four intramuscular injections per day of buprenorphine (0.01 mg/kg). The criteria for assessing the animal's clinical condition and possible pain were primarily behavioral: vocalizations (which do not occur after drinking), lordosis, prostration or apathy, absence or too frequent stands up, disinterest in the surroundings, low head carriage or low ears, etc. At every visit to feed them (3 h interval), the surgical wound, rectal temperature, quantity of ingested milk, stool and urine output were controlled. No additional enrichment was provided other than the presence of a littermate from the same litter and all physical contact with the animal technicians, particularly during feedings, which is particularly important at this physiological stage.

Sample recovery: After 24 h, the animal was injected intramuscularly with xylazine (0.5 mg/kg, ROMPUN®, Bayer, Leverkusen, Germany) and euthanized intravenously with pentobarbital (180 mg/kg, DOLETHAL®, Vetoquinol, Lure, France). The animal was then bled and the small and large intestine was harvested in a whole (duodenum to colon), then samples were taken from each intestinal loop. The unique 24 h-time point was selected to limit animal consumption based on the compromise between assessing the innate immune response to immunostimulants and allowing sufficient time to measure the effect on parasite invasion and development.

2.2. Use of the Gut-Loop Model for Investigating Immune Responses and Infection with C. parvum

2.2.1. Intestinal Stimulations

Intestinal responses to immunostimulants were evaluated with isolated loops in lambs and ex vivo with explants. Just after the end-to-end anastomosis, 150 µL of stimulants were injected into loops with an 8-mm 30G needle at the following concentrations: bacterial lipopolysaccharide (LPS) (10 µg/mL, LPS-EB, Invivogen, San Diego, CA, USA), R848 (10 µg/mL, Resiquimod, Invivogen, San Diego, CA, USA) and yeast cell wall fraction 1 (YCW1) or 2 (YCW2) (5 mg/mL, Phileo by Lesaffre, Marcq-en-Barœul, France). The two YCW were obtained from a *Saccharomyces cerevisiae* (*S. cerevisiae*) strain and differed in their polysaccharide composition. YWC1 is composed of an equilibrated content between β-glucans and mannoproteins (at minima 20% of each sugar compound), whereas YCW2 is enriched in β-glucans (50%). Twenty-four hours after surgery, animals were euthanized and intestinal tissues were collected either from the loops or from the ileum reconnected to the intestinal transit, and explants were generated with biopsy punches of 8 mm diameter. Tissue samples were incubated in culture medium (RPMI; 10% FBS; 100 U/mL Penicillin; 100 µg/mL Streptomycin; 100 µg/mL Normocin; 250 ng/mL Amphotericin B) alone or with LPS at a concentration of 10 µg/mL, for 4 h at +37 °C and 5% CO_2. Three explants per condition were used, referred as triplicates.

2.2.2. RNA Extraction and RT-qPCR

RNA extractions from explants or pieces of loop were performed after homogenization in TRIzol (Invitrogen, Waltham, MA, USA) with an Ultra-turrax and processed according to manufacturer's recommendations. Total RNA was reverse transcribed with the iScript RT SuperMix kit (Biorad, Hercules, CA, USA) and amplified by quantitative real-time PCR (RT-qPCR) in presence of EVA® Green using CFX96 Touch Real-Time PCR Detection System (Biorad, Hercules, CA, USA). Results were expressed as $2e^{-\Delta Ct}$, following normalization with three stable reference genes (*hprt*, *gapdh* and *actb*). mRNA was quantified for *il1α*, *il1β*, *mx1*, *cxcl1*, *cxcl2*, *cxcl8* and *tnfα* genes known to be upregulated by YCW extracts and/or TLR-agonists used in this study, LPS and R848 (see Table S1 for primer sequences).

2.2.3. *C. parvum* Infection

Intestinal loops were injected with 150 μL of physiological serum (0.9% NaCl) or with 150 μL of parasite solution containing 1.5×10^4 or 1.5×10^5 oocysts of *C. parvum* in presence or not of yeast cell wall fractions. The *nluc*-INRAE transgenic strain of *C. parvum* [16], which expresses the nanoluciferase enzyme was used in the study to quantify parasite load. All intestinal samples were collected after 24 h of infection with *C. parvum*.

2.2.4. Parasite Burden

The level of infection in the intestinal loops was determined by two complementary methods: The assessment of the luciferase activity of the *C. parvum nluc*-INRAE transgenic strain and the quantification of the *C. parvum 18S* (*Cp18S*) gene expression by RT-qPCR. For relative light unit (RLU) quantification, 500 μL of lysis buffer (50 mM Tris HCl; 2 mM DTT; 2 mM EDTA; 10% glycerol; 1% Triton) were added to biopsy punch, followed by a 30 min incubation at +4 °C. Samples were then vortexed during 1 min and centrifuged at 12,000× *g* for 1 min. Twenty-five μL of supernatant were mixed with 25 μL of Nano-Glo® substrate (1/50, Promega, Madison, WI, USA) and luminescence activity measured with a GloMax® plate reader (Promega, Madison, WI, USA). The values were expressed in RLU per cm^2 of intestinal tissue. *Cp18S* gene mRNA was quantified using the forward 5′-TAGAGATTGGAGGTTGTTCCT-3′ and reverse 5′-CTCCACCAACTAAGAACGGCC-3′ primers and results are expressed as $2e^{-\Delta Ct}$, after normalization with the expression of three reference genes (*hprt*, *gapdh* and *actb*).

2.2.5. Histology and Immunofluorescence

Following dissection and PBS wash, intestinal samples were immediately incubated in a 4% paraformaldehyde PBS solution during 24 h at +4 °C for tissue fixation. After 2 washes of 4 h at +4 °C in PBS, samples were incubated in a 30% sucrose PBS solution before inclusion in OCT (Cellpath, Newtown, UK) and conserved at −20 °C until histological analyses. Seven μm-thick histological sections were realized with the 3050S Leica cryotome (Leica, Wetzlar, Germany).

Intestinal sections were incubated with a rat polyclonal serum against oocyst antigens (as previously used [17]), overnight in the dark at +4 °C and then washed twice with PBS. The secondary antibody coupled with Alexa 594 (Invitrogen, Waltham, MA, USA) was added for 1 h in the dark and washed twice with PBS. Finally, cell nuclei were stained with Hoechst (Invitrogen, Waltham, USA) during 2 min and then washed twice with PBS. Immunofluorescent staining of *C. parvum* were analyzed by microscopy with the Eclipse 80i Nikon Microscope (Nikon, Tokyo, Japan).

2.2.6. Statistics

Statistical analyses were performed using the GraphPad Prism v6 software (GraphPad Software, San Diego, CA, USA). The Kruskal–Wallis non-parametric test and the Dunn's multiple comparison test were used to determine the significance of difference for parasite load determination by luminescence measure and RT-qPCR between the different conditions. *p*-values of less than 0.05 were considered statistically significant. The linear regression analysis was used to establish the correlation curve between parasite load evaluated by luminescence measure and *Cp18S* gene expression.

3. Results

3.1. Establishment of Gut-Loop Model on a Cesarean-Born Neonatal Lamb

In order to be able to compare the immunomodulatory properties of natural compounds in the intestinal response of newborn lambs without prior interference with natural microbial stimulation, we adapted the gut-loop model to cesarean-born lambs (Figure 1a).

All physiological parameters were constantly recorded during the surgery (SpO_2, $ETCO_2$, respiratory and heart rates, body temperature) and no loss was registered. The ligatures were made in the ileum of lambs since it is the main infected area in the case of natural infection by *C. parvum*. We performed set of triplicates for each treatment to be used, separated by an interloop region to avoid diffusion of the treatment to another adjacent treatment to be administered (Figure 1b). Leakage between loops were never observed confirming the effective sealing of each loop. At the end of the loop's generation, an end-to-end anastomosis was performed to reconnect the normal transit that will occur when lambs will first feed, and to allow the separation of the intestinal segment containing the isolated loops (Figure 1c). With the practice and experience gained over the first surgeries, the veterinary surgeon can now make around 25–30 intestinal loops (Figure 1d). The gut-loop lamb animals used to evaluate efficacy of immunostimulatory compounds were maintained with their littermates for 24 h before euthanasia and none presented abnormal behavior. Food intake was slightly lower in gut-loop lambs, but all animals were regularly fed on several occasions during the first 24 h (Figure 1e). In gut-loop animals, the body temperature drops before the beginning of the surgery and then regularly increases to reach similar temperature as littermate (Figure 1f), thanks to the heated neonatal resuscitation table, the heat lamp in the housing box and the early food intake following surgery.

3.2. Innate Immune Response to LPS Stimulation of Sterile Gut Loop

Intestinal immune tolerance initiates within the first hours of life in response to microbial colonization gained during vaginal delivery and subsequently via colostrum, milk and multiple contact with environment. This immune tolerance is characterized by a rapid hypo responsiveness to microbial antigens as demonstrated in mouse models [12,13]. We aim to stimulate the immune responses of animals from birth with colostrum supplemented with natural products and in particular YCW fractions that contain TLR2 and TLR4 ligands. This led us to use lambs born by cesarean-section and whose intestine had not yet been in contact with a microbiota. In order to validate the requirement of using cesarean-born lambs, we assessed LPS responsiveness of isolated ileal loops and of the ileum exposed to microbial colonization 24 h after birth. Chemokines (CXCL1, CXCL2, CXCL8) and TNFα, known to be induced after LPS stimulation, were compared by transcriptomic analysis in both conditions. Our results (Figure 2) demonstrated that in isolated loops LPS induced a chemokine response and TNFα expression while in the connected ileum exposed to microbial and alimentary antigens, these levels were already upregulated and LPS stimulation was without clear additive effect except for a slight increase in cxcl8 mRNA expression. In order to compare the potential of microbial productsm it is therefore necessary to use a cesarean-born lamb that remains fully responsive to microbial ligands.

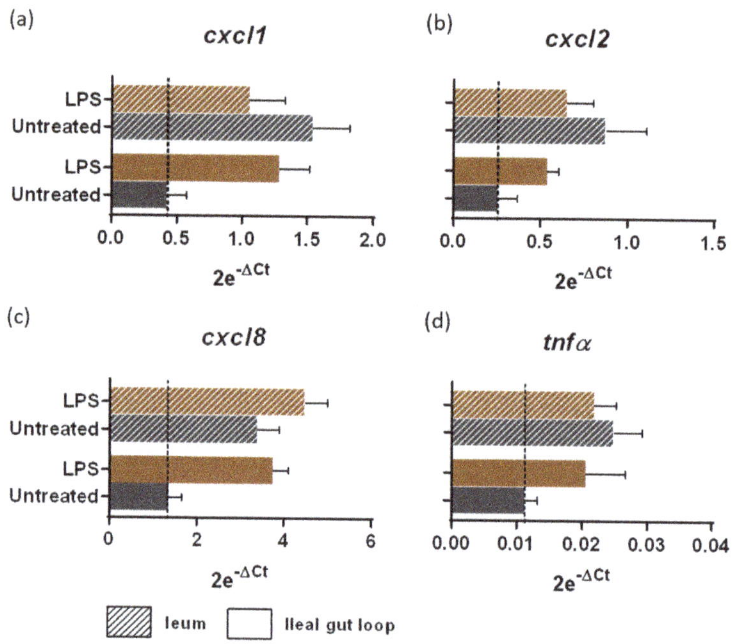

Figure 2. Intestinal responsiveness to LPS stimulation within the gut loop generated with a cesarean-born neonatal lamb. Intestinal tissues from the ileal gut loops and the ileum were recovered from the same neonatal lamb 24 h post-surgery. Small pieces of intestine, designed as "explants", were placed in a CO_2 incubator at 37 °C in culture medium alone (untreated) or with LPS at a concentration of 10 µg/mL for 4 h (n = 3 explants per condition). Explants were processed for RNA extraction and *cxcl1* (**a**), *cxcl2* (**b**), *cxcl8* (**c**) and *tnfα* (**d**) gene expressions were quantified by RT-qPCR. Results are expressed as $2e^{-\Delta Ct}$ (mean ± SEM), following normalization with the expression of three reference genes (*hprt; gapdh; actb*).

3.3. Innate Ileal Immune Response to TLR-Ligands and YCW Fractions in Isolated Loops

YCW principal components are β-glucans and mannoproteins, known to stimulate immune responses through multiple Pattern Recognition Receptors (PRR), mainly TLR2, TLR4 and Dectin-1 [18]. We investigated the innate immune response in ileal loops to two YCW from *S. cerevisiae*, differing in their polysaccharide composition. As controls, we used, *Escherichia coli* LPS, a TLR4-ligand, and R848, a synthetic TLR7-8 viral single-stranded ribonucleic acid mimic-ligand. We compared the responsiveness to these different agonists after 24 h of stimulation. YCW1 induced the mRNA expression of CXCL8 chemokine, IL1α, IL1β proinflammatory cytokines and the interferon-induced Mx1 while β-glucans enriched fraction only induced IL1β (Figure 3). LPS and R848 presented distinct cytokinic responses with R848 inducing the Mx1 upregulation while LPS was more prone to upregulate CXCL8 chemokine. These results confirm the potential of the neonatal gut-loop model to be used as a screening method to evaluate the immunostimulating properties of various compounds using a single animal.

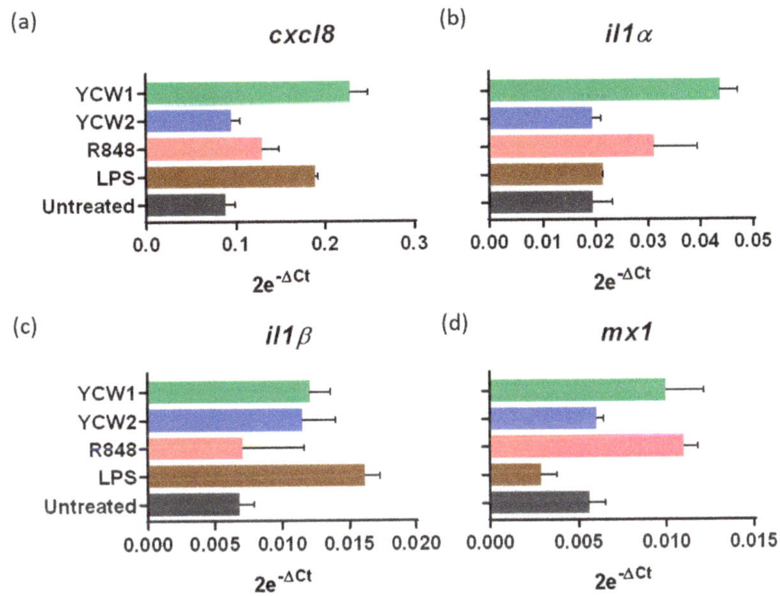

Figure 3. Innate ileal immune response to TLR-ligands and yeast cell wall fractions. Lamb's gut-loops were stimulated in vivo immediately after the surgery procedure by in situ injection of LPS (10 µg/mL), R848 (10 µg/mL), yeast cell wall fraction 1 (YCW1) or 2 (YCW2) (5 mg/mL) into distinct intestinal loops. Twenty-four hours later, intestinal loops were recovered and samples processed for RNA extraction. *cxcl8* (**a**), *il1α* (**b**), *il1β* (**c**) and *mx1* (**d**) gene expressions were quantified by RT-qPCR. Results are expressed as $2e^{-\Delta Ct}$ following normalization with the expression of three reference genes (*gapdh; hprt; actb*).

3.4. Evaluation of YCW Fractions on Early Cryptosporidium parvum Invasion and Development

To evaluate if immune stimulation with YCW fractions can reduce *C. parvum* infection with a limited number of animals, we used the benefit of the gut-loop model in providing multiple experimental conditions in a single animal. To evaluate *C. parvum* infection and development, we used a previously generated transgenic strain *Cp-Nluc*-INRAE allowing simple and accurate measurement of luciferase activity on tissue samples [16]. Two doses of *C. parvum* oocysts (1.5×10^4, 1.5×10^5) were inoculated in triplicate in independent loops and in both condition, parasite invasion and development were monitored by luciferase activity (Figure 4a). We confirmed the early *C. parvum* development within 24 h of infection by immunofluorescence microscopy, showing parasites developing into intestinal epithelial cells lining the ileal villi (Figure 4b). This result was also confirmed by RT-qPCR on *Cp18S* gene (Figure 4c) on tissue samples recovered from the same ileal loop as for luciferase activity. The correlation between the two variables presented a r^2 of 0.26 and a p value of 0.0044 (Figure 4d).

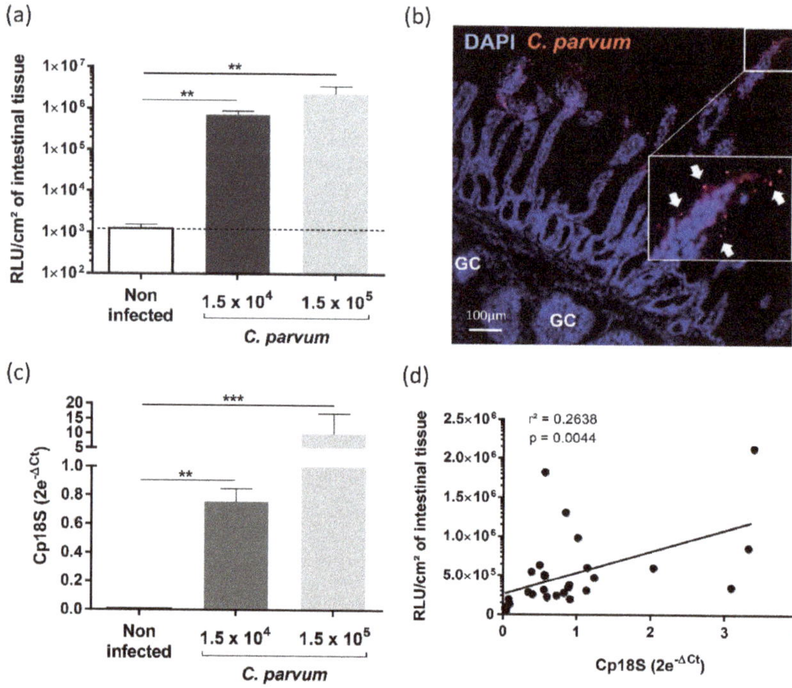

Figure 4. Establishment of *Cryptosporidium parvum* experimental infection in the gut-loop model. Data are cumulative results from three independent experimentations each performed with one newborn lamb. For each experimental condition, three loops were used; (**a**) Parasite load were determined by measuring luminescence activity in the ileal tissue from the loop 24 h after infection with oocysts of *C. parvum nluc*-INRAE transgenic strain ($n = 9$). Results are expressed as RLU per cm^2 of intestinal tissue (mean $\pm$ SEM); (**b**) Immunofluorescence microscopy of ileal tissue section from a *C. parvum* infected gut loop (1.5×10^5 oocysts) 24 h after infection. Parasites were stained in red with anti-*C. parvum* antibodies and nuclei in blue with DAPI. "GC" corresponds to the germinal centers of the ileal Peyer's patch. White arrows indicate parasites developing into intestinal epithelial cells lining the ileal villi; (**c**) Quantification of *Cp18S* gene expression in the intestinal loops was performed on the same samples as above. RNA was extracted from a piece of each intestinal loop after 24 h of infection with *C. parvum nluc*-INRAE transgenic strain, reverse transcribed and amplified by quantitative PCR. Results are expressed as $2e^{-\Delta Ct}$ (mean $\pm$ SEM) following normalization with the expression of three reference genes (*gapdh; hprt; actb*); (**d**) Correlation between parasite load evaluated by luminescence and *Cp18S* gene expression was determined by linear regression analysis. Statistical analyses in (**a**,**b**) were realized with the Kruskal–Wallis non-parametric test and the Dunn's multiple comparison test, and significative difference was determined by a *p*-value < 0.05 (** $p < 0.01$, *** $p < 0.001$).

The control of cryptosporidiosis is still limited and the search of natural alternative highly anticipated. With the same animal and benefiting from the multiple experimental possible conditions offered by the gut-loop model, we investigated the ability of two YCW fractions to limit *C. parvum* development. Both yeast fractions presented a limited but significant effect on *C. parvum* development (Figure 5), thus validating the gut-loop model to investigate new alternatives of treatment against this disease together with a limited use of experimental animals.

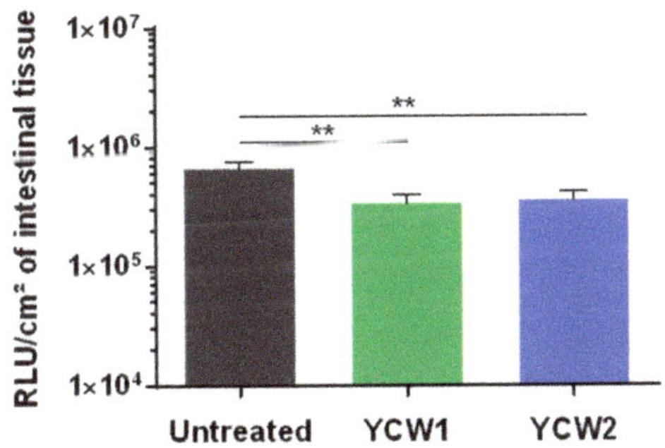

Figure 5. Evaluation of yeast cell wall fractions on early *Cryptosporidium parvum* invasion and development. Data are cumulative results from three independent experimentations each with one newborn lamb. For each experimental condition, three loops were used. Loops were infected by in situ injection of 1.5×10^4 oocysts of *C. parvum nluc-INRAE* transgenic strain alone (untreated), or associated with YCW1 or YCW2 at a concentration of 5 mg/mL. Parasite loads were evaluated by quantifying luminescence in intestinal tissues after 24 h of infection. Results are expressed as RLU per cm^2 of intestinal tissue with a logarithmic scale (mean ± SEM). Statistical analyses were performed with the Kruskal–Wallis non-parametric test followed by the Dunn's multiple comparison test (** $p < 0.01$).

4. Discussion

Neonates have an underdeveloped immune system at birth explaining their high susceptibility to infectious diseases. The high incidence of diarrhea in young ruminants must be addressed as soon as possible, as it results in the majority of mortality and morbidity [19,20]. Neonatal diseases, in particular those well controlled by neutralizing antibodies can often be prevented by maternal vaccination with vaccines administered before or during pregnancy. Multivalent vaccines against calf scours currently available are designed to prevent or limit infections with rotavirus, coronavirus, *E. coli* (K99) and *Salmonella* infections. There is no vaccine for *Cryptosporidium* to date which might be explained by the fact that antibodies seem to play a limited role in the protection process [21]. Therefore, alternative control strategies have to be investigated.

To promote gut health and improve growth, supplements containing immunoglobulins or mineral and vitamin complements, or with probiotics such as *Enterococcus faecium* can be orally given to young ruminants just after birth [22–24]. Live yeast and yeast-derived products are also extensively used as probiotic and prebiotic, respectively, for farm animals to reduce the severity of diarrhea by preventing pathogenic bacteria from binding to intestinal epithelial cells or by modulating gut mucosal immunity [25–27]. Indeed, YCW fractions are known for their immunostimulatory properties, they harbor ligands for various innate receptor such as TLR2, TLR4 and Dectin-1. They have previously shown to be effective against cryptosporidiosis and pathogenic bacterial colonization in young ruminants. Mammeri et al. identified the anti-cryptosporidial activities of chitosan, a natural polysaccharide present in YCW in in vitro and mouse studies [28]. In another study, yeast culture enriched with mannan-oligosaccharides were fed in milk to Holstein heifer calves enrolled at 4 to 12 h of age, and this led to the presence of fewer *Escherichia coli* and pathogenic *E. coli* compared with control calves [29].

In addition to their immunostimulatory properties, YCW products are suggested to be anti-adhesive agents able to reduce adhesion of intestinal pathogens [30,31]. *Cryptosporidium spp.* are known to express the lectin galactose/*N*-acetyl-D-galactosamine, which facilitates their adhesion to host epithelial cells via interaction with sulfated proteoglycans [32,33]. Therefore, further investigations are required regarding the precise mechanism of inhibition of *C. parvum* observed with YCW extracts while considering the role of polysaccharide composition.

We also wish to investigate if immunostimulatory components administered with the colostrum can strengthen immune system of neonatal ruminants and reduce the incidence of cryptosporidiosis. Comparing multiple derivates from yeast or other sources will require the use of many animals. In order to follow reduce animal consumption, we aimed to develop a gut-loop model adapted to caesarean-born animals. This surgical model, although invasive by definition, fits well with the 3Rs principle. With the practice and experience gained over the first surgeries, we can now make around 25–30 intestinal loops within the ileal Peyer's patch area which gives the opportunity to produce replicates and/or evaluate many anti-infectious products with just a single newborn animal. With this gut-loop model, the animal is its own control which reduces variability of responses especially in non-inbred animals that vary substantially in their genetic. We took a particular care to provide the best veterinary practices during surgery, anesthesia and analgesia, neonatal nursing, pre- and post-operative care. All of the contact related to the animals were optimized to avoid or limit pain and discomfort. After surgery, the animals fed naturally and did not show any particular behavior compared to its littermate.

The model is also flexible on the section of the intestine that could be investigated. Indeed, the small intestine contains distinct areas of organized lymphoid tissues at birth like Peyer's patches that are known to be the major inductive sites of immune responses. Ruminants possess jejunal Peyer's patches (JPP) that retain classical functions of intestinal Peyer's patches founds in mouse and human such as antigen sampling through Microfold cells and T- and B-cell activation, but also a peculiarly long ileal Peyer's Patch (IPP), which extends one meter along the terminal small intestine and which is known to be a primary lymphoid organ of B-cell development [34]. Since *C. parvum* infect primarily the distal small intestine, for our own investigations the intestinal loops were therefore generated in the ileum.

In order to compare various immunostimulant candidates containing TLR ligands we needed to use caesarean-section born lamb to have sterile intestinal environment mimicking the first encounter of the yeast derivates when administered with the first yeast-supplemented colostrum and putative interference with the presence of endotoxins that will raise rapidly in the gut lumen following microbiota installation. This process was previously reported in the mouse model by a mechanism that involves microRNA-146a-mediated translational repression and proteolytic degradation of the essential Toll-like receptor (TLR) signaling molecule interleukin-1 receptor-associated kinase 1 (IRAK1) [12]. This mechanism is sufficient to induce intestinal epithelial innate immune tolerance [13]. Although we did not demonstrate that similar mechanism occurs in lambs, we observed in this study that only in the gut loop explants, free of endotoxins, LPS induced chemokine production was preserved 5 h post-stimulation. Conversely, in explants generated with the ileum connected to the intestinal transit since 24 h, addition of LPS to these explants did not improve further chemokine upregulation with the exception of cxcl8 mRNA expression for which a slight increase was observed.

As a proof of concept, we next compared two YCW fractions for their immunostimulatory properties: one with a proportional content of β-glucans and mannoproteins and the second one enriched in β-glucans. We noticed that the YCW fractions enriched in β-glucans induced lower expression of pro-inflammatory cytokines in the gut-loop (CXCL8 and IL1α). Similar observation was previously made with macrophages cultured in vitro with *S. cerevisiae* extracts enriched in β-glucans that displayed weaker TLR2/4-related NFκB/AP-1 activity and less TNFα production [35]. We therefore can suspect that a similar mechanism may occur in gut lamb but this requires further investigations. When YCW1 and YCW2 were tested for their ability to reduce invasion and development of *C. parvum* in intestinal epithelial cells, they both limited in a modest but similar manner *C. parvum* early development despite different contents in β-glucans. One can conclude that the difference in ability to induce higher level of these immune effectors did not play a significant role in the protection process.

In addition to the evaluation of products for their immunostimulatory or their anti-infectious properties, the gut-loop in cesarean-born lambs can be used to further investigate host–microbial interactions in a controlled environment, and decipher immune feature of a specific area of the intestine. The gut-loop system is indeed suitable for investigations in various intestinal lymphoid and non-lymphoid segments by just making loops in the selected area as performed with jejunal loop made in a one-month-old piglet to study innate immune response to *Salmonella* [36]. Since our model does not require the use of antibiotics after the loop surgery, the neonatal gut-loop model can also be very useful to investigate the role of selected microbiota or probiotics that could be introduced in the "sterile loops" to evaluate their capacity to promote intestinal immune responses and mucosae maturation. For the later, maintenance of gut loop for long period would be required. A model of fetal lamb with a single 8–10 cm intestinal loop model was previously generated, and in this case, sterile intestinal loop constructed in utero retained functional GALT for as long as 6–7 months after birth [14]. This therefore demonstrates that even long-term slow progressive enteric diseases can be investigated with the gut-loop model. Our lamb gut-loop model could also be a model for human cryptosporidiosis to evaluate drug compounds and innate immune responses, considering the observed similarities between those of young ruminants and young children infected by *Cryptosporidium*. However, it has some limitations; for example, the large ileal Peyer's patch of young ruminants, which is not present in human.

Overall, this model paves the way for further new control method development such as immunostimulants, antimicrobial compounds and vaccines dedicated to the control of enteric infections in neonates.

Supplementary Materials: The following are available online at https://www.mdpi.com/article/10.3390/vetsci8090170/s1, Table S1: List of forward and reverse primers used for gene expression analysis by RT-qPCR.

Author Contributions: Conceptualization, F.L., N.K.-H. and S.L.-L.; methodology, A.B., C.B., F.L., J.C., N.K.-H. and S.L.-L.; formal analysis, A.B., F.L., N.K.-H. and S.L.-L.; investigation, A.B., A.P., C.B., J.C., J.P., N.K.-H. and S.L.-L.; writing—original draft preparation, A.B., F.L., J.C., J.S., N.K.-H., P.P.-P. and S.L.-L.; funding acquisition, F.L., J.S., P.P.-P. and S.L.-L. All authors have read and agreed to the published version of the manuscript.

Funding: A.B. benefited from a PhD grant from a CIFRE fellowship (Industrial Research Training Agreement) with Phileo by Lesaffre. This research was funded by both INRAE and PHILEO BY LESAFFRE funds.

Institutional Review Board Statement: The study was conducted according to the guidelines of the Declaration of Helsinki, and approved by the Institutional Review Board (or Ethics Committee) of Val de Loire (CEEA VdL n°19) (protocol code: APAFIS#16870-201809261558973 v2 and date of approval: 14 February 2019).

Informed Consent Statement: Not applicable.

Data Availability Statement: The data presented in this study are openly available in Zenodo at [10.5281/zenodo.5235377].

Acknowledgments: We would like to thank Léa Bouyonnet for generating the figure on the intestinal-loops with Illustrator® (Adobe, San José, CA, USA) and Eric Auclair and Christine Julien (Phileo by Lesaffre, Marcq-en-Baroeul, France) for their constructive discussions. We are also very grateful to Caroline Thérésine, Tiffany Pezier (ISP UMR, INRAE) and Fanny Faurie-Sarce (PFIE, INRAE) for the production of the *C. parvum nluc*-INRAE transgenic strain and to Aude Remot and Pierre Germon (ISP UMR, INRAE) for providing some primer sequences.

Conflicts of Interest: The authors declared no potential conflict of interest with respect to the research, authorship, and/or publication of this article.

References

1. Mayr, L.M.; Bojanic, D. Novel trends in high-throughput screening. *Curr. Opin. Pharmacol.* **2009**, *9*, 580–588. [CrossRef]
2. Brodin, P.; Christophe, T. High-content screening in infectious diseases. *Curr. Opin. Chem. Biol.* **2011**, *15*, 534–539. [CrossRef]
3. Li, V.S.W. Modelling intestinal inflammation and infection using 'mini-gut' organoids. *Nat. Rev. Gastroenterol. Hepatol.* **2021**, *18*, 89–90. [CrossRef] [PubMed]
4. Beaumont, M.; Blanc, F.; Cherbuy, C.; Egidy, G.; Giuffra, E.; Lacroix-Lamande, S.; Wiedemann, A. Intestinal organoids in farm animals. *Vet. Res.* **2021**, *52*, 33. [CrossRef]
5. Richmond, J. The 3Rs—Past, present and future. *Scand. J. Lab. Anim. Sci.* **2000**, *27*, 84–92.
6. Pinto, D.J.; Vinayak, S. Cryptosporidium: Host-Parasite Interactions and Pathogenesis. *Curr. Clin. Microbiol. Rep.* **2021**, *8*, 62–67. [CrossRef]
7. Gururajan, A.; Rajkumari, N.; Devi, U.; Borah, P. Cryptosporidium and waterborne outbreaks—A mini review. *Trop. Parasitol.* **2021**, *11*, 11–15. [CrossRef] [PubMed]
8. Khalil, I.A.; Troeger, C.; Rao, P.C.; Blacker, B.F.; Brown, A.; Brewer, T.G.; Colombara, D.V.; De Hostos, E.L.; Engmann, C.; Guerrant, R.L.; et al. Morbidity, mortality, and long-term consequences associated with diarrhoea from Cryptosporidium infection in children younger than 5 years: A meta-analyses study. *Lancet Glob. Health* **2018**, *6*, e758–e768. [CrossRef]
9. Striepen, B. Parasitic infections: Time to tackle cryptosporidiosis. *Nature* **2013**, *503*, 189–191. [CrossRef] [PubMed]
10. Lefay, D.; Naciri, M.; Poirier, P.; Chermette, R. Prevalence of Cryptosporidium infection in calves in France. *Vet. Parasitol.* **2000**, *89*, 1–9. [CrossRef]
11. Van Voorhis, W.C.; Hulverson, M.A.; Choi, R.; Huang, W.; Arnold, S.L.M.; Schaefer, D.A.; Betzer, D.P.; Vidadala, R.S.R.; Lee, S.; Whitman, G.R.; et al. One health therapeutics: Target-Based drug development for cryptosporidiosis and other apicomplexa diseases. *Vet. Parasitol.* **2021**, *289*, 109336. [CrossRef]
12. Lotz, M.; Gutle, D.; Walther, S.; Menard, S.; Bogdan, C.; Hornef, M.W. Postnatal acquisition of endotoxin tolerance in intestinal epithelial cells. *J. Exp. Med.* **2006**, *203*, 973–984. [CrossRef]
13. Chassin, C.; Kocur, M.; Pott, J.; Duerr, C.U.; Gutle, D.; Lotz, M.; Hornef, M.W. miR-146a mediates protective innate immune tolerance in the neonate intestine. *Cell Host Microbe* **2010**, *8*, 358–368. [CrossRef] [PubMed]
14. Mutwiri, G.; Watts, T.; Lew, L.; Beskorwayne, T.; Papp, Z.; Baca-Estrada, M.E.; Griebel, P. Ileal and jejunal Peyer's patches play distinct roles in mucosal immunity of sheep. *Immunology* **1999**, *97*, 455–461. [CrossRef] [PubMed]
15. Gerdts, V.; Uwiera, R.R.; Mutwiri, G.K.; Wilson, D.J.; Bowersock, T.; Kidane, A.; Babiuk, L.A.; Griebel, P.J. Multiple intestinal 'loops' provide an in vivo model to analyse multiple mucosal immune responses. *J. Immunol. Methods* **2001**, *256*, 19–33. [CrossRef]
16. Swale, C.; Bougdour, A.; Gnahoui-David, A.; Tottey, J.; Georgeault, S.; Laurent, F.; Palencia, A.; Hakimi, M.A. Metal-captured inhibition of pre-mRNA processing activity by CPSF3 controls Cryptosporidium infection. *Sci. Transl. Med.* **2019**, *11*, eaax7161. [CrossRef] [PubMed]
17. Lantier, L.; Lacroix-Lamande, S.; Potiron, L.; Metton, C.; Drouet, F.; Guesdon, W.; Gnahoui-David, A.; Le Vern, Y.; Deriaud, E.; Fenis, A.; et al. Intestinal CD103+ dendritic cells are key players in the innate immune control of Cryptosporidium parvum infection in neonatal mice. *PLoS Pathog.* **2013**, *9*, e1003801. [CrossRef]
18. Gow, N.A.; van de Veerdonk, F.L.; Brown, A.J.; Netea, M.G. Candida albicans morphogenesis and host defence: Discriminating invasion from colonization. *Nat. Rev. Microbiol.* **2011**, *10*, 112–122. [CrossRef]
19. Urie, N.J.; Lombard, J.E.; Shivley, C.B.; Kopral, C.A.; Adams, A.E.; Earleywine, T.J.; Olson, J.D.; Garry, F.B. Preweaned heifer management on US dairy operations: Part V. Factors associated with morbidity and mortality in preweaned dairy heifer calves. *J. Dairy Sci.* **2018**, *101*, 9229–9244. [CrossRef]
20. Scott, K.; Kelton, D.F.; Duffield, T.F.; Renaud, D.L. Risk factors identified on arrival associated with morbidity and mortality at a grain-fed veal facility: A prospective, single-cohort study. *J. Dairy Sci.* **2019**, *102*, 9224–9235. [CrossRef]
21. Chen, W.; Harp, J.A.; Harmsen, A.G. Cryptosporidium parvum infection in gene-targeted B cell-deficient mice. *J. Parasitol.* **2003**, *89*, 391–393. [CrossRef]
22. Bond, C. Evaluation of lamb colostrum supplements. *Vet. Rec.* **2020**, *187*, e100. [CrossRef]
23. Chang, M.N.; Wei, J.Y.; Hao, L.Y.; Ma, F.T.; Li, H.Y.; Zhao, S.G.; Sun, P. Effects of different types of zinc supplement on the growth, incidence of diarrhea, immune function, and rectal microbiota of newborn dairy calves. *J. Dairy Sci.* **2020**, *103*, 6100–6113. [CrossRef]
24. Renaud, D.L.; Kelton, D.F.; Weese, J.S.; Noble, C.; Duffield, T.F. Evaluation of a multispecies probiotic as a supportive treatment for diarrhea in dairy calves: A randomized clinical trial. *J. Dairy Sci.* **2019**, *102*, 4498–4505. [CrossRef]
25. Cangiano, L.R.; Yohe, T.T.; Steele, M.A.; Renaud, D.L. INVITED REVIEW: Strategic use of microbial-based probiotics and prebiotics in dairy calf rearing. *Appl. Anim. Sci.* **2020**, *36*, 630–651. [CrossRef]
26. Davis, M.E.; Maxwell, C.V.; Erf, G.F.; Brown, D.C.; Wistuba, T.J. Dietary supplementation with phosphorylated mannans improves growth response and modulates immune function of weanling pigs. *J. Anim. Sci.* **2004**, *82*, 1882–1891. [CrossRef] [PubMed]
27. Galvao, K.N.; Santos, J.E.; Coscioni, A.; Villasenor, M.; Sischo, W.M.; Berge, A.C. Effect of feeding live yeast products to calves with failure of passive transfer on performance and patterns of antibiotic resistance in fecal Escherichia coli. *Reprod. Nutr. Dev.* **2005**, *45*, 427–440. [CrossRef] [PubMed]

28. Mammeri, M.; Chevillot, A.; Thomas, M.; Polack, B.; Julien, C.; Marden, J.P.; Auclair, E.; Vallee, I.; Adjou, K.T. Efficacy of chitosan, a natural polysaccharide, against Cryptosporidium parvum in vitro and in vivo in neonatal mice. *Exp. Parasitol.* **2018**, *194*, 1–8. [CrossRef] [PubMed]
29. Lucey, P.M.; Lean, I.J.; Aly, S.S.; Golder, H.M.; Block, E.; Thompson, J.S.; Rossow, H.A. Effects of mannan-oligosaccharide and Bacillus subtilis supplementation to preweaning Holstein dairy heifers on body weight gain, diarrhea, and shedding of fecal pathogens. *J. Dairy Sci.* **2021**, *104*, 4290–4302. [CrossRef]
30. Ganner, A.; Schatzmayr, G. Capability of yeast derivatives to adhere enteropathogenic bacteria and to modulate cells of the innate immune system. *Appl. Microbiol. Biotechnol.* **2012**, *95*, 289–297. [CrossRef] [PubMed]
31. Posadas, G.A.; Broadway, P.R.; Thornton, J.A.; Carroll, J.A.; Lawrence, A.; Corley, J.R.; Thompson, A.; Donaldson, J.R. Yeast Pro- and Paraprobiotics Have the Capability to Bind Pathogenic Bacteria Associated with Animal Disease1. *Transl. Anim. Sci.* **2017**, *1*, 60–68. [CrossRef] [PubMed]
32. Joe, A.; Hamer, D.H.; Kelley, M.A.; Pereira, M.E.; Keusch, G.T.; Tzipori, S.; Ward, H.D. Role of a Gal/GalNAc-specific sporozoite surface lectin in Cryptosporidium parvum-host cell interaction. *J. Eukaryot. Microbiol.* **1994**, *41*, 44S. [PubMed]
33. Ludington, J.G.; Ward, H.D. The Cryptosporidium parvum C-Type Lectin CpClec Mediates Infection of Intestinal Epithelial Cells via Interactions with Sulfated Proteoglycans. *Infect. Immun.* **2016**, *84*, 1593–1602. [CrossRef] [PubMed]
34. Yasuda, M.; Jenne, C.N.; Kennedy, L.J.; Reynolds, J.D. The sheep and cattle Peyer's patch as a site of B-cell development. *Vet. Res.* **2006**, *37*, 401–415. [CrossRef]
35. Walachowski, S.; Tabouret, G.; Foucras, G. Triggering Dectin-1-Pathway Alone Is Not Sufficient to Induce Cytokine Production by Murine Macrophages. *PLoS ONE* **2016**, *11*, e0148464. [CrossRef]
36. Meurens, F.; Berri, M.; Auray, G.; Melo, S.; Levast, B.; Virlogeux-Payant, I.; Chevaleyre, C.; Gerdts, V.; Salmon, H. Early immune response following Salmonella enterica subspecies enterica serovar Typhimurium infection in porcine jejunal gut loops. *Vet. Res.* **2009**, *40*, 5. [CrossRef]

veterinary sciences

Article

Improved Cardiovascular Tolerance to Hemorrhage after Oral Resveratrol Pretreatment in Dogs

Jennifer Davis [1,*], Anthea L. Raisis [1], Claire R. Sharp [1,2], Rachel E. Cianciolo [3], Steven C. Wallis [4] and Kwok M. Ho [1,5,6]

1 School of Veterinary Science, Murdoch University, Murdoch, WA 6150, Australia; a.raisis@murdoch.edu.au (A.L.R.); C.Sharp@murdoch.edu.au (C.R.S.); kwok.ho@health.wa.gov.au (K.M.H.)
2 Centre for Terrestrial Ecosystem Science and Sustainability, Harry Butler Institute, Murdoch University, Murdoch, WA 6150, Australia
3 Department of Veterinary Biosciences, The Ohio State University, Columbus, OH 43210, USA; Cianciolo.14@osu.edu
4 University of Queensland Centre for Clinical Research, Brisbane, QLD 4029, Australia; s.wallis@uq.edu.au
5 Department of Intensive Care Medicine, Royal Perth Hospital, Perth, WA 6000, Australia
6 Medical School, University of Western Australia, Perth, WA 6009, Australia
* Correspondence: jen@vetanaesthesiatips.com

Citation: Davis, J.; Raisis, A.L.; Sharp, C.R.; Cianciolo, R.E.; Wallis, S.C.; Ho, K.M. Improved Cardiovascular Tolerance to Hemorrhage after Oral Resveratrol Pretreatment in Dogs. *Vet. Sci.* **2021**, *8*, 129. https://doi.org/10.3390/vetsci8070129

Academic Editors: Ana Faustino and Paula A. Oliveira

Received: 24 May 2021
Accepted: 10 July 2021
Published: 12 July 2021

Abstract: Resveratrol has been shown to preserve organ function and improve survival in hemorrhagic shock rat models. This study investigated whether seven days of oral resveratrol could improve hemodynamic response to hemorrhage and confer benefits on risk of acute kidney injury (AKI) without inducing coagulopathy in a canine model. Twelve greyhound dogs were randomly allocated to receive oral resveratrol (1000 mg/day) or placebo for seven days prior to inducing hemorrhage until a targeted mean blood pressure of $\leq$40 mmHg was achieved. AKI biomarkers and coagulation parameters were measured before, immediately following, and two hours after hemorrhage. Dogs were euthanized, and renal tissues were examined at the end of the experiment. All investigators were blinded to the treatment allocation. A linear mixed model was used to assess effect of resveratrol on AKI biomarkers and coagulation parameters while adjusting for volume of blood loss. A significant larger volume of blood loss was required to achieve the hypotension target in the resveratrol group compared to placebo group (median 64 vs. 55 mL/kg respectively, $p = 0.041$). Although histological evidence of AKI was evident in all dogs, the renal tubular injury scores were not significantly different between the two groups, neither were the AKI biomarkers. Baseline (pre-hemorrhage) maximum clot firmness on the Rotational Thromboelastometry (ROTEM®) was stronger in the resveratrol group than the placebo group (median 54 vs. 43 mm respectively, $p = 0.009$). In summary, seven days of oral resveratrol did not appear to induce increased bleeding risk and could improve greyhound dogs' blood pressure tolerance to severe hemorrhage. Renal protective effect of resveratrol was, however, not observed.

Keywords: acute kidney injury; biomarkers; coagulation; ischemia-reperfusion injury; shock

1. Introduction

Perioperative hemorrhage is a major complication in surgical patients, resulting in increased morbidity and mortality [1–4]. Severe hemorrhage can lead to severe hypotension resulting in reduced vital organs perfusion and oxygenation [5]. Ischemia followed by reperfusion—also known as ischemia-reperfusion (I-R) injury—will induce formation of free radical oxygen species, inflammatory mediators, and toxic metabolites which can cause organ injury [6–8]. Indeed, I-R injury is one of the main causes of perioperative acute kidney injury (AKI) [9,10]. AKI is common in patients undergoing major surgery and can contribute to increased perioperative mortality [11–15]. As such, strategies capable of attenuating I-R injury due to severe hemorrhage are both clinically relevant and desirable.

Resveratrol is a naturally occurring polyphenol and administrating resveratrol in large doses parenterally has been shown to preserve organ function and improve survival in hemorrhagic shock rat models [6]. Resveratrol can activate expression of the Silent Information Regulator (SIRT1) gene to produce situin-1 protein which can inhibit pro-inflammatory mediators, enhance anti-oxidant pathways, and improve mitochondrial function. In addition, it also has an agonist effect at estrogen receptors [6–8,16]. These effects may explain why administering intraperitoneal or parenteral resveratrol as a resuscitation agent could improve blood pressure in rats with decompensated hemorrhagic shock [6,8]. Although rodent models are appropriate to investigate the mechanistic benefits of resveratrol, confirmation of resveratrol's benefits in large-animal models is needed before it can be safely tested on humans in a clinical setting [17,18]. Previous experimental studies have primarily used a single large dose of resveratrol during or after hemorrhage [6,8]. Because sterile preparation of parenteral resveratrol is not commercially available, translating this treatment strategy to patients at risk of hemorrhage remains difficult.

It is possible that oral resveratrol, taken for a prolonged period of time prior to hemorrhage, may achieve a similar benefit as parenteral resveratrol in hemorrhage. If this is the case, this strategy would be more clinically applicable for patients who are at risk of massive blood loss in elective major surgery. Another concern about the use of resveratrol in perioperative patients is its potential antiplatelet action [19,20]. Currently no safety data exists to guide clinicians whether resveratrol should be ceased prior to major surgery to avoid increased risk of bleeding.

We hypothesized that seven days of oral resveratrol treatment would improve hemodynamic tolerance to induced bleeding. Using a canine model, we aimed to investigate whether resveratrol, administered orally for seven days prior to anesthesia, could improve blood pressure tolerance due to induced hemorrhage. Our secondary objective was to investigate whether resveratrol could protect against AKI without inducing adverse effect on the coagulation.

2. Materials and Methods

2.1. Animals and Selection Criteria

Twelve donated adult entire male dogs were included in the study. The dogs were retired racing greyhounds, surrendered by their owners to be used as terminal blood donors. Physical examination, renal ultrasonography, urinalysis, complete blood count, serum creatinine (SCr), blood urea nitrogen, serum albumin concentration, platelet closure time (PCT), and Rotational Thromboelastometry (ROTEM® delta, Tem International GmbH, Munich, Germany) for all dogs were within reference intervals for adult Greyhounds [21]. Ethics approval was granted by the Murdoch University Animal Ethics Committee (permit number R2726/15) and the dogs cared for in accordance with the Australian code for the care and use of animals for scientific purposes.

2.2. Resveratrol Administration

Dogs were randomly allocated to receive either no resveratrol supplementation (C-control group; n = 6) or seven days of 1000 mg (>10 mg/kg) per day of micronized trans-resveratrol (Micronized Resveratrol Micro500, Harmoni-T, Las Vegas, NV 89119, USA) orally (R-resveratrol group; n = 6), via a computer-generated random number sequence (Excel, Microsoft Corporation, Redmond, WA, USA). The final dose of resveratrol was administered in the morning of anesthesia. Serum and urine supernatant collected during the enrolment screening process for all dogs was divided into aliquots and stored at −80 °C as "baseline" samples for subsequent analysis of creatinine, protein, gamma-glutamyl transferase (GGT), and AKI biomarker concentrations. Animals in both groups were cared for in the same premises, under the same conditions, for seven days prior to anesthesia. Throughout this period, water was provided ad lib, and dogs were fed an intestinal health diet (Hills I/D) twice daily. All dogs received a deworming prophylactic medication (Popantel®, Jurox Animal Health, Rutherford, NSW, Australia) at the time of enrolment into the study.

Dogs in both groups were examined daily by a veterinarian to ensure maintenance of good health, and any abnormalities detected on physical examination recorded.

2.3. Anesthesia

For all dogs, food was removed at least eight hours prior to the procedure but access to water provided until premedication. Dog were anesthetized by a veterinary anesthesiologist (AR) using a standardized protocol. Premedication with methadone 0.3 mg/kg (Ilium Methadone 10 mg/mL, Troy Laboratories, Glendenning, NSW, Australia) IM was performed 30 min prior to induction of general anesthesia with alfaxalone (Alfaxan injection 10 mg/mL, Jurox, Glendinning, NSW, Australia) 2.2–3.2 mg/kg intramuscularly. Endotracheal intubation was used to maintain anesthesia with isoflurane (I.S.O., Veterinary Companies of Australia Pty Ltd, Kings Park, NSW, Australia) in 30% oxygen through a circle rebreathing system. End tidal isoflurane was maintained at 1.3–1.4%. Infusion of a balanced isotonic crystalloid solution (Compound Sodium Lactate, Baxter Healthcare, Toongabbie, NSW, Australia) was provided at 10 mL/kg/hour IV throughout anesthesia, along with IV fentanyl (Fentanyl injection 50 μg/mL, AstraZeneca, Macquarie Park, NSW, Australia) 2 μg/kg/hour. Arterial carbon dioxide tension was maintained between 35 and 45 mmHg by provision of mechanical ventilation by an anesthesia workstation (Datex-Ohmeda S/5 Aespire Anaesthesia Machine, GE Healthcare, Chicago, IL, USA). Active warming maintained esophageal temperature between 36.0 and 38.0 °C.

2.4. Instrumentation Procedures

Instrumentation procedures were performed within the first 60 min of anesthesia. Dogs were positioned in dorsal recumbent position during this period. The cranial vena cava was cannulated via the right jugular vein, using a 14-gauge 13 cm cannula, to facilitate collection of venous blood and injection of lithium chloride. To allow measurement of arterial blood pressure and cardiac output (Qt) via the lithium dilution technique, and to facilitate controlled hemorrhage, a 14-gauge 9 cm cannula was placed in the left femoral artery. Prior to surgical exposure of the femoral artery and cannula placement, a femoral nerve block was performed using 0.1 mL/kg bupivacaine (Bupivacaine hydrochloride 5.0 mg/mL, Pfizer, Sydney, NSW, Australia). Non-distensible tubing filled with heparinized saline connected the arterial cannula to an electronic pressure transducer (DTX Plus, Argon Critical Care Systems, Singapore) for measurement of arterial blood pressure using a multi-parameter monitor (Surgivet V9203; Smiths Medical, Minneapolis, MN, USA). Before anesthesia of each animal, the transducer was calibrated using a mercury manometer (at 40, 80 and 120 mmHg). Once connected, the transducer was placed level with the manubrium of the sternum (i.e., approximate level of the right atrium) and zeroed to atmospheric pressure. Hourly during the study (just prior to each data collection interval), damping factor was subjectively assessed via a rapid flush test. Lack of baseline drift was confirmed at the end of each anesthetic by ensuring pressure read zero upon re-opening the transducer to the atmosphere. An 8 Fr 55 cm Foley urinary catheter was inserted into the bladder for collection of urine samples and measurement of urine output (UOP).

2.5. Experimental Design

This study utilized a predefined blood pressure (mean arterial pressure-MAP of $\leq$40 mmHg) target to assess whether the volume of blood loss needed to achieve such target was different between the two groups (see Figure 1). Following instrumentation and collection of baseline samples (T0), bleeding was commenced by removing blood from the femoral artery cannula until MAP fell to $\leq$40 mmHg and remained at that level for 60 mins. After the 60-min hypotensive period was completed, intravenous fluid resuscitation was initiated using 20 mL/kg of succinylated gelatine solution 4% (Gelofusine®, B. Braun, Bella Vista, NSW, Australia) at 1200 mL/h (without crystalloid boluses) until MAP was $\geq$60 mmHg. The dogs then underwent a further three hours of anesthesia with their MAP maintained at $\geq$60 mmHg, by infusing more Gelofusine® if required. Approximately about five hours following induction of anesthesia, the animals were euthanized using pentobarbitone (Lethabarb Euthanasia Injection 300 mg/mL, Virbac, Milperra, NSW, Australia) 150 mg/kg IV. The total amount of blood volume needed to achieve and maintain the 60 min hypotensive period and the amount of colloid fluid needed to achieve normotension subsequently were recorded.

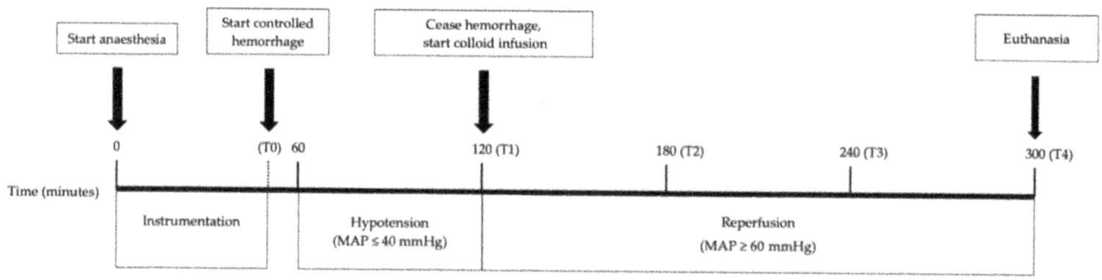

Figure 1. Timeline of the pressure-targeted hemorrhage model performed in 12 adult male Greyhounds. MAP = mean arterial pressure. T0–T4 were time points at which cardiac output was measured, and blood (arterial and venous) and urine samples were collected.

2.6. Data Collection

Data collection was performed at T0 (baseline: 60 min after induction of anesthesia, following instrumentation, prior to ischemia), T1 (60 min after MAP first reached $\leq$40 mmHg, prior to reperfusion), T2 (60 min after commencing reperfusion), T3 (120 min after reperfusion commenced), and T4 (180 min after reperfusion commenced).

2.6.1. Cardiovascular Parameters

At each time point, the mean of five consecutive MAP measurements was calculated. A blood gas analyzer and co-oximeter (ABL 725, Radiometer, Copenhagen) was used to measure blood gas variables (oxygen tension, carbon dioxide tension, bicarbonate concentration, calculated base excess, oxygen saturation) on arterial and central venous blood collected into heparinized syringes (PICO syringe, Radiometer, Copenhagen). These samples were analyzed immediately following collection. The same analyzer was used to measure hemoglobin and sodium concentration in central venous blood, necessary for the Qt measurement using lithium dilution. The lithium dilution technique, previously validated for Qt measurement in dogs, was performed at each time interval [22]. The mean of two consecutive Qt measurements that were not different by >10% was used for analysis. The calculation of the oxygen extraction ratio (OER) is described in Table 1. OER reflects the balance between systemic oxygen delivery and demand, and an OER greater than 0.5 in anesthetized dogs implied the presence of anaerobic metabolism according to previous studies [23–25].

Table 1. Calculation of oxygen extraction ratio (OER).

Calculated Variable	Formula
BSA (m^2)	$10.1 \times$ (bodyweight in grams$^{2/3}$) $\times 10^{-4}$
Cardiac index (L minute^{-1} $(m^2)^{-1}$)	Qt $\div$ BSA
CaO_2 (mL L^{-1})	$(1.36 \times SaO_2 \times Hb) + (0.003 \times PaO_2) \times 10$
DO_2I (mLO_2 min^{-1} $(m^2)^{-1}$)	Cardiac index $\times$ CaO_2
$CcvO_2$ (mL L^{-1})	$(1.36 \times ScvO_2 \times Hb) + (0.003 \times PcvO_2) \times 10$
VO_2I (mLO_2 min^{-1} $(m^2)^{-1}$)	Cardiac index $\times$ $(CaO_2 - CcvO_2)$
OER	$VO_2I \div DO_2I$

BSA = body surface area; CaO_2 = arterial oxygen content; $CcvO_2$ = central venous oxygen content; DO_2I = oxygen delivery index; Hb = hemoglobin concentration; PaO_2 = arterial oxygen tension; $PcvO_2$ = central venous oxygen tension; Qt = cardiac output; SaO_2 = arterial oxygen saturation; $ScvO_2$ = central venous oxygen saturation; VO_2I = oxygen consumption index.

2.6.2. Renal Parameters

Urine Sediment Examination

At each time point urine bag was emptied into a sterile collection pot for UOP measurement, and urine specific gravity and dipstick analyses (Multistix®, Siemens, Erlangen, Germany). Urine samples were centrifuged at 339 g for five minutes. At each time interval, urine sediment in the urine was examined using light microscopy at low (×100) and high (×400) power. A minimum of ten high power fields were examined for presence of blood cells, epithelial cells, and granular and hyaline casts. Urine supernatant was divided into multiple aliquots and stored at −80 °C for later analysis of creatinine, protein, GGT, and biomarker concentrations.

Kidney Light Microscopy

Kidneys were removed immediately following euthanasia, sectioned and stored in 10% formalin. Additionally, small cubes of cortex (approximately 2 to 3 mm^3) were placed into 3% glutaraldehyde (Koch-Light, Johannesburg, South Africa) for transmission electron microscopic (TEM) evaluation. Samples were processed routinely by the Comparative Pathology and Mouse Phenotyping Shared Resource (The Ohio State University, Columbus, OH, USA). Samples were embedded in paraffin, sectioned at 3µm thickness and stained with hematoxylin and eosin and periodic acid-Schiff. One veterinary pathologist, blinded to group allocation and AKI biomarker concentrations, examined sections of both kidneys by light microscopy and counted the number of injured tubules per 200× field. Tubular injury was defined as loss of the apical brush border, denudation of tubular basement membranes, singly dead tubular epithelial cells and tubules with intra-luminal detached cells/cellular debris. Twenty randomly selected cortical fields were assessed, and the average number of injured tubules per field was calculated for each specimen. Based on these quantitative data, the samples were then categorized into groups of normal (no injured tubules), minimal (mean tubular injury ≤0.5), mild (mean tubular injury >0.6 but ≤1) and moderate (mean tubular injury >1 but ≤2) histologic injury.

Kidney Transmission Electronic Microscopy (TEM)

Due to resources constraint, the glutaraldehyde-fixed tissue from only 4 dogs (2 control and 2 treated) were processed for electronic microscopy. After post-fixation in 1% osmium tetroxide, the specimens were serially dehydrated, infiltrated in an acetone/epoxy plastic, and embedded in plastic. The plastic blocks were sectioned to a silver-grey interference color (55–60 nm) and placed on copper mesh grids. The sections were stained with filtered lead citrate/sodium citrate solution (Electron Microscopy Sciences, Inc., Hatfield, PA, USA). Grids were imaged on a JEOL JEM-1400 TEM (JEOL USA, Inc., Peabody, MA, USA) and representative images were photographed with an Olympus SIS Veleta 2K camera (Olympus Soft Imaging Solutions GmbH, Muenster, Germany.

Renal Biomarkers

At baseline, T0, T1, and T3; aliquots of plain serum and urinary supernatant were stored at −80 °C for measurements of SCr and AKI biomarkers within three months.

Urinary creatinine, protein, and GGT concentrations; and SCr concentration, were measured using the same biochemistry analyzer (Cobas Integra 400 plus, Roche Diagnostics) that was calibrated prior to each use, with two control samples included in each run. Concentrations of AKI biomarkers; clusterin, cystatin C, kidney injury molecule 1 (KIM-1), monocyte chemoattractant protein 1 (MCP-1), and neutrophil gelatinase-associated lipocalin (NGAL), were measured in the urine and serum samples from each dog at baseline, T0, T1, and T2 using a bead-based multiplexed immunoassay (MILLIPLEX™ MAP Canine Kidney Toxicity Expanded Magnetic Bead Panel 1, Merck Millipore, Burlington, MA, USA). Urinary samples were diluted to 1:2, and serum samples to 1:10, using assay buffer prior to analysis. All samples were analyzed in duplicate. The assays were performed according to manufacturer instructions. Briefly, samples were incubated with a solution of antibody-immobilized magnetic beads (for each of the five biomarkers previously listed) for two hours at room temperature. The plate was then washed using a magnetic plate washer (Bio-Plex® Pro II Wash Station, Bio-Rad, Hercules, CA, USA), detection antibodies added, and incubated at room temperature for one hour. Streptavidin-Phycoerythrin was added to the detection antibodies, prior to a further 30 min period of incubation at room temperature. The plate was then washed, and beads re-suspended with drive fluid (MAGPIX® Drive Fluid, Merck Millipore, Burlington, MA, USA). The plate was analyzed five minutes later using a multiplex reader (Bio-Plex® MAGPIX™ Multiplate Reader, Bio-Rad, Hercules, CA, USA) with xPONENT® software. The Median Fluorescent Intensity data was analyzed using a 5-parameter logistic curve to calculate analyte concentrations in each sample. If the coefficient of variance between two duplicates was >15% the sample was re-analyzed on a subsequent plate. If the analyte concentration of a sample was above the upper limit of quantification (ULOQ) the sample was diluted until the biomarker concentration fell within the measurable range of the assay. If the analyte concentration of a sample was below the lower limit of quantification (LLOQ) the sample was reanalyzed on a subsequent plate using undiluted sample. Any result still falling below the LLOQ was allocated the same value as the LLOQ in order to allow statistical analysis. When serum samples were analyzed using the multiplex assay, a serum diluent (Serum Matrix, Merck Millipore, Burlington, MA, USA) was added to wells containing the Standards and Controls in place of assay buffer in order to mitigate any matrix effects.

2.6.3. Coagulation Parameters

Central venous blood was obtained from the jugular vein catheter at each time point for assessment of coagulation. At T0, T1, and T2, blood samples for coagulation assay were transferred into blood tubes containing 3.2% buffered sodium citrate with a blood: anticoagulant ratio of 9:1. Tubes were gently agitated to ensure even mixing of anticoagulant. Platelet closure time (PCT) was measured immediately, and blood was stored at room temperature for Rotational Thromboelastometry analysis (ROTEM® delta, Tem International GmbH, Munich, Germany) 30 min after collection.

The Platelet Function Analyzer-100 (Dade Boehring Inc., Deerfield, IL, USA), with collagen and adenosine diphosphate cartridges, was used to measure PCT in duplicate within 10 min of sample collection. Analysis was immediately repeated if the coefficient of variation was greater than 15%. ROTEM® was performed according to the manufacturer's instructions using both InTEM and ExTEM. Each profile was run for at least one hour following initiation. Data recorded for the InTEM and ExTEM profiles included clotting time (CT), clot formation time (CFT), alpha angle, maximum clot firmness (MCF) and lysis index at 60 min (LI60). Peak thrombin generation (normal range 243–507 nM) and area under the thrombin generation curve (AUC: normal range 1500–2835 nM × min) were measured by the Calibrated Automated Thrombogram using Thrombinoscope® (Stago™, Asnières sur Seine, France).

Total concentrations of resveratrol in plasma and qualitative investigation of its metabolites (glucuronide and sulphate) both prior to the administration and after 7 days of resveratrol treatment (or control) were measured by the UHPLC-MS/MS on a Shimadzu Nexera2 UHPLC system coupled to a Shimadzu 8030+ triple quadrupole mass spectrometer (Kyoto, Japan) [26].

2.7. Data Analysis

A sample size calculation was performed using data from a previous study by Holthoff et al. [27] in which the benefits of resveratrol on septic AKI were quantified by histological changes in a murine model ($n = 6$). A sample size of 6 dogs per group would be needed assuming a difference in histology score of 1.2 between the treatment and control groups, with a standard deviation in the scores of 0.6, in order to achieve 80% power while taking an alpha-value <0.05 as significant. Due to the small sample size of this study, a non-parametric Mann–Whitney test was used to compare the outcomes between the two treatment groups. The changes in MCF on the ROTEM between the two groups was analyzed by adjusting for volume of blood loss in each animal using a linear mixed model. All analyses were two tailed using SPSS for Windows (version 23, IBM, Armonk, NY, USA) and MedCalc Statistical Software (version 18.11.3, Ostend, Belgium). A p-value < 0.05 was taken as significant, and no adjustment was made for multiple statistical testing in this study.

3. Results

Six dogs were included in each group. Exact age of each dog was unknown, but veterinary examination suggested all dogs were adult (2–7 years). The body weights of the dogs were not statistically different between the two groups (median [range] for the control group was 30.9 kg (30.1–32.5), and for treatment group was 32.6 kg (27.9–34.0); $p = 0.441$). All dogs had a lean body condition (score 3/9), appropriate for trained racing dogs of this breed. No adverse effects of oral resveratrol were noted, and all dogs completed the study. There was no significant difference in renal and coagulation parameters between the two groups prior to initiation of resveratrol treatment (Tables 2 and 3), but there was a suggestion that seven days of resveratrol treatment could improve clot strength compared to the control prior to induction of hemorrhage (InTEM MCF 54 vs. 43 mm respectively; $p = 0.009$)(Table 3). No resveratrol and its metabolites were detected in the control dogs. Among the six dogs treated with resveratrol, only one had a detectable level of resveratrol at a concentration < 50 mg/mL.

Table 2. Renal parameters of anesthetized greyhound dogs before and after induced hemorrhage ($n = 11$).

Variables	Resveratrol Treated ($n = 5$)	Control ($n = 6$)	p-Value #
Serum creatinine, µmol/L:*			
-Pre-treatment	129 (121–146)	118 (105–128)	0.126
-Pre-hemorrhage	125 (113–136)	105 (99–120)	0.126
-1 h post-hemorrhage	145 (133–155)	133 (128–151)	0.329
-2 h post-resuscitation	155 (145–185)	154 (130–167)	0.537
Serum cystatin:			
-Pre-treatment	49 (36–56)	45 (36–58)	0.999
-Pre-hemorrhage	52 (41–63)	46 (41–54)	0.537
-1 h post-hemorrhage	64 (46–69)	56 (49–72)	0.931
-2 h post-resuscitation	47 (32–55)	39 (31–47)	0.537
Serum NGAL:			
-Pre-treatment	8.4 (4.6–12.8)	8.8 (4.8–12.9)	0.999
-Pre-hemorrhage	9.4 (7.7–20.2)	10.8 (8.5–13.1)	0.931
-1 h post-hemorrhage	12.4 (11–22.9)	14.9 (12.3–22.8)	0.662
-2 h post-resuscitation	12.1 (7.8–19.5)	11.2 (9.9–16.9)	0.999
Serum KIM-1:			
-Pre-treatment	59 (51–94)	75 (42–101)	0.73
-Pre-hemorrhage	71 (54–87)	74 (53–90)	0.999
-1 h post-hemorrhage	54 (46–68)	64 (48–75)	0.537
-2 h post-resuscitation	39 (35–45)	42 (32–47)	0.931
Urine clusterin:			
-Pre-treatment	219 (146–617)	250 (83–921)	0.999
-Pre-hemorrhage	299 (134–348)	72 (24–186)	0.082
-1 h post-hemorrhage	318 (147–1315)	368 (49–554)	0.999
-2 h post-resuscitation	2242 (1275–3769)	1381 (727–4443)	0.537
Urine cystatin:			
-Pre-treatment	3.6 (2.9–4.3)	3.2 (2.1–3.8)	0.42
-Pre-hemorrhage	3.8 (2.5–5.4)	3.7 (1.8–5.7)	0.999
-1 h post-hemorrhage	4.0 (2.7–5.9)	5.0 (1.7–7.5)	0.999
-2 h post-resuscitation	8.7 (8.0–10.8)	9.1 (7.3–10.6)	0.792
Urine NGAL:			
-Pre-treatment	0.5 (0.3–3.7)	1.5 (0.1–6.2)	0.999
-Pre-hemorrhage	1.2 (0.3–3.0)	0.9 (0.6–3.3)	0.931
-1 h post-hemorrhage	1.1 (0.3–2.3)	0.8 (0.6–2.8)	0.841
-2 h post-resuscitation	25.6 (18.1–30.1)	19.7 (16.2–24.7)	0.429

Table 2. *Cont.*

Variables	Resveratrol Treated (n = 5)	Control (n = 6)	p-Value #
Urine KIM-1:			0.052
-Pre-treatment	50 (8–119)	139 (109–279)	
-Pre-hemorrhage	22 (11–138)	82 (44–184)	0.247
-1 h post-hemorrhage	18 (12–223)	137 (38–415)	0.286
-2 h post-resuscitation	49 (38–75)	57 (36–115)	0.662
Urine GGT to creatinine ratio:			
-Pre-treatment	21 (4–224)	13 (9–21)	0.841
-Pre-hemorrhage	25 (21–145)	51 (29–80)	0.548
-1 h post-hemorrhage	76 (24–306)	39 (26–285)	0.999
-2 h post-resuscitation	1190 (842–12996)	1299 (420–2500)	0.69
Glomerular injury, %	3.9 (2.4–15.1)	0.6 (0–3.7)	0.082
Tubular injury score	0.30 (0.28–0.89)	0.30 (0.11–0.49)	0.429

p-value by non-parametric Mann–Whitney test. All values are median (with interquartile range –IQR unless stated otherwise). Except for the urine KIM-1 (p = 0.536), all kidney injury parameters listed in this table differed significantly (p-values < 0.05) during the time course of the experiment. * The number of animals developed stage-1 acute kidney injury–with an increment in serum creatinine > 26.4 µmol/L-between the resveratrol (n = 2, 40%) and control (n = 6, 100%) groups appeared different but this was not statistically significant (p = 0.061). GGT = gamma-glutamyl transferase; KIM-1 = kidney injury molecule 1; NGAL = neutrophil gelatinase-associated lipocalin.

Table 3. Rotational Thromboelastometry (ROTEM®) viscoelastic and thrombin generation parameters of the anesthetized greyhound dogs before and after induced hemorrhage (n = 9).

Variables	Resveratrol Treated (n = 5)	Control (n = 4)	p-Value #
CT-InTEM:			
-Pre-treatment	198 (167–233)	194 (169–209)	0.73
-Pre-hemorrhage	168 (137–192)	133 (80–169)	0.177
-1 h post-hemorrhage	142 (130–165)	168 (140–174)	0.548
-2 h post-resuscitation	160 (142–200)	183 (168–193)	0.286
CT-ExTEM:			
-Pre-treatment	71 (40–90)	72 (49–98)	0.556
-Pre-hemorrhage	40 (35–60)	86 (31–155)	0.429
-1 h post-hemorrhage	62 (38–76)	101 (65–112)	0.082
-2 h post-resuscitation	76 (69–88)	113 (100–188)	0.016
MCF-InTEM:			
-Pre-treatment	51 (44–56)	48 (43–50)	0.556
-Pre-hemorrhage	54 (47–63)	43 (38–47)	0.009
-1 h post-hemorrhage	38 (33–45)	40 (32–44)	0.841
-2 h post-resuscitation	33 (30–41)	33 (28–37)	0.556
MCF-ExTEM:			
-Pre-treatment	41 (38–51)	40 (33–47)	0.73
-Pre-hemorrhage	52 (47–63)	45 (33–50)	0.052
-1 h post-hemorrhage	38 (34–45)	27 (25–36)	0.052
-2 h post-resuscitation	35 (30–39)	21 (14–32)	0.063
PFA-closure time:			
-Pre-hemorrhage	78 (71–93)	87 (70–131)	0.914
-1 h post-hemorrhage	111 (106–115)	95 (79–95)	0.057
-2 h post-resuscitation	91 (75–91)	89 (79–120)	0.857
Peak thrombin generation, nM:			
-Pre-treatment	66 (47–93)	101 (56–121)	0.286
-1 h post-hemorrhage	93 (61–137)	41 (19–56)	0.032
AUC thrombin generation, nM × min:			
-Pre-treatment	329 (252–329)	406 (254–406)	0.999
-1 h post-hemorrhage	495 (495–495)	294 (160–427)	0.4

p-value by non-parametric Mann–Whitney test. All values are median (with interquartile range –IQR unless stated otherwise). CT-INTEM (p = 0.008), MCF-INTEM (p = 0.001), and MCF-EXTEM (p = 0.001) all significantly differed with time. AUC = area under curve; CT = clotting time; MCF = maximum clot firmness; PFA = platelet function analyzer.

3.1. Cardiovascular Results

MAP over time are summarized in Figure 2. The median time to induce hemorrhage to achieve a MAP $\leq$ 40 mmHg was within 4 min (interquartile range [IQR] 2.75–8.75) in the treatment group and 5 (3–7.75) minutes in the control group which were not statistically different (p = 0.939). A significantly larger amount of blood was, however, withdrawn to achieve and maintain this MAP target for 60 min for the dogs treated with resveratrol compared to control (median 63.8 vs. 54.9 mL respectively, p = 0.041), with a slightly lower MAP at end of one-hour of hemorrhage in the resveratrol group (33 vs. 39 mmHg respectively, p = 0.041). Arterial and venous pH were also significantly lower in the treatment group versus control group at end of one-hour of hemorrhage (arterial 7.24 vs. 7.35, p = 0.011; venous 7.23 vs. 7.32, p = 0.047 respectively). The other blood gas and hemodynamic parameters between the two groups were not significantly different (Tables 4 and 5). Time for the MAP to return to $\geq$ 60 mmHg following cessation of induced bleeding and initiation of Gelofusine® fluid resuscitation was not different between the two groups (R group: median 17, IQR 12.5–25 vs. C group: 14, IQR 11.3–17.3; p = 0.571). No additional fluid administration was required in any dog in order to maintain MAP > 60 mmHg during the subsequent three-hour reperfusion period. There were also no significant differences in the arterial and venous blood gases including OER between the two groups during the experiment.

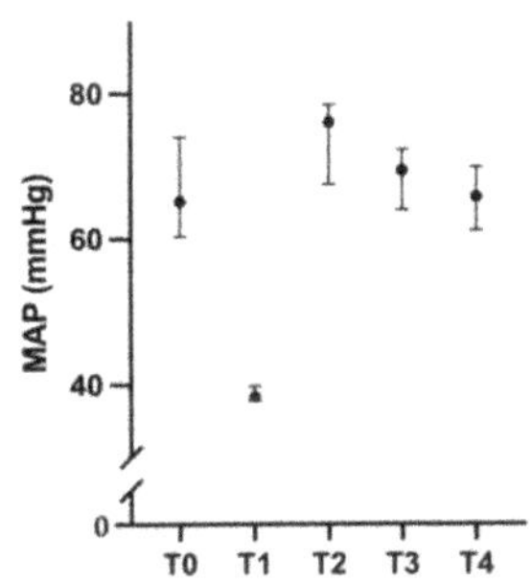

(A)

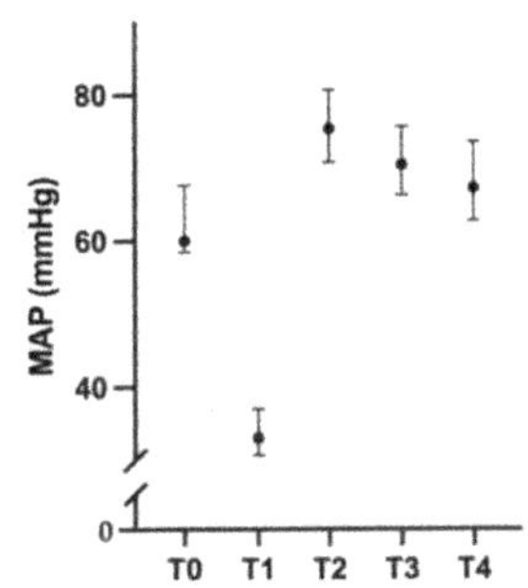

(B)

Figure 2. Changes in mean arterial pressure (MAP) over time in anesthetized greyhounds subjected in a pressure-targeted hemorrhage model. Values are median and interquartile range. (**A**) Control group (n = 6). (**B**) Resveratrol pretreated group (n = 6).

Table 4. Arterial and venous blood gas parameters during the experiment (n = 12).

Variables	Resveratrol Treated (n = 6)	Control (n = 6)	p-Value [#]
Arterial pH:			
-Pre-hemorrhage	7.44 (7.41–7.48)	7.46 (7.40–7.52)	0.696
-1 h post-hemorrhage	7.24 (7.21–7.33)	7.35 (7.33–7.36)	0.011
-1 h post-resuscitation	7.33 (7.32–7.38)	7.35 (7.31–7.39)	0.913
-2 h post-resuscitation	7.35 (7.34–7.39)	7.36 (7.32–7.42)	0.853
P_aO_2 (mmHg):			
-Pre-hemorrhage	187 (178–191)	196 (179–2.8)	0.329
-1 h post-hemorrhage	133 (124–153)	147 (130–170)	0.485
-1 h post-resuscitation	170 (146–207)	150 (132–163)	0.147
-2 h post-resuscitation	159 (152–174)	157 (138–172)	0.699
P_aCO_2 (mmHg):			
-Pre-hemorrhage	33.1 (28.4–38.5)	32.1 (26.7–36.0)	0.662
-1 h post-hemorrhage	45.9 (36.9–51.6)	37.9 (36.1–40.7)	0.162
-1 h post-resuscitation	43.2 (40.4–46.4)	44.7 (38.9–48.0)	>0.999
-2 h post-resuscitation	41.7 (37.0–43.7)	41.3 (37.0–46.2)	>0.999
Arterial HCO_3 (mmol/L)			
-Pre-hemorrhage	22.1 (20.8–23.7)	22.3 (21.2–23.3)	0.892

Table 4. *Cont.*

Variables	Resveratrol Treated (*n* = 6)	Control (*n* = 6)	*p*-Value [#]
-1 h post-hemorrhage	19.5 (17.1–20.9)	20.8 (18.3–22.3)	0.225
-1 h post-resuscitation	22.5 (21.2–23.6)	23.6 (22.9–24.2)	0.121
-2 h post-resuscitation	21.8 (20.9–23.4)	23.2 (22.7–23.9)	0.178
Arterial Base excess (mmol/L):			
-Pre-hemorrhage	−1.7 (−2.3–(−0.4))	−1.5 (−2.3–(−0.2))	0.792
-1 h post-hemorrhage	−6.9 (−8.3–(−5.0))	−4.0 (−6.7–(−2.9))	0.132
-1 h post-resuscitation	−2.4 (−3.7–(−1.6))	−1.5 (−2.2–(−0.5))	0.143
-2 h post-resuscitation	−2.9 (−3.6–(−1.5))	−1.7 (−2.3–(−0.7))	0.102
S_aO_2 (%):			
-Pre-hemorrhage			
-1 h post-hemorrhage	All values 100%, at all time points		
-1 h post-resuscitation			
-2 h post-resuscitation			
Central venous pH:			
-Pre-hemorrhage	7.41 (7.39–7.46)	7.43 (7.38–7.51)	0.792
-1 h post-hemorrhage	7.23 (7.16–7.28)	7.32 (7.29–7.32)	0.047
-1 h post-resuscitation	7.29 (7.28–7.34)	7.32 (7.30–7.36)	0.29
-2 h post-resuscitation	7.32 (7.30–7.35)	7.34 (7.29–7.39)	0.771
P_vO_2 (mmHg):			
-Pre-hemorrhage	46.1 (39.9–51.2)	46.5 (43.8–50.4)	0.931
-1 h post-hemorrhage	25.4 (19.2–29.3)	25.9 (23.8–27.7)	0.784
-1 h post-resuscitation	51.4 (45.1–59.6)	50.4 (46.4–53.0)	0.732
-2 h post-resuscitation	50.3 (44.9–59.0)	46.9 (41.9–49.9)	0.24
P_vCO_2 (mmHg):			
-Pre-hemorrhage	38.3 (33.8–42.1)	33.4 (29.3–40.0)	0.247
-1 h post-hemorrhage	59.3 (47.4–70.0)	49.1 (45.6–51.2)	0.18
-1 h post-resuscitation	50.4 (48.4–51.7)	48.5 (43.2–52.0)	0.589
-2 h post-resuscitation	47.1 (44.8–50.4)	48.0 (42.3–53.2)	0.181
Central venous HCO_3 (mmol/L)			
-Pre-hemorrhage	24.0 (23.0–25.1)	23.4 (20.8 - 24.6)	0.699
-1 h post-hemorrhage	23.0 (21.7–25.0)	24.1 (21.4–25.4)	0.667
-1 h post-resuscitation	23.3 (22.9–25.0)	24.5 (23.3–24.9)	0.407
-2 h post-resuscitation	23.2 (22.6–24.5)	24.7 (23.1–25.2)	0.37
Venous Base excess (mmol/L):			
-Pre-hemorrhage	0.9 (−0.8–1.4)	−0.5 (−3.1–0.8)	0.306
-1 h post-hemorrhage	−3.8 (−5.5–(−1.2))	−1.6 (−4.8–(−0.1))	0.589
-1 h post-resuscitation	−2.5 (−2.8–(−0.5))	−1.3 (−1.8–(−0.4))	0.234
-2 h post-resuscitation	−2.2 (−2.6–(−0.1))	−1.1 (−1.7–(−0.5))	0.31
S_vO_2 (%):			
-Pre-hemorrhage	88.2 (76.3–90.8)	91.7 (85.2–92.6)	0.162
-1 h post-hemorrhage	29.3 (22.9–38.5)	41.0 (34.1–49.9)	0.064
-1 h post-resuscitation	84.0 (59.4–86.2)	85.1 (80.1–90.0)	0.589
-2 h post-resuscitation	84.2 (82.1–88.0)	79.6 (72.4–86.9)	0.589
Hemoglobin-venous (g/dL):			
-Pre-hemorrhage	13.5 (12.7–13.7)	15.8 (14.5–16.1)	0.262
-1 h post-hemorrhage	17.2 (14.9–19.5)	17.5 (15.8–18.1)	0.965
-1 h post-resuscitation	8.7 (7.1–10.7)	9.2 (9.0–9.5)	0.662
-2 h post-resuscitation	8.8 (7.3–10.8)	9.3 (8.8–9.5)	0.79
Sodium (mmol/L):			
-Pre-hemorrhage	145 (145–147)	146 (145–148)	>0.999
-1 h post-hemorrhage	144 (141–145)	145 (143–146)	0.298
-1 h post-resuscitation	144 (143–146)	145 (145–146)	0.141
-2 h post-resuscitation	144 (143–145)	145 (144–147)	0.175

[#] *p*-value by non-parametric Mann–Whitney test. All values are median (with interquartile range –IQR unless stated otherwise).

3.2. Renal Outcomes

Renal data from one dog in the resveratrol group were excluded from further analysis due to identification of pre-existing renal inflammatory disease on histological examination of the kidney tissue. Except for the urine KIM- 1 ($p = 0.536$), all kidney injury parameters increased significantly (p-values < 0.05) within 2 h after induced hemorrhage. The risks of

AKI—defined according to the IRIS criteria [22]—were not significantly different ($p = 0.844$) between the resveratrol (no AKI = 3 dogs, AKI grade 2 = 3 dogs) and control groups (no AKI = 1 dog, AKI grade 1 = 1 dog, AKI grade 2 = 4 dogs). All serum and urinary AKI biomarkers were also similar between the two groups (Table 2).

Table 5. Hemodynamic parameters, bleeding volume and intravenous fluid administered during the experiment ($n = 12$).

Variables	Resveratrol treated ($n = 6$)	Control ($n = 6$)	p-Value #
Bleeding volume, mL	63.8 (57.8–78.6)	54.9 (42.5–58.9)	0.041
Total intravenous fluid volume, mL:			
- Compound Sodium Lactate	978 (924.8–1004.3)	925 (013.5-942)	0.31
- Colloid	652 (616.5-669.5)	617 (609-628)	0.31
Heart rate, beats/min:			
-Pre-hemorrhage	86 (52–122)	104 (93–118)	0.485
-1 h post-hemorrhage	204 (179–221)	182 (156–208)	0.31
-1 h post-resuscitation	172 (153–184)	150 (134–171)	0.18
-2 h post-resuscitation	176 (156–189)	154 (126–167)	0.093
Mean arterial pressure, mmHg:			
-Pre-hemorrhage	60 (59–68)	65 (60–74)	0.31
-1 h post-hemorrhage	33 (31–37)	39 (38–40)	0.041
-1 h post-resuscitation	76 (71–81)	76 (68–79)	0.589
-2 h post-resuscitation	71 (67–76)	69 (64–73)	0.394
Cardiac index, mL/min/m^2:			
-Pre-hemorrhage	95 (90–104)	86 (74–93)	0.082
-1 h post-hemorrhage	36 (32–40)	41 (33–49)	0.18
-1 h post-resuscitation	118 (111–130)	134 (120–170)	0.093
-2 h post-resuscitation	121 (115–139)	143 (119–152)	0.31
Oxygen extraction ratio, %:			
-Pre-hemorrhage	13 (12–23)	11 (10–17)	0.24
-1 h post-hemorrhage	71 (62–77)	60 (50–66)	0.065
-1 h post-resuscitation	19 (16–20)	17 (12–22)	0.699
-2 h post-resuscitation	18 (14–20)	22 (15–29)	0.589

p-value by non-parametric Mann–Whitney test. All values are median (with interquartile range –IQR unless stated otherwise).

3.3. Coagulation Outcomes

The clotting time became prolonged ($p = 0.008$) and MCF was weakened in both InTEM ($p = 0.001$) and ExTEM ($p = 0.001$) in the study dogs after hemorrhage. In addition to having a better clot firmness (or MCF) prior to hemorrhage, dogs treated with resveratrol also had a higher peak thrombin generation potential at 1-h after hemorrhage (Table 3). MCF on the InTEM appeared to remain better in the resveratrol group during the period of hemorrhage after adjusting for volume of blood withdrawn in each animal in the linear mixed model analysis (Table 6).

Table 6. Linear mixed models showing the associations between resveratrol treatment with maximum clot firmness on the Rotational Thromboelastometry (ROTEM®) of the anesthetized greyhound dogs before and after induced hemorrhage ($n = 12$) after adjusting for volume of blood removed using unstructured covariance structure. MCF = maximum clot firmness.

Predictors	Coefficient (95% Confidence Interval)	p-Value
MCF-InTEM		
Resveratrol treatment	10.6 (1.0–20.3)	0.034
Volume of blood removed	0.1 (0.3 to −0.3)	0.673
Time points	−4.7 (−3.3 to −6.1)	0.001
Resveratrol × time	1.8 (−0.3 to 3.7)	0.078
MCF-ExTEM		
Resveratrol treatment	5.5 (−2.8 to 13.8)	0.158
Volume of blood removed	−0.4 (−0.1 to −0.7)	0.028
Time points	−0.8 (−2.3 to 0.7)	0.260
Resveratrol × time	2.9 (0.7 to 5.1)	0.017

3.4. Renal Histopathology

Notably, severe tubular injury (evidence of ischemic coagulative necrosis) was not identified in any specimen. In the 20 randomly selected 200× fields of cortex (due to the lack of lesions in the deeper regions of the renal parenchyma) for each kidney examined, there was evidence of injury (loss of brush border, singly dead cells, sloughed cellular debris) in the study animals. Both control and resveratrol treated dogs had normal to mildly injured tubules (Figure 3A,B). One resveratrol-treated dog had evidence of moderate tubular injury (Figure 3C).

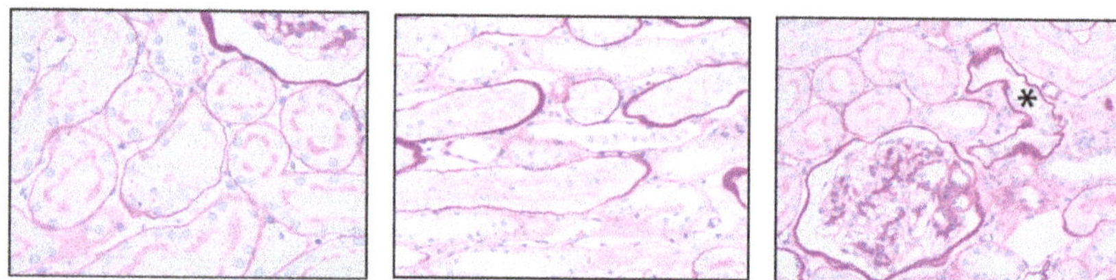

(A) (B) (C)

Figure 3. (**A**) Photomicrograph from a control dog with mild acute tubular epithelial injury, characterized by loss of the brush border and simplification of the tubular epithelium. There are 2 sloughed necrotic cells within the lumen. (**B**) Photomicrograph from a treated dog with mild acute tubular epithelial injury characterized by the presence of many sloughed epithelial cells within a tubular lumen. (**C**) Photomicrograph from a treated dog that had moderate acute tubular epithelial injury, characterized by simplification and attenuation of tubular epithelium (asterisk) that was more frequently observed than in dogs with mild injury. Periodic acid Schiff, 40 × magnification.

3.5. Renal Transmission Electron Microscopy (TEM)

Samples from two control dogs and two treated dogs were further evaluated using TEM. All four dogs had evidence of loss of the apical brush border, cellular swelling and intraluminal cellular fragments (Figure 4). *Proximal tubules* were most frequently damaged but distal convoluted tubules were occasionally affected. Ultra-structurally, there were no obvious differences between the two dogs in each study group (although this was not formally tested statistically because of the small sample size of this sub-group). Dogs that had minimal injury in the histology specimen had similar ultrastructural lesions on the TEM as dogs that had mild and moderate histologic lesions.

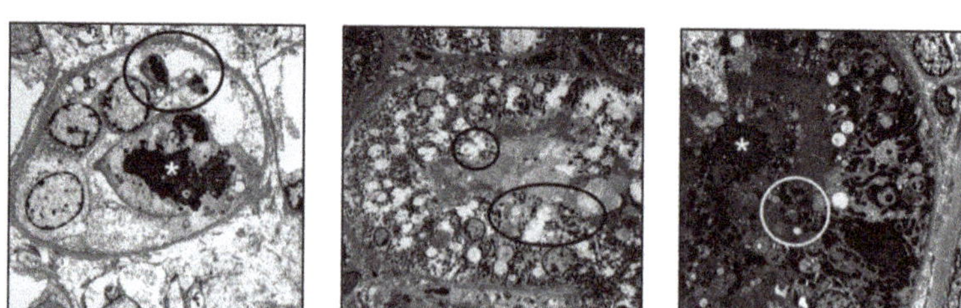

(A) (B) (C)

Figure 4. (**A**) Transmission electron micrograph from a control dog that had minimal acute tubular epithelial injury in the histology sample. The sample submitted for TEM showed sloughed cellular debris (circled) and electron dense cytoplasmic material (mineralization) (asterisk). There is denuded tubular basement membrane. Bar is 2 micron. (**B**) Transmission electron micrograph from a treated dog that had mild acute tubular epithelial injury in the histology sample. There is loss of the apical brush border of multiple cells (circled) whereas neighboring cells maintain intact brush borders. Bar is 20 micron. (**C**) Transmission electron micrograph from a control dog that had minimal injury in the histology sample. There is loss of the apical brush border (circled) as well as sloughed cells within the tubular lumen (asterisk). Bar is 2 micron.

4. Discussion

This study showed that seven days of oral resveratrol treatment in greyhound dogs improved their blood pressure tolerance to induced hemorrhage; that is, treated animals required a larger amount of blood loss to develop the same degree of hypotension. There was also a signal to suggest that resveratrol might improve clot strength and thrombin generation. The potential renal protective effect of resveratrol was, however, not observed both in terms of biochemical and histological assessments. These findings are clinically relevant and require further discussion.

First, a blood pressure targeted hemorrhage model was used in this study in an attempt to assess whether resveratrol could improve the maintenance of blood pressure during hemorrhage. Whether this was a better model than a model that induces a fixed amount of blood loss remains uncertain [28]. The volume of blood loss to achieve the predetermined MAP in our control animals (median 54.9, range 42.5–58.9 mL/kg) was consistent with the results of another study (53 mL/kg, 95% confidence interval 48–57) [29]. As such, the greater amount of blood loss needed to achieve and maintain the same degree of hypotension (and OER in the central venous blood) in the resveratrol-treated dogs (median 63.8, range 57.8–78.6 mL/kg) suggests that oral resveratrol pretreatment may have genuinely assisted the dogs in maintaining a better blood pressure during acute bleeding. This beneficial effect of resveratrol may theoretically, at least in part, be related to activation of SIRT1 gene, its estrogen agonist effect, or interactions with other targets or receptors that are relevant to pathogenic process of I-R injury [6–8]. Because we could not detect resveratrol levels in most blood samples of our resveratrol-treated dogs, the precise mechanisms how resveratrol can help to maintain the blood pressure remains unclear, but modification of gut microbiota is one possibility [30]. Nonetheless, the dogs in the resveratrol group did have a lower MAP and pH in the arterial and venous blood at the end of 60-min of hemorrhage. Repeating the study with a larger sample size, using a volume-guided hemorrhagic model using preset volume of blood loss, and extending the study duration to include survival time and effect on gut microbiota would be helpful to confirm the benefits of resveratrol, and clarify the mechanisms through which oral resveratrol pretreatment improves cardiovascular tolerance to hemorrhage.

Second, all study dogs had an increase in many serum and urinary AKI biomarkers induced hemorrhage; and these changes were associated with histopathological evidence of AKI. This was not surprising as a reduction in blood volume and pressure would certainly compromise renal blood flow, triggering a reduction in glomerular filtration rate and induction of renal ischemia [31]. As such, the hemorrhagic model used in our study was sensitive enough to induce bleeding-related AKI [32]. Despite this, there were no significant difference in all the renal biomarkers and histological changes between the resveratrol and control groups. This negative result could be explained by a number of reasons. First of all, a much lower dose (per body weight) of resveratrol was used in this study and outcomes were assessed in a shorter time frame than in the rat models in previous studies [6–8,16,27,33]. Furthermore, a small sample size could undermine the statistical power of this study. Of the dogs that were treated with resveratrol, 40% (2/6 dogs) developed stage-1 AKI (defined by an increment in SCr > 26.4 μmol/L) compared to 100% of the dogs in the control group, even though this difference did not reach statistical significance. Another possibility is that resveratrol might have different renal effects in different animal species. To the best of our knowledge, using resveratrol to prevent AKI during hemorrhage in dogs (in contrast to rats [16]) has not been assessed previously. Some types of intravenous fluid used for resuscitation can also have an adverse effect on the kidneys. Although the same types and similar quantities of intravenous fluid (Compound Sodium Lactate for maintenance at 10 mL/kg/hour and Gelofusine® to restore normal blood pressure: see Table 4) were used for both groups in this study, it is possible that resveratrol's renal protective effect—if there is one—was confounded or overwhelmed by a potential harmful or beneficial effect of Gelofusine® on the kidneys. Previous studies have shown that using Gelofusine® for resuscitation was less likely to induce AKI compared to the older forms of intravenous starches in humans [34], but recent animal studies have shown that Gelofusine® may induce AKI compared to crystalloids or fresh whole blood [35]. As such, it would be better to use crystalloid fluid alone during the reperfusion period in similar experiments in the future.

An interesting and important finding of this study was the positive association between resveratrol, clot firmness and thrombin generation. A previous study has showed that resveratrol could help to preserve platelet function in stored blood [36]. Conversely, resveratrol has also been shown to induce platelet dysfunction and coagulation derangements in a few other experimental studies [19,20,37,38]. Given hemostasis is a paramount outcome in any hemorrhagic situations, the effect of resveratrol on bleeding tendency requires further investigation.

Finally, we need to acknowledge the limitations of this study. The circumstances of how our study animals were recruited had limited our ability to address the important issue of survival time after hemorrhage. It is important to note that race-trained greyhounds are highly adapted to anaerobic tissue metabolism during exercise, modulating their cardiovascular, hemostatic and homeostatic responses to hemorrhage [39,40]. Prolonged anesthesia (5 h in this study) might have interfered with the normal cardiovascular compensatory responses to bleeding. In addition, pure μ-agonist opioids, such as the fentanyl used in this study, are known to cause venous and arterial blood wall relaxation, via endothelium and opioid-receptor independent mechanisms in various species [41,42]. Similarly, isoflurane also has a dose-dependent vasodilatory and negative inotropic effect, rendering our study animals more likely to develop hypotension after hemorrhage [43,44]. Nonetheless, a parallel-arm randomized-controlled design should have balanced the confounding effects of these factors.

5. Conclusions

This exploratory study showed that seven days of oral resveratrol treatment prior to bleeding improved greyhound dogs' blood pressure stability in response to severe hemorrhage and possibly also coagulation profiles compared to no resveratrol treatment. A renal protective effect of resveratrol in hemorrhage was, however, not observed. An adequately-powered study using a volume-targeted hemorrhage model assessing the benefits of resveratrol, including survival time, is needed before this dietary supplement can be recommended prior to exposure to hemorrhage.

Author Contributions: Conceptualization, A.L.R. and K.M.H.; Methodology, J.D., A.L.R. and K.M.H.; Investigation, J.D., A.L.R., C.R.S., R.E.C. and S.C.W.; Formal Analysis, K.M.H.; Resources, J.D., A.L.R., C.R.S., R.E.C., S.C.W. and K.M.H.; Writing—Original Draft, J.D. and A.L.R.; Writing—Review and Editing, J.D., A.L.R., C.R.S., R.E.C., S.C.W. and K.M.H.; Visualization, J.D., R.E.C. and K.M.H.; Supervision, A.L.R. and K.M.-H.; Project Administration, J.D. and A.L.R.; Funding Acquisition, K.M.-H. All authors have read and agreed to the published version of the manuscript.

Funding: This research was funded by the Australian Society of Anaesthetists Jackson Rees Research Award in 2015.

Institutional Review Board Statement: This study was conducted according to the guidelines of the Declaration of Helsinki, and approved by the Animal Ethics Committee of Murdoch University (R2726/15).

Informed Consent Statement: Written informed consent was obtained from the animal owners to publish this paper.

Data Availability Statement: The data presented in this study are available on request from the corresponding author.

Acknowledgments: The authors thank Alan Flechtner and Anne Saulsbery from the Comparative Mouse Phenotyping Shared Resource (Cancer Center Support Grant P30 CA016058) for histopathology and electron microscopy specimen preparation, and Jeffrey Lipman from the University of Queensland Centre for Clinical Research for measurement of plasma resveratrol concentration.

Conflicts of Interest: The authors declare no conflict of interest. The funders had no role in the design of the study; in the collection, analyses, or interpretation of data; in the writing of the manuscript, or in the decision to publish the results.

References

1. Hardy, J.F.; De Moerloose, P.; Samama, M. Massive transfusion and coagulopathy: Pathophysiology and implications for clinical management. *Can. J. Anaesth.* **2004**, *51*, 293–310. [CrossRef]
2. Pham, H.P.; Shaz, B.H. Update on massive transfusion. *Br. J. Anaesth.* **2013**, *111* (Suppl. S1), i71–i82. [CrossRef]
3. Fredericks, C.; Kubasiak, J.C.; Mentzer, C.J.; Yon, J.R. Massive transfusion: An update for the anesthesiologist. *World J. Anesthesiol.* **2017**, *6*, 14–21. [CrossRef]
4. Muirhead, B.; Weiss, A.D.H. Massive hemorrhage and transfusion in the operating room. *Can. J. Anaesth.* **2017**, *64*, 962–978. [CrossRef]
5. Gaunt, C.; Woolley, T. Management of haemorrhage in major trauma. *Contin. Educ. Anaesth. Crit. Care Pain* **2014**, *14*, 251–255. [CrossRef]
6. Sims, C.A.; Baur, J.A. The grapes and wrath: Using resveratrol to treat the pathophysiology of hemorrhagic shock. *Ann. N. Y. Acad. Sci.* **2017**, *1403*, 70–81. [CrossRef] [PubMed]
7. Liu, F.C.; Tsai, Y.F.; Tsai, H.I.; Yu, H.P. Anti-Inflammatory and Organ-Protective Effects of Resveratrol in Trauma-Hemorrhagic Injury. *Mediat. Inflamm.* **2015**, *2015*, 643763. [CrossRef] [PubMed]
8. Ayub, A.; Poulose, N.; Raju, R. Resveratrol Improves Survival and Prolongs Life Following Hemorrhagic Shock. *Mol. Med.* **2015**, *21*, 305–312. [CrossRef] [PubMed]
9. Goren, O.; Matot, I. Update on perioperative acute kidney injury. *Curr. Opin. Crit. Care* **2016**, *22*, 370–378. [CrossRef]
10. Moore, E.M.; Bellomo, R.; Nichol, A.D. The meaning of acute kidney injury and its relevance to intensive care and anaesthesia. *Anaesth. Intensive Care* **2012**, *40*, 929–948. [CrossRef]
11. Hobson, C.; Ruchi, R.; Bihorac, A. Perioperative Acute Kidney Injury: Risk Factors and Predictive Strategies. *Crit. Care Clin.* **2017**, *33*, 379–396. [CrossRef]
12. Calvert, S.; Shaw, A. Perioperative acute kidney injury. *Perioper Med.* **2012**, *1*, 6. [CrossRef] [PubMed]
13. Susantitaphong, P.; Cruz, D.N.; Cerda, J.; Abulfaraj, M.; Alqahtani, F.; Koulouridis, I.; Jaber, B.L. World incidence of AKI: A meta-analysis. *Clin. J. Am. Soc. Nephrol. CJASN* **2013**, *8*, 1482–1493. [CrossRef] [PubMed]
14. Lee, Y.J.; Hu, Y.Y.; Lin, Y.S.; Chang, C.T.; Lin, F.Y.; Wong, M.L.; Kuo-Hsuan, H.; Hsu, W.L. Urine neutrophil gelatinase-associated lipocalin (NGAL) as a biomarker for acute canine kidney injury. *BMC Vet. Res.* **2012**, *8*, 248. [CrossRef]
15. Legatti, S.A.M.; El Dib, R.; Legatti, E.; Botan, A.G.; Camargo, S.E.A.; Agarwal, A.; Barretti, P.; Paes, A.C. Acute kidney injury in cats and dogs: A proportional meta-analysis of case series studies. *PLoS ONE* **2018**, *13*, e0190772. [CrossRef] [PubMed]
16. Wang, H.; Guan, Y.; Karamercan, M.A.; Ye, L.; Bhatti, T.; Becker, L.B.; Baur, J.A.; Sims, C.A. Resveratrol Rescues Kidney Mitochondrial Function Following Hemorrhagic Shock. *Shock* **2015**, *44*, 173–180. [CrossRef] [PubMed]
17. Fulop, A.; Turoczi, Z.; Garbaisz, D.; Harsanyi, L.; Szijarto, A. Experimental models of hemorrhagic shock: A review. *Eur. Surg. Res.* **2013**, *50*, 57–70. [CrossRef]
18. Majde, J.A. Animal models for hemorrhage and resuscitation research. *J. Trauma* **2003**, *54*, S100–S105. [CrossRef]
19. Olas, B.; Wachowicz, B.; Saluk-Juszczak, J.; Zielinski, T. Effect of resveratrol, a natural polyphenolic compound, on platelet activation induced by endotoxin or thrombin. *Thromb. Res.* **2002**, *107*, 141–145. [CrossRef]
20. Zbikowska, H.M.; Olas, B.; Wachowicz, B.; Krajewski, T. Response of blood platelets to resveratrol. *Platelets* **1999**, *10*, 247–252. [CrossRef]
21. Zaldivar-Lopez, S.; Marin, L.M.; Iazbik, M.C.; Westendorf-Stingle, N.; Hensley, S.; Couto, C.G. Clinical pathology of Greyhounds and other sighthounds. *Vet. Clin. Pathol.* **2011**, *40*, 414–425. [CrossRef]
22. Mason, D.; O'Grady, M.; Woods, J.; McDonell, W. Assessment of lithium dilution cardiac output as a technique for measurement of cardiac output in dogs. *Am. J. Vet. Res.* **2001**, *62*, 1255–1261. [CrossRef]
23. Cilley, R.E.; Polley, T.Z., Jr.; Zwischenberger, J.B.; Toomasian, J.M.; Bartlett, R.H. Independent measurement of oxygen consumption and oxygen delivery. *J. Surg. Res.* **1989**, *47*, 242–247. [CrossRef]
24. Samsel, R.W.; Nelson, D.P.; Sanders, W.M.; Wood, L.D.; Schumacker, P.T. Effect of endotoxin on systemic and skeletal muscle O_2 extraction. *J. Appl. Physiol.* **1988**, *65*, 1377–1382. [CrossRef]
25. Gutierrez, G.; Warley, A.R.; Dantzker, D.R. Oxygen delivery and utilization in hypothermic dogs. *J. Appl. Physiol. (1985)* **1986**, *60*, 751–757. [CrossRef] [PubMed]
26. Muzzio, M.; Huang, Z.; Hu, S.C.; Johnson, W.D.; McCormick, D.L.; Kapetanovic, I.M. Determination of resveratrol and its sulfate and glucuronide metabolites in plasma by LC-MS/MS and their pharmacokinetics in dogs. *J. Pharm. Biomed. Anal.* **2012**, *59*, 201–208. [CrossRef] [PubMed]
27. Holthoff, J.H.; Wang, Z.; Seely, K.A.; Gokden, N.; Mayeux, P.R. Resveratrol improves renal microcirculation, protects the tubular epithelium, and prolongs survival in a mouse model of sepsis-induced acute kidney injury. *Kidney Int.* **2012**, *81*, 370–378. [CrossRef]
28. Tsukamoto, T.; Pape, H.C. Animal models for trauma research: What are the options? *Shock* **2009**, *31*, 3–10. [CrossRef]

29. Davis, J.; Raisis, A.L.; Cianciolo, R.E.; Miller, D.W.; Shiel, R.E.; Nabity, M.B.; Hosgood, G.L. Urinary neutrophil gelatinase-associated lipocalin concentration changes after acute haemorrhage and colloid-mediated reperfusion in anaesthetized dogs. *Vet. Anaesth. Analg.* **2016**, *43*, 262–270. [CrossRef]
30. Anhê, F.F.; Choi, B.S.Y.; Dyck, J.R.B.; Schertzer, J.D.; Marette, A. Host-Microbe Interplay in the Cardiometabolic Benefits of Dietary Polyphenols. *Trends Endocrinol. Metab.* **2019**, *30*, 384–395. [CrossRef]
31. Basile, D.P.; Anderson, M.D.; Sutton, T.A. Pathophysiology of Acute Kidney Injury. *Compr. Physiol.* **2012**, *2*, 1303–1353. [CrossRef]
32. Tirapelli, L.F.; Barione, D.F.; Trazzi, B.F.; Tirapelli, D.P.; Novas, P.C.; Silva, C.S.; Martinez, M.; Costa, R.S.; Tucci, S.; Suaid, H.J.; et al. Comparison of two models for evaluation histopathology of experimental renal ischemia. *Transpl. Proc.* **2009**, *41*, 4083–4087. [CrossRef]
33. Powell, R.D.; Swet, J.H.; Kennedy, K.L.; Huynh, T.T.; McKillop, I.H.; Evans, S.L. Resveratrol attenuates hypoxic injury in a primary hepatocyte model of hemorrhagic shock and resuscitation. *J. Trauma Acute Care Surg.* **2014**, *76*, 409–417. [CrossRef]
34. Saw, M.M.; Chandler, B.; Ho, K.M. Benefits and risks of using gelatin solution as a plasma expander for perioperative and critically ill patients: A meta-analysis. *Anaesth. Intensive Care* **2012**, *40*, 17–32. [CrossRef] [PubMed]
35. Boyd, C.J.; Claus, M.A.; Raisis, A.L.; Cianciolo, R.; Bosio, E.; Hosgood, G.; Nabity, M.; Mori, T.; Barden, A.; Sharp, C.R.; et al. Evaluation of biomarkers of kidney injury following 4% succinylated gelatin and 6% hydroxyethyl starch 130/0.4 administration in a canine hemorrhagic shock model. *J. Vet. Emerg. Crit. Care* **2019**, *29*, 132–142. [CrossRef] [PubMed]
36. Ekaney, M.L.; Grable, M.A.; Powers, W.F., 4th; McKillop, I.H.; Evans, S.L. Cytochrome c and resveratrol preserve platelet function during cold storage. *J. Trauma Acute Care Surg.* **2017**, *83*, 271–277. [CrossRef] [PubMed]
37. Shahidi, M.; Parhizkary, F.; Sharifi, R.; Ghotaslou, A.; Barati, M. Effects of resveratrol on coagulative, fibrinolytic, and inflammatory marker expression and secretion by endothelial cells (human umbilical vein endothelial cells). *Blood Coagul. Fibrinolysis* **2020**, *31*, 207–212. [CrossRef] [PubMed]
38. Marumo, M.; Ekawa, K.; Wakabayashi, I. Resveratrol inhibits Ca(2+) signals and aggregation of platelets. *Environ. Health Prev. Med.* **2020**, *25*, 70. [CrossRef] [PubMed]
39. Horvath, S.J.; Couto, C.G.; Yant, K.; Kontur, K.; Bohenko, L.; Iazbik, M.C.; Marín, L.M.; Hudson, D.; Chase, J.; Frye, M.; et al. Effects of racing on reticulocyte concentrations in Greyhounds. *Vet. Clin. Pathol.* **2014**, *43*, 15–23. [CrossRef]
40. Cox, R.H.; Bagshaw, R.J.; Detweiler, D.K. Baroreceptor reflex cardiovascular control in mongrel dogs and racing greyhounds. *Am. J. Physiol.* **1985**, *249*, H655–H662. [CrossRef]
41. Sasmaz, S.; Sahin, A.; Duman, I. Inhibitory Effect of Fentanyl on Phenylephrine-Induced Contraction on Rabbit Aorta. *Pharmacol. Pharmacy* **2011**, *2*, 141–145. [CrossRef]
42. Sahin, A.S.; Duman, A.; Atalik, E.K.; Ogün, C.O.; Sahin, T.K.; Erol, A.; Ozergin, U. The mechanisms of the direct vascular effects of fentanyl on isolated human saphenous veins in vitro. *J. Cardiothorac. Vasc. Anesth.* **2005**, *19*, 197–200. [CrossRef] [PubMed]
43. Pagel, P.S.; Kampine, J.P.; Schmeling, W.T.; Warltier, D.C. Evaluation of myocardial contractility in the chronically instrumented dog with intact autonomic nervous system function: Effects of desflurane and isoflurane. *Acta Anaesthesiol. Scand.* **1993**, *37*, 203–210. [CrossRef] [PubMed]
44. Merin, R.G.; Bernard, J.M.; Doursout, M.F.; Cohen, M.; Chelly, J.E. Comparison of the effects of isoflurane and desflurane on cardiovascular dynamics and regional blood flow in the chronically instrumented dog. *Anesthesiology* **1991**, *74*, 568–574. [CrossRef]

veterinary sciences

MDPI

Review

Animal Models of Colorectal Cancer: From Spontaneous to Genetically Engineered Models and Their Applications

Elisabete Nascimento-Gonçalves [1,†], Bruno A.L. Mendes [1,†], Rita Silva-Reis [1,†], Ana I. Faustino-Rocha [1,2,*], Adelina Gama [3,4] and Paula A. Oliveira [1,4,*]

1 Center for the Research and Technology of Agro-Environmental and Biological Sciences (CITAB), University of Trás-os-Montes and Alto Douro (UTAD), 5000-801 Vila Real, Portugal; elisabete.nascimento.g@gmail.com (E.N.-G.); b.alex.mendes@hotmail.com (B.A.L.M.); ritareis96@hotmail.com (R.S.-R.)

2 Department of Zootechnics, School of Sciences and Technology, University of Évora, 7000-812 Évora, Portugal

3 Department of Veterinary Sciences, University of Trás-os-Montes and Alto Douro (UTAD), 5000-801 Vila Real, Portugal; agama@utad.pt

4 Animal and Veterinary Research Center (CECAV), University of Trás-os-Montes and Alto Douro (UTAD), 5000-801 Vila Real, Portugal

* Correspondence: anafaustino@uevora.pt (A.I.F.-R.); pamo@utad.pt (P.A.O.)

† Authors contributed equally to the work.

Abstract: Colorectal cancer is one of the most common gastrointestinal malignancies in humans, affecting approximately 1.8 million people worldwide. This disease has a major social impact and high treatment costs. Animal models allow us to understand and follow the colon cancer progression; thus, in vivo studies are essential to improve and discover new ways of prevention and treatment. Dietary natural products have been under investigation for better and natural prevention, envisioning to show their potential. This manuscript intends to provide the readers a review of rodent colorectal cancer models available in the literature, highlighting their advantages and disadvantages, as well as their potential in the evaluation of several drugs and natural compounds' effects on colorectal cancer.

Keywords: spontaneous models; induced models; genetically engineered models

Citation: Nascimento-Gonçalves, E.; Mendes, B.A.L.; Silva-Reis, R.; Faustino-Rocha, A.I.; Gama, A.; Oliveira, P.A. Animal Models of Colorectal Cancer: From Spontaneous to Genetically Engineered Models and Their Applications. *Vet. Sci.* **2021**, *8*, 59. https://doi.org/10.3390/vetsci8040059

Academic Editor: Patrick Butaye

Received: 5 February 2021
Accepted: 29 March 2021
Published: 5 April 2021

1. Introduction

Worldwide, colorectal cancer is the third most common cancer in men and second in women [1]. Many risk factors have been considered for the development of colorectal cancer, such as the ingestion of processed meat, alcoholic drinks, body fatness, low intake of vegetables and fruits, smoking, and other concomitant diseases, such as inflammatory bowel disease (IBD), Crohn's disease, and ulcerative colitis [2,3].

Colorectal cancer is characterized by the invasion of neoplastic epithelial cells below the muscularis mucosae of the colorectal wall [4]. Its evolution is slow and characterized by different stages. Progressive changes in the amount or activity of proteins that regulate cell proliferation, differentiation, and cell survival occur, leading to a disorder in cell replication that contributes to the development of proliferative lesions, such as adenoma [5]. Subsequently, the intestinal epithelium undergoes a malignant transformation to invasive carcinoma [4,5]. Besides adenomas, hyperplastic polyps, serrated adenomas, flat adenomas, and dysplastic lesions are also observed in the colon as other types of preneoplastic lesions [5]. In humans, colorectal cancer is histologically classified as an adenocarcinoma [4,6]. In Figure 1 we can observe the progression from normal intestinal epithelium to carcinoma.

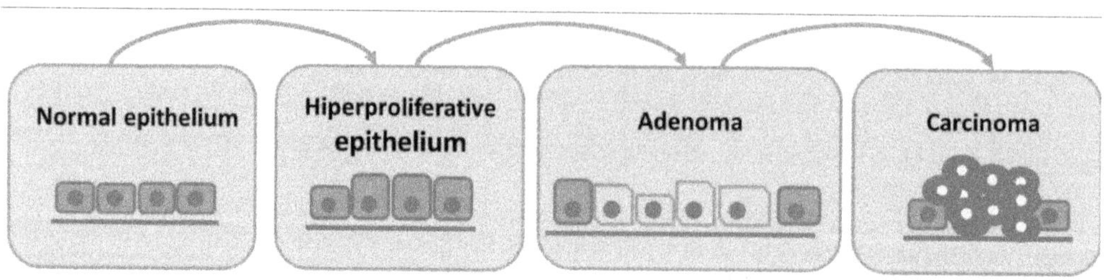

Figure 1. Schematic representation of adenoma–carcinoma multistep model. The normal cells of colon epithelium progress sequentially to a hyperproliferative epithelium, premalignant adenoma, and then carcinoma.

About 97% of colorectal cancers are spontaneous, and the remaining are due to one of two autosomal dominant inherited diseases: hereditary non-polyposis colorectal cancer (HNPCC) and familial adenomatous polyposis (FAP) [4,5]. The genetic mechanisms of spontaneous CRC are present in the adenoma–carcinoma sequence. Carcinogenesis is initiated with inactivating mutations in the tumor suppressor adenomatous polyposis coli (APC) gene, followed by an accumulation of mutations in the genes K-RAS, PI3K, DCC, SMAD2, SMAD4, and lastly the mutation in the tumor suppressor gene TP53 that determines the progression from the non-invasive to the invasive CRC [7].

Laboratory rodents are commonly used as animal models in experimental research because they are easy and cheap to maintain, their physiology and genetics are well studied, and they are mammals like humans [8]. They allow us to understand and follow the progression of diseases, enable the discovery and development of new preventive strategies, which can be later used in clinical trials. An ideal animal model of human disease should be simple, not expensive, and mimic the disease in terms of morphology, biochemical alterations, and biological behavior [4,9]. Several works have reviewed the use of animal models of CRC [10]. However, this manuscript not only intends to augment the information on rodent models of CRC, highlighting their advantages and disadvantages, but also to review their applications and how they can be used to evaluate natural compounds, nutrition habits, and drugs.

2. Rat and Mouse Colon and Rectum: Anatomy and Histology

The rat and mouse intestine are similar to that of humans concerning development, structure, and functions [9]. The large intestine comprises the cecum, the colon, the rectum, and anus, and it is responsible for the absorption of water and salt from feces [11] (Figure 2).

The cecum is a curved blind sac responsible for bacterial fermentation and empties into the proximal/right colon. Even though the rodents' colon and rectum represent a percentage of the total size of the large intestine similar to the humans, the cecum is much bigger in rats, which may be attributed to the high fiber content of their diet [12]. The colon continues toward the pyloric region of the stomach and has the same histological structure of the gastrointestinal tract: mucosa, submucosa, inner circular and outer longitudinal tunica muscularis, and serosa [1]. Despite histological similarities, rats and mice do not have adipose tissue in the submucosa, unlike humans who have it in abundance. The colon can be divided into ascending (it leads cranially to the thoracic cavity), transverse colon (from the left to the right side), and descending colon (on the right side of the abdominal cavity). The rodents' middle and distal colon corresponds to the human left colon [13]. The rectum is relatively short and indistinct from the distal colon. The anorectal junction has no stratified columnar epithelium, and the anal canal is lined by keratinized stratified squamous epithelium [11].

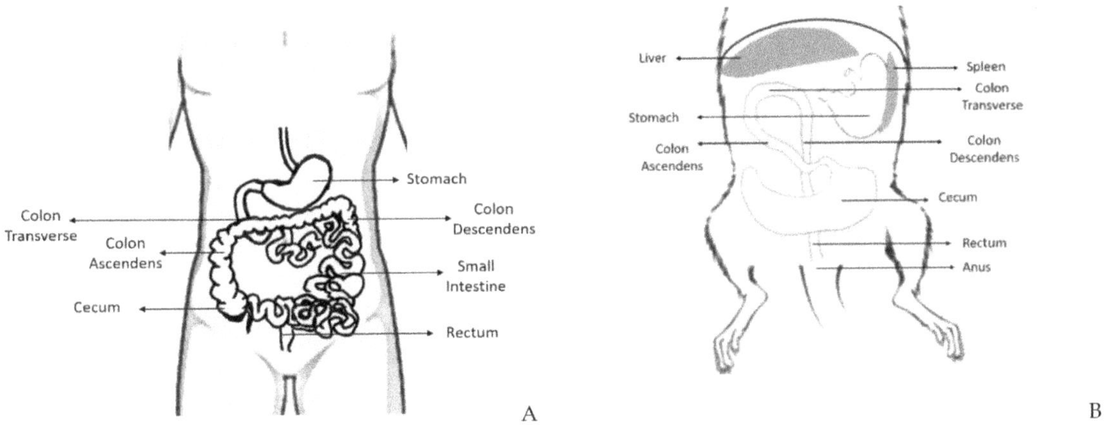

Figure 2. Schematic representation of some parts of the human (**A**) and rodent (**B**) digestive system, where is possible to observe the distinct portions of the large intestine: cecum, colon (ascendant, transverse, and descendent), rectum and anus, and its topographic anatomy. Human and rodent in supine position.

3. Rodents as Models of Colorectal Cancer

Although there is no ideal animal model that replicates all human disease aspects, the rodents are accepted as good models to study colorectal carcinogenesis because of their physiological similarity with humans, reproducible tumor induction, and the possibility to study the disease biopathology and test strategies for cancer prevention and treatment [4].

An ideal rodent model of colorectal cancer should develop carcinomas in the colon and rectum, with a high incidence in a short period, allow non-invasive monitoring of disease progression, and follow the histological and molecular characteristics of human colorectal cancer [8,13]. The models available to study colorectal cancer include spontaneous, induced, genetically engineered, xenograft, and syngeneic models (Figure 3).

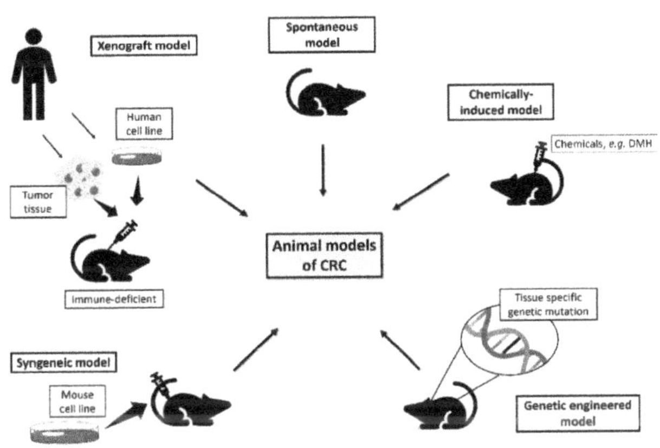

Figure 3. Rodent models available to study human colorectal cancer.

3.1. Spontaneous Models

Spontaneous development of colorectal cancer in rats and mice is rare, although some cases were reported in the literature. In 1969 it was reported that C57BL mice developed adenomas in the colon [9], and in 1975 Miyamoto and colleagues showed that 30–40% of animals from the Wistar-Furth/Osaka strain developed adenocarcinomas [14]. More recently, in 2009, Newark and colleagues showed that C57BL/6J developed cancer in the large intestine with an incidence of 1% [15]. These models are not very used due to unpredictability and low reproducibility [4]. In 40% of the spontaneous rat models, the period of latency is approximately eight months [16,17].

3.2. Induced Models

Colorectal tumors can be induced in rodents through the administration of chemical carcinogens alone or in combination [5]. There are two types of chemical carcinogenic agents: direct and indirect. Direct carcinogens do not need to be metabolized to induce cancer, while the indirect agents are administered in their inactive form and only acquire carcinogenic activity when biotransformed and converted into their active form in the liver [5,6].

3.2.1. Chemically Induced Models

In 1941, Lorenz and Stewart were the first to induce intestinal mouse tumors by feeding them with dibenzanthracene or methylcholanthrene [4], leading to the development of adenocarcinoma of the small intestine [4]. Later, in 1947, Lisco and colleagues induced carcinomas in the rat colon through feeding with radioactive yttrium [17]. Some years later, in 1963, Laquer and colleagues stated hydrazines are colonic carcinogens. Rats developed adenocarcinomas after feeding with a large quantity of cycad flour, which have hydrazine called cycasin, a form of methylazoxymethanol (MAM) [18,19].

Over the years, experimental research was conducted to discover chemical carcinogens specific for colorectal cancer such as 3,2′-dimethyl-4-aminobiphenyl (DMAB), alkylnitrosamines such as N-methyl-N-nitrosourea (MNU) and N-methyl-N-nitrosoguanidine (MNNG), 1,2-dimethylhydrazine (DMH), azoxymethane (AOM), and 2-amino-1-methyl-6-phenylimidazo (4,5-b) pyridine (PhIP) [20].

These carcinogens can be indirect-acting agents (DMAB, DMH, AOM, and PhiP), which need an enzymatic reaction to be converted into an active form, or direct-acting carcinogens (MNU and MNNG) that do not need biological catalysis [5].

3,2′-Dimethyl-4-Aminobiphenyl (DMAB)

Walpole and colleagues, in 1952, described the first induction of intestine cancer in male rats by subcutaneous administration of DMAB dissolved in arachis oil, at a mean total dose of 2.8 g/kg, for 141 days [21]. After that, other researchers evaluated the carcinogenic potential of this compound [22]. Of these works, the one made by Reddy et al. [23] stands out; they showed that 30% of F344 rats fed with a low-fat diet and 75% of animals fed with a high-fat diet developed colon cancer after being injected with DMAB (50 mg/kg), once a week, for 20 weeks. DMAB forms carcinogenic DNA adducts through the N-hydroxylation by cytochrome P450, followed by O-acetylation and hydrolysis, reacting with DNA [23].

Nevertheless, this model has some disadvantages because multiple DMAB administrations are needed [23–25] and it has low specificity, leading to the development of tumors in various other tissues, such as salivary glands, mammary glands, urinary bladder, ear, and skin [6]. The studies performed using this model may be consulted in Table 1.

Table 1. Studies using the DMAB model to study different therapeutic approaches for colorectal cancer.

Animal Strain and Gender	Carcinogenic Administration Route	Drugs or Compounds Evaluated (Classification)	Dose/Treatment	Therapeutic Effects (Ref)
F344 male rats	s.c. 100 mg/kg b.w.	Copper-zinc (CU), manganese (Mn), and iron (Fe)	p.o. (0.8 or 5.1 μg CU/g diet; 0.6 or 17 μg Mn/g diet and 37 or 140 μg Fe/g diet) 3.5 wks before DMAB and for 8 wks	Increased neoplastic lesions by low doses of copper and manganese relative to iron [25]
	s.c. 100 mg/kg b.w. 1/wk for 2 wks	Selenium (nutritionally essential trace element)	p.o. (0, 0.1 or 2.0 mg selenium/kg diet as selenite, selenate or selenomethionine) 4 wks before DAMBP for 12 wks	Dietary administration of selenium in the form of the inorganic salts selenite and selenate reduced colon ACF [26]
	s.c. 100 mg/kg b.w.	Celecoxib (selective cyclooxygenase-2 inhibitor)	Diet supplemented (0, 500, 1000, or 1500 ppm celecoxib) 2 wks before DMABP and for 2 days	Chemopreventive effect for colorectal cancer in a dose-response manner [27]
	Gavage 50 or 5 mg/kg b.w. 1/wk for 4 wks	Acetaminophen	Diet supplemented (1000 ppm) 2 wks before DMAB and for 6 wks	Protective effect on the development of colorectal carcinogenesis [28]

ACF: aberrant crypt foci; b.w.: body weight; p.o.: per os; s.c.: subcutaneous injection; wk: week; wks: weeks.

N-Methyl-N-Nitrosourea (MNU) and N-Methyl-N-Nitrosoguanidine (MNNG)

Since 1967, after discovering that oral administration of alkylnitrosamide induced adenocarcinomas in the glandular stomach in rats, other works were performed envisioning to address the carcinogenic potential of MNNG and MNU [4,13]. MNU and MNNG are direct DNA alkylating agents; they transfer a methyl group to nucleobases leading to the accumulation of genetic mutations [13]. Intra-rectal instillation of MNNG during 20 weeks at a dose of 1–3 mg/rat/week induced colon cancer in 100% of F344 rats [13,18,20]. Female ICR/Ha Swiss mice instilled with 0.3 mg of MNU intrarectally, three times a week for 10 weeks, developed tumors in the distal colon, rectum, and anus with an incidence of 78% [29]. The intrarectal administration allows a more selective induction in the distal colon and rectum, which is a huge advantage of this model. However, a precise technique is needed, and the quantification of drug volume is difficult [18,20]. In addition, the animals need to be kept in an inverted position for one minute after administration to prevent the return of the compound to the anus [19,20].

This model can be used to evaluate the therapeutic effects of several compounds on colorectal cancer development. More details about colorectal cancer studies using the MNU model to evaluate the influence of diet, drugs, and natural compounds can be consulted in Table 2.

Table 2. Studies using the MNU model to evaluate the effects of different therapeutic strategies for colorectal cancer.

Animal Strain and Gender	Carcinogenic Administration Route	Drugs or Compounds Evaluated (Classification)	Dose/Treatment	Therapeutic Effects [Ref]
F344/NSlc female rats	i.r. 2 mg 3x/wk for 3 wks	Ursodeoxycholic acid (UDCA) and 5-aminosalicylic acid (5-ASA) (non-steroid anti-inflammatory drug)	p.o. (0%, 0.11 or 0.02%) for 30 wks	Inhibited colorectal cancer development [29]
Sprague Dawley female rats	i.r. 10 mg/Kg 3x/wk for 4 wks	Omega 3 polyunsaturated fatty acid (Omega-3PUFA)	i.g. (2 g/kg) daily for 4 wks	Attenuated CRC by blocking PI3K/AKT/Bcl-2 [30]
F344 male rats	i.r. 2 mg/rat 3x/wk for 4 wks	4-[3,5-Bis(trimethylsilyl)benzamido]benzoic acid (TAC-101) (retinobenzoic acid derivative)	p.o. (0.8 or 8 mg/kg for 1 or 4 wks) for 20 wks	Inhibited colorectal cancer development [31]
	i.r. 4 mg on days 1 and 4	Calcium and cholic (bile acid)	d.t. (0.2% cholic acid or 1.6% calcium) for 28 wks	Increased colorectal tumor development by bile acid and no protective effect of calcium [32]

b.w.: body weight; d.t.: diet supplement. i.g.: intra-gastrically; i.r.: intra-rectal administration; p.o.: per os; wks: weeks.

1,2-Dimethylhydrazine (DMH)

DMH is an alkylating agent that needs liver metabolic activation to become a carcinogen. Therefore, DMH is oxidized in the liver into azoxymethane and is then hydroxylated to form methylazoxymethanol (MAM). MAM is converted to formalin and methyldiazonium ion that are responsible for DNA, RNA, and protein alkylation [4,33].

The induction of colon cancer in rats with this compound was described for the first time in 1967 by Druckrey and colleagues, through its subcutaneous administration, at a dose of 21 mg/kg [20,34]. They showed that DMH cancer induction in the distal portion of the colon is histopathologically similar to humans [16,33]. These results were later confirmed by other authors [6,13,19].

DMH can be administered through different routes, including subcutaneous, intraperitoneal, oral, and intrarectal [4]. For example, a subcutaneous injection of 20 mg/kg DMH, once a week, for 20 weeks induces colonic adenomas in about 60% of male F344 rats [21]. Oral administration of 20 mg/kg showed a lower tumor incidence in male Wistar rats, depending on the nature of the diet [35]. Intrarectal administration of 250 mg/kg of DMH in Sprague-Dawley rats induced multiple colorectal adenocarcinomas with a latency period of 34 weeks [36]. Of the routes of administration presented, subcutaneous seems to be the one that leads to high incidence and consequently the most used in chemopreventive studies [4]. More information about other studies with this model may be consulted in Table 3.

Although DMH-induced colon tumors in rodents are similar to human colon tumors [21], this model has disadvantages, e.g., multiple injections of DMH are necessary to induce tumors, it is characterized by at least six months of the latency period, and no hepatic metastases were observed until now (Figure 4C) [20].

Azoxymethane (AOM)

In 1970, Druckrey and colleagues showed for the first time the ability of azoxymethane (AOM) to induce intestinal tumors. Other works were then published confirming that AOM is a potent inducer of carcinomas of the large intestine in various strains of rats, such as F344, and mice, such as C57BL/6J and SWR/J, among others [21,37,38].

AOM is a metabolite of DMH that has been more frequently used in the induction of colon tumors than DMH, given some of its advantages over the original compound, such as its increased efficacy and greater chemical stability [37]. Like DMH, AOM is also an indirect carcinogenic compound, and it is activated in the liver by N-oxidation through cytochrome P450 2E1, producing metabolites such as methylazoxymethanol and methyl-diazoxide, which induce inflammation [39]. AOM seems to be a more effective carcinogen than DMH because it requires fewer reactions to be activated [6].

AOM induces rodent colon carcinogenesis when administered over 6–8 weeks via subcutaneous or intraperitoneal injection, with a latency period ranging from 20 to 30 weeks [13]. The distribution of tumors developed in the small intestine and colon (predominantly in the distal colon) is similar to that observed in the human colon [21,40]. Histological and histochemical properties of AOM-induced tumors are similar to those described in humans, being classified as adenomas and adenocarcinomas. Using this induction model, it was possible to identify metastases in lymph nodes and the liver similar to those described in humans [6,38].

Details concerning studies using AOM induction model to understand the influence of diet, drugs, or natural compounds in colorectal cancer can be consulted in Table 4.

Table 3. Studies using the DMH model to study different therapeutic approaches for colorectal cancer.

Animal Strain and Gender	Carcinogenic Administration Route	Drugs or Compounds Evaluated (Classification)	Dose/Treatment	Therapeutic Effects (Ref)
Wistar male rats	i.p. 40 mg/kg b.w. 2x/wk for 2 wks	Hyperbaric oxygen (HBO2)	HBO2 alone or DMH + HBO2; 15 daily 90 min HBO2 sessions every 24 h at 2.0 atm absolute pressure	HBO2 had a protective effect in colorectal cancer, demonstrated by the decrease in COX-2 [41]
	s.c. 40 mg/kg b.w. 2x/wk for 2wks	Astaxanthin	p.o. (15 mg/kg b.w.) 1 wk before and after DMH for 16 wks	Positive effects against colorectal cancer [42]
	s.c. 30 mg/kg 1x/week for 18 wks	Aspirin (a non-steroidal anti-inflammatory drug)	Gavage (0, 5, 30 or 60 mg/kg diet) daily for 18 wks	Reduced tumor incidence [43]
	s.c. 50 mg/kg b.w. 1 wk after diet supplemented and physical activity	Probiotic soy product and physical exercise	Gavage (3 mL/kg b.w./day fermented or unfermented soy products) and t.r. (60 min/day at 3–5% inclination at 355 m/min or 17–20 m/min) alone or in combination for 6 wks	No inhibition of colorectal cancer by the ingestion of fermented soy products or physical activity or by a combination of both [44]
	s.c. 40 mg/Kg b.w. for 8 wks	Epigallocatechin gallate (EGCG)	p.o. (50, 100 or 200 mg/Kg b.w.; once daily) for 8 wks	EGCG inhibited the formation of DMH-induced CRC by regulating key pathways, namely p53 and PI3K-Akt signaling pathways and I-kappaB kinase/NF-kappaB signal pathways, apoptosis signal pathways and MAPK cascades, involved in tumorigenesis [29]
Wistar female rats	s.c. 20 mg/kg b.w. for 20 wks	High fiber diet and aspirin	Exp1: gavage (10 or 30 mg/kg/day b.w aspirin) Exp.2: diet supplemented with high fiber (16% crude fiber) from the beginning and for 32 wks	Protective effects of high fiber diet and aspirin. The aspirin effect is dose-related [45]
	s.c. 20 mg/Kg b.w. for 5 wks	Methanolic extract of Muntingia calabura L. leaves (MEMC)	p.o. (100 or 200 mg/Kg b.w.) all days till the 15 week	MEMC offered a protective role against experimentally induced CRC via suppressing hyperproliferation and inflammation [46]

Table 3. *Cont.*

Animal Strain and Gender	Carcinogenic Administration Route	Drugs or Compounds Evaluated (Classification)	Dose/Treatment	Therapeutic Effects (Ref)
F344 male rats	Exp.1: s.c.20 mg/kg b.w. daily for 16 wks; Exp.2: s.c. 20 mg/kg b.w. daily for 12 wks	Cellulose, calcium and folic acid	p.o. (Exp.1: 10% cellulose for 30 wks; Exp.2: 250 or 500 mg/100 g diet calcium + 0 or 0.1 mg/100 g folic acid for 22 wks)	Protective effects of cellulose and reduced tumor number and multiplicity of calcium [47]
	i.p. 40 mg/kg b.w. once weekly for 4 wks	Adlay bran ethanol extract (ABE-Ea)	p.o. (8.64, 17.28 or 34.56 mg/day ABE-Ea) 1 wk before 1st DMH and for 9 or 18 wks	Inhibited preneoplastic lesions [48]
	i.p. 30 mg/kg weekly for 10 wks	Calcium and vitamin 3 in low or high-fat diet	p.o. (0.5 or 1%supplemental calcium and 1000 or 2000 IU/kg diet vitamin 3 in combination with a low-fat diet, 0.5% corn oil, or high-fat diet, 20% corn oil) 2 wks before DMH and for 20 wks	Preventive effects only in a high-fat diet [49]
	s.c. 100 mg/kg b.w. 2x/wk for 2 wks	Arabinoxylan-oligosaccharides (AXOS) (prebiotic)	p.o. (60 g/kg diet) 10 days before DMH and for 13 wks	Chemopreventive effect [50]
Sprague-Dawley male rats	Gavage 30 mg/kg	Aspirin (non-steroidal anti-inflammatory drug)	Exp.1: s.c.(100 mg/kg/day aspirin) 1 wk before and after DMH and for 1 wk; Exp.2: s.s. (50 mg/kg/day aspirin) 4 wks after DMH and for 36 wks)	Reduced tumor incidence when aspirin was administered 1 wk before or after DMH but no effect when administered 4 wks after [51]
	s.c. 30 mg/kg b.w. for 6 wks	etoricoxib (selective cycloxygenase-2 inhibitor) and diclofenac (a preferential cycloxygenase-2 inhibitor)	Gavage (0.6 mg/kg b.w. ectoricoxib and 8 mg/kg b.w. diclofenac) daily for 6 wks	Chemopreventive effect of both compounds [52]
	s.c. 20 mg/kg b.w. 2x/wk for 4 wks	Soy isoflavones	p.o. (1, 10, 50, 150 or 500 mg/kg diet) 1 wks after DMH and for 12 wks	Inhibited colorectal cancer in dose-independent manner [53]
	s.c. 20 mg/kg b.w. weekly from day 3 and for 12 wks	β-carotene (derived from carrots) sodium ascorbate (L-ascorbic acid) and cellulose	Diet supplemented (0.005% β-carotene or 0.02% sodium ascorbate or 1.5% cellulose) 12 wks before and for 28 wks	Only β-carotene showed an inhibitory effect of carcinogenesis [54]
	s.c. 20 mg/kg 1x/wk 13 wks	Wheat bran	Diet supplemented (fiber-free diet or 20% wheat bran supplement) for 31 wks	Increased colorectal carcinogenesis [55]
	s.c. 20 mg/kg b.w. 6x/wk	Milk and calcium	Diet supplemented (37 g/kg diet of milk and 40 mg/kg rat/day of calcium carbonate)	Protective activity by milk supplementation [56]
Sprague-Dawley male and female rats	s.c. 21 mg/kg 1x/week for 18 wks	Calcium	d.w.(3.2 g/L calcium lactate) daily from the start until 25–34 wks	Inhibited colorectal cancer [57]

b.w.: body weight; d.w.: drinking water; i.p.: intraperitoneal injection; p.o.: per os; s.c.: subcutaneous injection; t.r.: treadmill running; wk: week; wks: weeks.

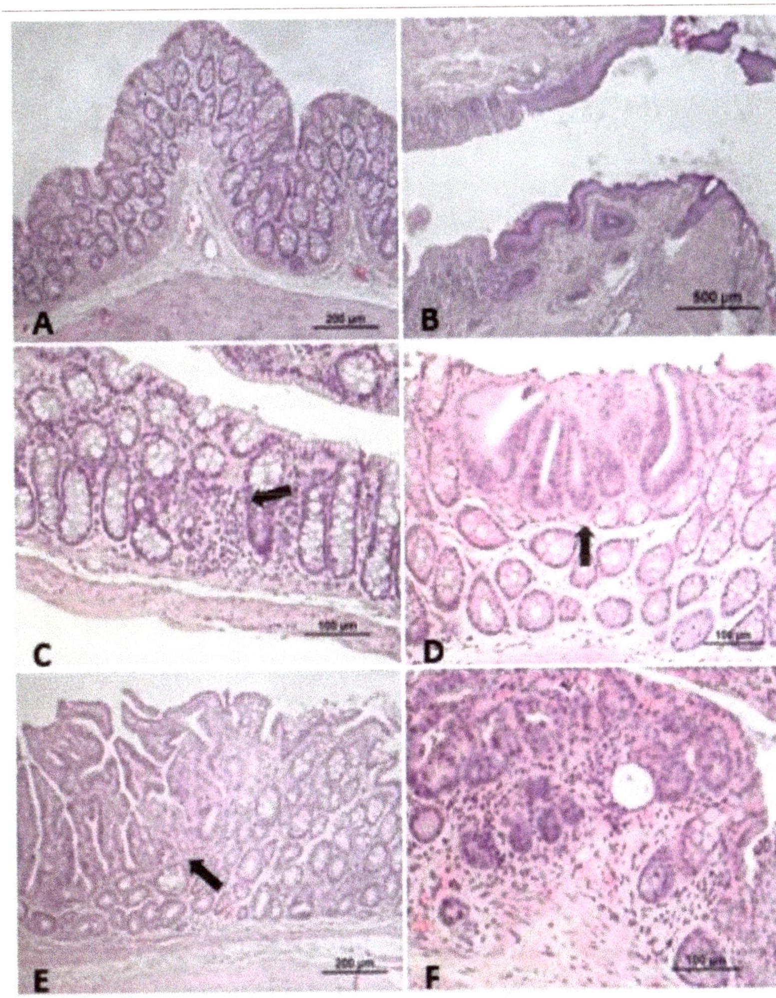

Figure 4. Representative photomicrographs of mouse/rat colon and rectum. (**A**) Rat colon with no alterations. (**B**) Mouse anorectal junction, with no alterations. (**C**) Mouse colon showing mild inflammatory infiltrate at the mucosa (DMH induction CRC model). (**D**) Rat colon with the presence of focal mild epithelial dysplasia (DMH induction CRC model). (**E**) Rat colon adenoma, characterized by a benign epithelial tubulopapillary neoplastic proliferation, non-invasive (DMH induction CRC model). (**F**) Mouse rectum, adenocarcinoma, characterized by a carcinomatous proliferation, associated with stromal invasion and inflammation (DMH induction CRC model). HE staining.

Table 4. Studies using the AOM model to evaluate several therapeutic strategies for colorectal cancer.

Animal Strain and Gender	Carcinogenic Administration Route	Drugs or Compounds Evaluated (Classification)	Dose/Treatment	Therapeutic Effects (Ref)
F344 male rats	s.c. injections 15 mg/kg b.w./wk once a week for 2 wks	Ursodeoxycholic acid and cholic acid (bile acids)	p.o.(0.2% or 0.4% cholic acid, 0.2% or 0.4% ursodeoxycholic acid, 0.2% cholic acid + 0.2% ursodeoxycholic acid) for 30 wks	Higher dose of ursodeoxycholic acid reduced the incidence of colorectal tumors [58]
	s.c. 15 mg/kg bw once weekly for 2 wks	Celecoxib (a non-steroidal anti-inflammatory drug)	p.o. (500, 1000 or 1500 ppm) before exposure to AOM, during treatment, and until termination of the study at 52 wks	Chemopreventive activity in all tumor stages [59]
	s.c. 15 mg/kg b.w. 1x/wk for 2 wks	iNOS inhibitor L-N^6 -(1-iminoethyl) lysine tetrazole-amide (SC-51), celocoxib (nonsteroidal anti-inflammatory)	p.o. (10, 30 or 100 ppm SC-51; 500 ppm celocoxib; 30 or 100 ppm SC-51 + 500 ppm celocoxib) for 8 wks	The combination of SC-51 with celocoxib was more effective in colorectal cancer prevention than the compounds alone [60]
	s.c. 15 mg/kg b.w. 1x/wk for 3 wks	Rebaudioside A, oleanolic acid, costunolide and soyasionin A2 (terpenoids), liquiritin (flavonoid), phyllodulcin and hydrangenol (isocumarins)	p.o. (200 ppm of each) for 5 wks	Costunolide is the most effective chemopreventive agent [61]
	s.c. 29.6 mg/kg b.w.	Piroxicam (a non-steroidal anti-inflammatory drug) and D, L-α-difluoromethylornithine (DFMO)	p.o. (25, 75 and 150 ppm piroxicam or 400, 1000 and 4000 ppm DFMO) 1 wk after AOM for 26 wks	A combination of piroxicam and DFMO was more effective in the inhibition of colorectal cancer than compounds alone [62]
	s.c. 15 mg/kg b.w. 1x/wk for 2 wks	Phenylethyl-3-methylcaffeate (PEMC)	p.o. (750 ppm) 2 wks before AOM for 52 wks	Inhibited colonic tumors [63]
	s.c. 15 mg/kg b.w. 1x/wk for 2 wks; start 2 wks after diet	Celocoxib (COX-2 inhibitor)	p.o. (1500 ppm) for 50 wks	Chemopreventive activity [64]
	s.c. 15 mg/kg b.w. at 7 and 8 wks of rat age	S-methylmethane thiosulfonate (S-MMTS) (isolate from cauliflower) and sulindac	p.o. (80 ppm S-MMTS, 160 ppm sulindac or 40 ppm S-MMTS + 160 ppm sulindac) 14 wks after AOM for	A combination of S-MMTS and sulindac was more effective in the inhibition of colorectal cancer than compounds alone [65]
	s.c. 15 mg/kg 1x/wk for 2 wks	Naproxen and NO-naproxen (nonspecific nonsteroidal anti-inflammatory drugs)	p.o. (200 or 400 ppm naproxen and 300 or 600 ppm nitric oxide-naproxen) 3 days after AOM for 8 wks	Chemopreventive effects [66]
	15 mg/kg i.p. 1x/wk for 2 wks	Lovastatin (statin) and exisulind (selective apoptotic antineoplastic drug)	p.o. (50 ppm lovastatin, 100, 250 or 1000 ppm exisulind alone or in combination with 50 ppm lovastatin) for for 4 wks	Chemopreventive effects of lovastatin but not exisulind [67]
	s.c. 15 mg/kg b.w. 1x/wk for 2 wks	CP-31398 (p53-modulating agent) and celocoxib (non-steroidal anti-inflammatory drug)	Diet supplemented (1, 150 or 300 ppm CP-31398, 300 ppm celecoxib or 1500 ppm CP-31398 + 300 ppm celecoxib) 2 wks after AOM and for 48 wks	A combination of compounds enhanced colorectal cancer chemopreventive efficacy [68]
	s.c. 15 mg/kg b.w. 1x/wk for 2 wks	Aspirin (a non-steroidal anti-inflammatory drug)	p.o. (0, 200 or 400 ppm) daily 2 wks before AOM and for 52 wks	Inhibited incidence and multiplicity of colorectal carcinomas [69]

Table 4. *Cont.*

Animal Strain and Gender	Carcinogenic Administration Route	Drugs or Compounds Evaluated (Classification)	Dose/Treatment	Therapeutic Effects (Ref)
	s.c. injection 15 mg/kg b.w. 1x/wk for 2 wks	Prebiotic germinated barley foodstuff (a mixture of insoluble protein and dietary fiber)	Diet supplemented with prebiotic germinated barley foodstuff for 4 wks	Anti-tumorigenicity activity [70]
	i.p. 15 mg/kg b.w.	Aspirin (a non-steroidal anti-inflammatory drug) and α-Difluoromethylornithine (DFMO) (ornithine decarboxylase inhibitor)	p.o. (Exp1.: 0, 200, 600 or 1800 mg/kg/diet of aspirin or 1000 mg/kg diet of DFMO; 8 days before 1st AOM; Exp.2: 200, 600, 1800 mg/kg/diet aspirin or 1000 or 3000 mg/kg/diet of DFMO or 1000 mg/kg/diet DFMO + 200 or 600 mg/kg/diet aspirin; 8 days before 1st) for 43 wks after last AOM	The combination of aspirin and DFMO after AOM reduced colorectal tumors [71]
	s.c. 15 mg/kg b.w 1x/wk for 2 wks	Vitamin D, acetylsalicylic acid (a non-steroid anti-inflammatory drug) and calcium	Diet supplemented (0, 2500, 5000 or 7500 ppm calcium; 0 or 300 ppm acetylsalicylic acid alone or combination with 0 or 0.02 µg/kg diet vitamin D) 20 days before AOM and for 18 wks	Increased incidence of tumors with high levels of calcium alone or in combination with vitamin D; Vitamin D with acetylsalicylic acid also increased tumor incidence [72]
	s.c. 8 mg/kg b.w./wk for 10 wks	Dietary wheat bran and dehydrated citrus fiber (in form of orange peel)	Diet supplemented (0 or 15% wheat bran or citrus fiber) for 20 wks	Reduced the risk of colorectal tumors [24]
	s.c. 15 mg/kg 1x/wk for 2 wks)	Tea extracts, Polyphenols and epigallocatechin gallate (EGCG)	d.w. (360 or 3600 ppm black and green tea extracts; 360 or 1800 ppm EGCG; 360 or 1800 black tea polyphenols and 360 or 3600 green tea polyphenols) at 6 wks and for 43 wks	No effect in tumor incidence [73]
	s.c. 15 mg/kg b.w. 1x/wk for 2 wks)	Aspirin, celecoxib, (cyclooxygenase-2 inhibitor), and atorvastatin (3-hydroxy-3-methylglutaryl CoA reductase inhibitors)	Diet supplemented (150 ppm atorvastatin, 600 pp celecoxib, 400 ppm aspirin, 100 ppm atorvastatin + 300 ppm celecoxib or 100 ppm atorvastatin + 200 ppm aspirin) one day after AOM and for 42 wks	Inhibited the incidence and multiplicity of colorectal carcinomas alone or in combination [74]
	s.c. 15 mg/kg b.w. 1x/wk for 2 wks	Grape seed extract (GSE)	Diet supplemented (0.25 or 0.5% (*w/w*) GSE) 1 wk before AOM, 4 wks last AOM or during all study and for 16 wks	Chemopreventive efficacy against early steps of colorectal carcinogenesis [75]
	s.c. 15 mg/kg b.w. 1x/wk for 2 wks)	Celecoxib (cyclooxygenase-2 inhibitor) in diets high in mixed lipids (HFML) or fish oil (HFFO)	Diet supplemented (0, 250, 500, or 1000 ppm celecoxib with HFML or HFFO diet) one day after AOM and for 26 wks	Preventive effect of low doses of celecoxib in HFFO diet [76]
F344 female rats	i.p. 20 mg/kg b.w.	Polyethylene-glycol (PEG) (non-fermented polymer)	Diet supplemented (3 g/kg b.w/day) 7 days after AOM and for 105 days	Chemopreventive effects [77]
	i.p. 20 mg/kg b.w.	Heme in food (in form of chicken, beef, black pudding)	Diet supplemented (600 g/kg diet chicken, beef and black pudding) 7 days after AOM and for 100 days	Increased colorectal carcinogenesis for all compounds [78]
	s.c. 8 mg/kg b.w./wk for 10 wks	Alfalfa, pectin and wheat bran	Diet supplemented (0 or 15% alfalfa, pectin and wheat bran) for 40 wks after 1st AOM	Inhibited colorectal tumor incidence, especially by pectin or wheat bran [79]
BALB-c female mice	i.p. 15 mg/kg1x/wk for 2 wks	Kefir (a probiotic fermented milk product)	p.o. (5 mL/kg b.w. fermented kefir milk) for 8 wks	Decreased and prevented the growth of colorectal tumors [80]

Table 4. *Cont.*

Animal Strain and Gender	Carcinogenic Administration Route	Drugs or Compounds Evaluated (Classification)	Dose/Treatment	Therapeutic Effects (Ref)
Sprague-Dawley male rats	s.c. 15 mg/kg 1x/wk for 2 wks, 28 days after diet supplementation	Amylose maize starch and butyrylated high-amylose maize starch	Diet supplemented (10% of high-amylose maize starch or 10% butyrylated alone or in combination) start at day 0 until euthanasia	The compound combination reduced the risk of developing colorectal cancer [81]
	i.p. 15 mg/kg 1x/wk for 4 wks	indomethacin and copper-indomethacin (non-steroidal anti-inflammatory drug)	i.p. (3.0 mg/kg indomethacin or 3.8 mg/kg copper-indomethacin) daily	Both compounds showed chemopreventive activity, but indomethacin was more effective [82]
	s.c. 15 mg/kg b.w. 1x/wk for 2 wks	R-Flurbiprofen (non-steroidal anti-inflammatory drug)	Gavage (30 mg/kg b.w./per day) 6 days a week, 1 wk before AOM and for 30 wks	Protective effects against colorectal cancer development [83]
	s.c. 15 mg/kg b.w. 1x/wk for 2 wks, at day 45 of rat's life	Soy isoflavones	p.o. (0, 40 100 mg/kg diet) from birth, including pregnancy and lactation, until 26 wks of life. AOM at day 45	Lifetime exposure suppressed colon tumors growth [84]
	s.c. 15 mg/kg b.w. 1x/wk for 2 wks	Probiotic bacteria "*bifidobacterium lactis*" (*B. lactis*) and carbohydrate "resistant starch" (from a commercial source called Hi-maize 958 or Hi-maize S260)	Diet supplemented (100 g/kg/diet of Hi-maize 958 or Hi-miaze 260 and 1% lyophilized culture of *B. lactis*)	Protective effects by the combination of the two products [85]
	s.c. 15 mg/kg b.w. for three weekly doses	Xanthohumol (a polyphenol isolated from *Humulus lupulus L.*)	Gavage (5 mg/kg b.w.) every alternate day for 8 wks	Inhibited cell proliferation and induced apoptosis [86]
Wistar rats	i.p. 15 mg/kg	L-lysine, propolis, or gum arabic	Gavage water (150 mg/kg L.-lysine, 100 mg/5 mL/kg propolis or 5 mL/kg gum arabic) daily for 16 wks	Gum arabic and propolis reduced the total number of aberrant crypt foci, L-lysine neither protected against nor enhanced colorectal cancer [87]

b.w.: body weight; d.w.: drinking water; i.p.: intraperitoneal injection; p.o.: per os; s.c.: subcutaneous injection; wk: week; wks: weeks.

Azoxymethane (AOM) and Dextran-Sodium Sulfate Model (DSS)

Because colon cancer is associated with long-standing IBD, such as ulcerative colitis and Crohn's disease, the risk of colorectal cancer development increases with the extent and duration of disease [19]. Chronic and repeated mucosal inflammation may result in tumors through several mechanisms, such as induction of genetic mutations, increased cryptal cell proliferation, changes in crypt cell metabolism, changes in bile acid, and alterations in the bacterial flora [5]. In 2003, Tanaka and colleagues developed a colitis-related mouse model of colorectal cancer initiated with AOM and promoted by dextran-sodium sulfate (DSS) [88]. DSS is an inflammatory compound that causes damages to the epithelial lining of the colon and induces colitis. Using this model, male Crj: Cd-1 (IRC) mice were intraperitoneally injected with AOM (10 mg/kg of body weight) and, one week later, received 2% of DSS in drinking water for seven days. Twenty weeks later, 88% of animals had colonic dysplasia, and the incidence of adenoma and adenocarcinoma was 100% [88]. This model allows the reduction in the number of AOM administrations, avoiding prolonged exposure to this compound, and still allows a reduction in the latency period. After this finding, other researchers have associated DSS with other compounds (PhIP and DMH), with tumor induction in a shorter period than the AOM/DSS model [19,89,90]. Concerning the AOM/DSS-induced model, different mice strains present diverse sensitivity; for example, the incidence and multiplicity of adenocarcinomas appear to be higher in the BALB/c mouse strain [91].

The AOM/DSS model mimetics human colorectal cancer pathogenesis, with a similar location (distal colon) and initiation by a polypoid growth. However, this model has a very low tendency to metastasize, which constitutes a limitation [92]. This model has been used in several chemopreventive studies of colitis-related colon carcinogenesis [88] (Table 5).

2-Amino-1-Methyl-6-Phenylimidazo (4,5-b) Pyridine (PhIP)

The PhIP is a heterocyclic amine isolated from cooked fish and meat, which can be used to induce tumors in the colon, prostate, and mammary gland [93]. After administration, it is rapidly absorbed by the gastrointestinal tract and widely distributed through the body [19]. Then it is bio-transformed by the liver cytochrome P450 s, being converted in the amino group to a hydroxyamino group, which is then activated by forming esters with acetic acid, sulfuric acid, and proline. These esters are responsible to induce carcinogenic DNA adducts and genetic alterations leading to colorectal cancer [94]. The work performed by Ito and colleagues was the first to induce colon tumors in rats with this compound [95]. F344 rats from both sexes were fed with 400 ppm PhIP for up to 52 weeks and presented a high incidence of colon carcinomas [95]. PhIP did not induce colon cancer in mice, it just induced the formation of colonic aberrant crypt foci and lymphomas [4,9]. The mechanisms responsible for the non-induction of cancer in mice are not well understood [94]. This model has been used to evaluate the therapeutic effects of several compounds on colorectal cancer. Detailed information concerning mouse and rat models' studies using PhIP induction can be consulted in Table 6.

Table 5. Colorectal cancer chemopreventive studies using the AOM/DSS model.

Animal Strain and Gender	Carcinogenic Administration Route	Drugs or Compounds Evaluated (Classification)	Dose/Treatment	Therapeutic Effects (Ref)
CF-1 male mice	s.c. AOM 10 mg/kg body wt + 1 wk later d.w. 1.2% DSS for 7 days	Aspirin (acetylsalicylic acid)	Diet supplemented (0.02% aspirin) 1 wk before AOM and for 20 wks	Suppressed inflammatory colitis symptoms and tumor multiplicity [96]
C57BL/6 male mice	i.p. AOM 10 mg/kg + 1 wk later d.w. 2% DSS for 1 wk	*Asther glehni Franchet et Sckmidt* (common Korean dietary edible herb)	p.o. (25 mg/kg/day) 1 wk after AOM + DSS and for 1 wk	Inhibited colitis-associated colon carcinogenesis [97]
	i.p. AOM 10 mg/kg b.w. + d.w. 2% DSS for 5 days	DA-6034 (7-Carboxymethyloxy-39,49,5-trimethoxyflavone) (synthetic derivative of flavonoid eupatilin)	Gavage (30 mg/kg) from day 7 to the end	Reduced the number of colon tumors [98]
	i.p. AOM 10 mg/kg b.w. + 5 days later d.w. 2.5% DSS followed by 14 days of normal water x 3 cycles	Pristimerin (a naturally triterpenoid)	i.p. (125 ng/kg) every 2 days for 80 days	Reduced the number and size of the tumors [99]
	i.p. AOM 10 mg/kg + 1 wk after d.w. 2% DSS for 1 wk	Chitooligosaccharides (oligomers that are depolymerized from chitosan)	i.g. (300 mg/kg) once a day and 6x/wk	Prevented colorectal cancer through regulating the gut microbiota and mycobiota [100]
C57BL/6 female mice	i.p. AOM 10 mg/kg + 1wk later d.w. 2% DSS for 1x/wk for 3 wks	Conjugated linoleic acid (CLA)	Diet supplemented (1% CLA) 3 wk before AOM + DSS and for 13 wks	Increased colorectal cancer [101]
	i.p. AOM 10 mg/kg b.w. + d.w. DSS 2.5% 1 wk after AOM for two cycle of 7 days	Licorice flavonoids (LFs) (Chinese herbal medicine)	Gavage (0, 50 or 100 mg/kg) once a day for 10 wks	Reduced tumorigenesis [102]
BALB/c female mice	i.p. AOM 10 mg/kg + 1 wk after d.w.DSS 2.5% 2.5% of 3 cycles of 1 wk	*Aloe vera* gel	p.o. (200 or 400 mg/kg/day) 1 wk before AOM and for 13 wks	Reduced the multiplicity of colorectal adenomas and adenocarcinomas [103]
BALB/c male mice	i.p. AOM 12.5 mg/kg + 1 wk after dDSS 2.5% in d.w. for 3 cycles of 5 days	*Triticum aestivum* sprouts ethanol extract (TAEE)	Gavage (100 or 200 mg/kg/day) for 40 days	Inhibited colon inflammation and neoplasm formation [104]
CD-1 (ICR) male mice	i.p. AOM 10 mg/kg b.w. + 1 wk after d.w. DSS 1.5% for 7 days	Zerumbone (tropical ginger sesquiterpene)	Diet supplemented (100, 250, or 500 ppm) for 17 wks	Zerumbone suppresses mouse colon carcinogenesis through mechanisms of growth, apoptosis, inflammation that are involved in carcinogenesis in the colon [105]
	i.p. AOM 10 mg/kg b.w. + 1 wk after d.w. DSS 1% for 7 days	Prenyloxycoumarins, auraptene and collinin nonsteroidal anti-inflammatory drugs)	Diet supplemented (0.01 and 0.05% of all compounds) 1 wk after DSS and for 17 wks	Chemopreventive activity [106]

Table 5. *Cont.*

Animal Strain and Gender	Carcinogenic Administration Route	Drugs or Compounds Evaluated (Classification)	Dose/Treatment	Therapeutic Effects (Ref)
	i.p. AOM 10 mg/kg b.w. + 1 wk after d.w. DSS 1% for 7 days	Ursodeoxycholic Acid (UDCA) and Sulfasalazine (anti-inflammatory agents)	Diet supplemented (0.016, 0.08 or 0.4% UDCA, 0.05% sulfasalazine or 0.5% UDCA + 0.05% sulfasalazine) 1 wk after DSS and for 20 wks	UDCA showed more suppressing effects on colorectal cancer [107]
	i.p. AOM 10 mg/kg b.w. + d.w. DSS 2.5% 1 wk after AOM and for 7 days	Dried açaí berry powder	Diet supplemented (2.5 or 5%) for 14 wks	Reduced the incidence of colorectal cancer [108]
	i.p. AOM (10 mg/kg b.w.) + d.w. DSS 1.5% 1 wk after AOM and for 7 days	Fucoxanthin (a xanthophyll present in marine brown algae)	Gavage (6 or 30 mg/kg)	Reduced the number of colorectal polyps [109]
CD-1 (ICR) female mice	i.p. AOM 10 mg/kg b.w. + 1 wk after d.w. DSS 2% for 7 days	Nimesulide (a cyclooxygenase-2 inhibitor), troglitazone and bezafibrate (ligands for peroxisome proliferator-activated receptors)	Diet supplemented (0.04% nimesulide, 0.05% troglitazone and 0.05% bezafibrate) 1 wk after DSS and for 14 wk	Suppressed development of colorectal cancer [110]
129SvJxC57BL6 male and female mice	i.p. AOM 12.5 mg/kg + 5 days later d.w. DSS 2% for 5 days followed by a 2-wk rest period and again 5 days of DSS	Chalcone lonchocarpin isolated from Lonchocarpus sericeus	i.p. (2.5 mg/mL) 4 wks after the last DSS cycle and for 4 days	Reduced tumor proliferation [111]

b.w.: body weight; d.w.: drinking water; i.g.: intragastrically; i.p.: intraperitoneal injection; p.o.: per os; s.c.: subcutaneous injection; wk: week; wks: weeks.

Table 6. Studies using PhIP model to study several strategies for colorectal cancer.

Animal Strain and Gender	Carcinogenic Administration Route	Drugs or Compounds Evaluated (Classification)	Dose/Treatment	Therapeutic Effects (Ref)
F344 male rats	p.o. 200 ppm for the first 20 wks	Tomato + broccoli powder in AIN93G diet	control, PhIP alone, PhIP + diet with 10% of tomato and broccoli powder for 20 wks and without PhIP for 32 wks	A diet rich in tomato + broccoli can reduce or prevent dietary carcinogens-induced cancer. Tomato + broccoli group reduced incidence and/or severity of cancer lesions [112]
	Gavage 75 mL/kg b.w. 5 times a week for 2 wks	Yogurt powder (milk fermented by *Lactobacillus delbrueckii* subsp. *Bulgaricus strain 2038* and *Streptococcus salivariu* subsp. *thermophilus* strain 1131)	Diet supplemented (10.4646% yogurt powder) 14 days before PhIP and for 14 days	Yogurt appears to have tumor-suppressing properties [113]
	Gavage daily 100 mg/kg b.w. for 2 wks	White tea, green tea, epigallocatechin-3-gallate (EGCG) and caffeine	d.w. (2% white tea, 2% green tea, 0.5 mg/mL EGCG or 9.5 mg/mL caffeine) 1 wk after last PhiP and for 16 wks	Inhibition of tumor initiation mostly by white tea, caffeine and EGCG [114]
	i.g. 100 mg/kg2x/wk for 10 wks	Nobiletin (5,6,7,8,3,4 -hexamethoxy flavone) (polymerthoxy-flavonoid extracted from citrus fruits)	Diet supplemented (0.05% nobiletin) for 50 wks	Chemopreventive activity of early carcinogenesis changes [115]
	Gavage 150 mg/kg for 5 alternate days	White tea	d.w. (2% wt/vol white tea) for 2 wks	Inhibition of preneoplastic lesion development [116]
	i.g. 200 mg/kg 2x/wk for 10 wks)	Fujiflavone (a commercial isoflavone supplement)	Diet supplemented (0.25% fujiflavone) for 50 wks	Preventive effects on colorectal cancer [117]
	Gavage 50 mg/kg b.w.	Clorophyllin (CHL) indole-3-carbinol (I3C)	p.o. (0.1% I3C and 0.1% CHL), before and during PhiP exposure or 1 wk after PhIP and for 16 wks	Protective effects for CHL and I3C on colorectal carcinogenesis [118]
F344 female rats	Diet supplemented 0.02%	caffeine, α-tocopherol (lipophilic antioxidant), and n-tritriacontane-16,18-dione (TTAD) (β-diketone derivative)	p.o. (0.1% caffeine, 0.5% α-tocopherol or 0.1% TTAD) for 54 wks	Increase the incidence of colorectal tumors by caffeine; α-tocopherol and TTAD had no effect on colorectal tumors [119]
Sprangue-Dawley male rats	Gavage 10 mg/kg b.w.	Chinese cabbage (*Brassica chinensis*)	p.o. (20% freeze-dried cabbage powder) 10 days before PhIP and for 20 h	Preventive effect on initiated colorectal tumors [120]

b.w.: body weight; d.w.: drinking water; h: hours; i.g.: intragastrically; p.o.: per os; wk: week; wks: weeks.

3.3. Genetically Engineered Models

Genetically engineered models allow the study of genetic predisposition to colorectal cancer development, and its interaction with environmental and modifying factors. These models mimic the genetic alterations of spontaneous and hereditary forms of colorectal cancer [9]. Through the study of hereditary colorectal syndromes, such as FAP and HNPCC, it was possible to discover the mutations under colonic carcinogenesis and replicate genetic lesions in mice and rats by developing genetically engineered models [9].

3.3.1. Adenomatous Polyposis Mouse Models (APC)

It was demonstrated that human colorectal cancer is a multi-step genetic process and that the mutation of the APC gene occurs at the beginning of the carcinogenesis process. The APC gene is responsible for the regulation of β-catenin, cytoskeleton organization, cell cycle regulation, apoptosis, and cell adhesion. When mutated in the germline, this gene is associated with FAP [5,121].

Moser and colleagues discovered that C57BL/6 mice treated with ethylnitrosourea developed a mutation that predisposes to spontaneous development of intestinal cancer, naming the model as ApcMin mouse [122]. The Min mouse model is the only animal model of cancer that contains a single genetic alteration capable of producing a fully penetrating, consistent, and organ-specific tumor phenotype. The adenomas developed rapidly, with lesions identified within 60 days, and high tumor multiplicity. This model allows the study of multiple pathways impacting tumorigenesis and enables numerous entry points for basic or applied studies [10]. Over the years, this model has been tested, improved, characterized, and used to understand the role of the APC gene in colorectal cancer and also in chemopreventive studies [14] (Table 7).

Other mouse models with target genetic modifications at different locations on the APC gene have been generated, such as ApcMin/850, ApcΔ716, Apc1638N, Apc1638T, ApcΔ468, and ApcΔ474 [10], that allow the study of the colorectal cancer mechanisms, create models more similar to those changes found in humans, and test the role of specific regions in the APC gene on cancer development [9,10].

Table 7. Studies using the $APC^{Min/+}$ model to evaluate the effect of several compounds in colorectal cancer.

Animal Strain/Gender	Drugs or Compounds Evaluated (Classification)	Dose/Treatment	Therapeutic Effects (Ref)
Min mice/n.d.	α-phenyl-tert-butyl nitrone (PBN) and 4-hydroxyl-PBN (4-O-PBN) (nitrones)	d.w. (100 ng/kg/day PBN or 4-O-PBN) for 3–4 months	Anti-cancer activity of PBN more significant than 4-O-PBN [123]
	Bilberry (Vaccinium myrtillus), lingonberry (Vaccinium vitis-idaea), cloudberry (Rubus chamaemorus), cloudberry seeds or cloudberry pulp, or pure ellagic acid	p.o. (1564 mg/kg of each) for 10 wks.	Chemopreventive potential [124]
	Atorvastatin (hydroxy-3- methylglutaryl CoA reductase (HMGR) inhibitor) and celecoxib (cyclooxygenase-2 (COX-2) inhibitors)	p.o. (0 or 100 ppm atorvastatin or 300 celecoxib alone or in combination) for 80 days	Inhibited intestinal tumorigenesis by atorvastatin and increased chemopreventive activity in combination with celecoxib [125]
	Piroxicam (a non-steroidal anti-inflammatory drug)	p.o. (200 ppm piroxicam) for 100 or 200 days of rat age	Reduced tumor multiplicity [126]
	Annurca Apple polyphenol extract	d.w. (60 µmol/L) in combination with a western diet or balanced diet for 12 wks	Chemopreventive potential [127]
	Fermented brown rice and rice bran (FBRA)	Exp.1: diet supplemented (5 or 10% FBRA) for 20 wks; Exp2: diet supplemented (5 or 10% FBRA and 2% DSS in d.w. for 1 wk) for 12 wks; Exp3: diet supplemented (10% FBRA and 1.5% DSS in d.w. for 1 wk) for 7 wks	No effect on tumor development by FBRA alone but in combination with DSS suppressed the multiplicity of colon tumors [128]
	Sulforaphane (SFN) (isothiocyanate)	Diet supplemented (600 ppm SFN) for 1, 3, or 5 days	Chemopreventive potential [129]
	Bilirubin, bovine serum albumin (BSA) and sodium taurocholate (NaTC)	p.o (0.5 mM BSA alone or in combination with 0.25 mM bilirubin or 5 mM NaTC) for 8 wks	Reduced intestinal adenomas by NaTC [130]
	Metformin (biguanide derivative)	p.o. (250 mg/kg/day) for 10 wks	Chemopreventive potential [131]
	Silibinin	Gavage (750 mg/kg b.w.) for 5 days a wk for 13 wks	Chemopreventive potential [132]
Min mice/female and male	Aspirin	Diet supplemented (250 or 500 ppm) for 7 wks	Chemopreventive potential [133]
	Curcumin	Diet supplemented (2% curcumin) from 4 to 18 wks of age	Chemopreventive potential [134]
	Anthocyanin-rich tart cherry extract and sulindac (a nonsteroidal anti-inflammatory drug)	p.o. (0, 375, 750, 1500 or 3000 mg anthocyanin-rich tar cherry extract/kg if diet with 100 mg sulindac/kg diet) for 19 wks	The combination of both compounds had a more protective effect than compounds alone [135]
	Physical activity	t.r. (18 m/min, 60 min/day, 6 days/wk or voluntary wheel running) for 9 wks	Reduced number and size of intestinal polyps, dependent on exercise mode and gender [136]
	DMU-135 (3,4-Methylenedioxy-3,4,5 -trimethoxy chalcone) (anticancer prodrug)	Diet supplemented (0.2% w:w) from 4–18 wks	Chemopreventive activity [137]
	MCC-555 (peroxisome proliferator-activated receptor (PPAR) ligand)	Gavage (30 mg/kg/day 5 days/wk) for 4 wks	Suppressed activity [138]
	Soy isoflavones	Diet supplemented (low-isoflavone: 11.5 genistein, 2.3 daidzein and 2.3 mg of glycitein/kg diet, rich isoflavone diet: 280.6 genistein, 147.2 daidzein and 48.3 mg of glycitein/kg diet) for 107 days	No inhibition of colorectal tumor development [139]

Table 7. *Cont.*

Animal Strain/Gender	Drugs or Compounds Evaluated (Classification)	Dose/Treatment	Therapeutic Effects (Ref)
	Orange peel extract (OPE)	Diet supplemented (0.25 or 0.5% OPE) for 9 wks	Inhibited colorectal tumorigenesis [140]
	Physical activity	t.r.(18 m/min, 60 min, 6 days/wk, 5% grade) for 9 wks	Reduced the overall tumor burden (size and number) [141]
Min mice/male	Guanidinoethyldisulfide (GED) [14–17], peroxynitrite decomposition catalyst, FP 15 and poly(ADPribose) synthetase (PARP) inhibitor, N-(6-oxo-5,6-dihydrophenanthridin- 2-yl)-N, N-dimethylacetamide hydrochloride (PJ 34) (specific inhibitors of inducible nitric oxide synthase)	Gavage (10 or 30 mg/kg/day GED, 1 or 3 mg/kg/day FP15 and 3 or 10 mg/kg/day PJ34) twice a day from 5 wks of age until 12 wks	Chemopreventive activity of all compounds [142]
Min mice/female	(–)-epigallocatechin-3-gallate (EGCG) and fish oil	Diet supplemented (0.16% EGCG alone or in combination with high-fat fish oil diet, 20% w:w) for 9 wks	Inhibited tumor multiplicity by a combination of low doses of EGCG and fish oil [143]
Apc1638N mice/male and female	Aspirin and α-amylase resistant starch (RS)	Diet supplemented (125 g/kg diet RS or 0.3 g/kg aspirin alone or in combination) from 6 wks	The combination of two compounds showed more preventive activity than compounds alone [144]

d.w.: drinking water; n.d.: no data; p.o.: per os; t.r.: treadmill running; wk: week; wks: weeks.

3.3.2. F344-Pirc Rat Model

In 2007, Landgraf and colleagues developed a rat model carrying a knockout allele in the APC gene on an inbred F344/NTac genetic background rat [145]. To distinguish it from the APC mouse, they called this strain Pirc form (polyposis in the rat colon). The Pirc rats developed adenomas similar to those found in humans, showed the same progression to invasive carcinomas, and dependence on gender was observed, with males more prone to develop tumors in the intestinal tract than female rats [145]. In comparison to APC mouse models, this model takes some advantages due to the rat's size, the facility of diagnostic imaging, colonic predisposition, and longevity [10].

3.3.3. Hereditary Nonpolyposis Colon Cancer Mouse Models (HNPCC)

HNPCC is an inherited disease characterized by inactivated DNA mismatch repair genes, such as MLH1, MSH2, MSH6, and PMS2, that leads to the development of a variety of cancers, including colorectal cancer [15]. Mice with a targeted inactivating mutation in the mismatch repair genes are used to study these genes and evaluate their contribution to carcinogenesis [14,146]. Developed tumors are not specific to the colon-rectum; they occur in other organs such as the skin, lung, lymphatic system, stomach, and small intestine [15].

3.4. Xenograft and Syngeneic Models

Colorectal cancer cells or grafts of tissue can be implanted into animals to evaluate tumor development and to analyze the effects of several chemicals and natural compounds (Figure 5).

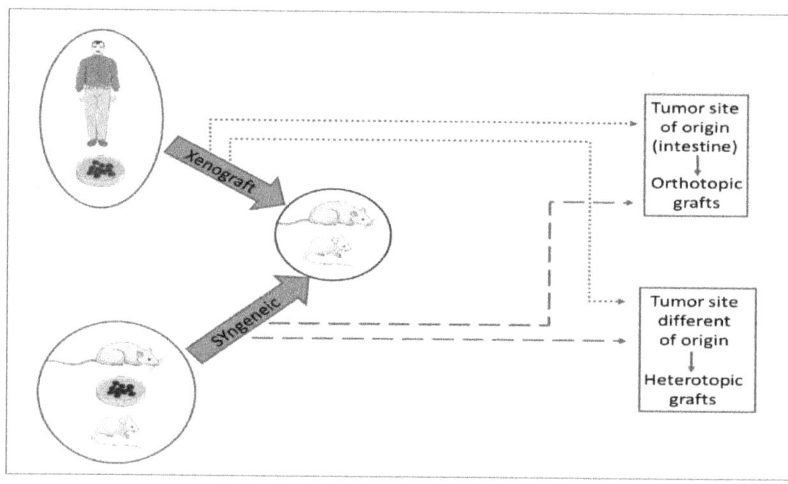

Figure 5. Schematic representation of xenograft and syngeneic models. In both models, the cells may be implanted in tumor site origin (orthotopic grafts) or a site different from tumor origin (heterotopic grafts).

In xenograft models, human tumor cells or tumor fragments are implanted into immunocompromised animals. Syngeneic tumor models are characterized by the use of animal tumor cell lines obtained from chemically induced rodent colorectal cancer and are implanted into animals with the same genetic background as the cell line [13]. These models can be used, among others, to study the effects of treatment on colorectal cancer metastases [4,147,148].

The cells may be implanted subcutaneously, intrasplenically into the renal capsule (heterotopic models), or directly in the colon or rectum (orthotopic models) [13,14]. The subcutaneous inoculation (heterotopic model) is one of the most used methods due to the simplicity, easy access, and high tumor growth [4]; however, the tumor microenvironment is different from the colon [13], and the metastases do not develop [149]. Comparing with the subcutaneous heterotopic model, the intrasplenically and renal capsule heterotopic models and the orthotopic model are technically more advanced and more difficult to work with, requiring the animals' anesthesia and use of imaging modalitics (e.g., ultrasonography) to implant the cells specifically in the spleen, the renal capsule or colon, respectively [150].

Orthotopic implantation refers to cells or tumor fragments implanted in the tumor site of origin, i.e., colon or rectum [13]. These models allow replication of tumor invasion, vascular spread, mimic the progression to advanced colorectal cancer in humans, and metastasize to distal organs [12]. For example, MCA-38 cells were intramurally injected into the cecum of C5BL/6J mice, and 40–65% of them developed metastases eight weeks later [151]. In another study, CT26 cells were transanal rectal injected in Balb/c mice with uptake rates of 65%, but only 3.3% developed metastases [152].

These models can be used to evaluate some therapeutic drugs. For example, Tao and colleagues used a commercial human colon cancer cell line, HCT-116, to evaluate the anti-colorectal cancer activity of Weichang'an, a Chinese herbal medicine, with 5-fluorouracil. The cells were injected subcutaneously in male BALB/C mice axilla, and after tumor growth, they were transplanted into the cecum. The group concluded that the compounds evaluated inhibited both colon tumor growth and hepatic metastases [153].

4. Conclusions

Experimental data concerning dietary, drugs, and natural compounds' effects on colon cancer models were reviewed in this work. Although several animal models are available to study colon cancer, there is no perfect model; all constitute an important tool to study human and animal colon carcinogenesis and to evaluate the potential effects of preventive and therapeutic strategies.

Whereas the AOM/DSS model mimetics the pathogenesis observed in human colorectal cancer, others like genetically engineered models allow studying genetic predisposition to the development of this type of cancer. The model selection should consider the studies' goals, the costs, and the advantages and disadvantages of each model, animal, strain, and gender.

Considering dietary patterns and natural products used as chemoprevention or chemotherapy, some like soy isoflavones, β-carotene, dried plums, fuji flavone, and Chinese cabbage showed an inhibitory effect on colorectal carcinogenesis, and adlay bran ethanol extract, grape seed extract, and pomegranate peel extract decreased the development of colonic premalignant lesions. However, groups that studied the effects of wheat bran and heme groups (in form of chicken, beef, black pudding) in the mice diet have concluded the opposite, observing a higher incidence of colorectal carcinogenesis.

In some cases, natural compounds, several drugs, and dietary patterns results are inconsistent and depend on multiple factors, and the best way to obtain better results is to select the most appropriate model and try to reduce most of the external factors. To achieve this goal, more research with controlled parameters is warranted. Moreover, ideally, the studies to evaluate the effects of natural compounds in CRC should not only evaluate the whole compound, but also each active substance in an isolated way. However, these studies imply the use of a higher number of animals, and consequently, higher costs for researchers, which may constitute a limitation.

Author Contributions: Writing—original draft preparation, E.N.-G., B.A.L.M., R.S.-R.; writing—review and editing, A.I.F.-R., A.G., P.A.O.; supervision, P.A.O. All authors have read and agreed to the published version of the manuscript.

Funding: This work was supported by National Funds by FCT—Portuguese Foundation for Science and Technology, under the projects UIDB/04033/2020 and UIDB/CVT/00772/2020, and post-graduation grant SFRH/BD/136747/2018.

Institutional Review Board Statement: Not applicable.

Informed Consent Statement: Not applicable.

Data Availability Statement: Data sharing not applicable.

Conflicts of Interest: The authors declare no conflict of interest.

References

1. Bray, F.; Ferlay, J.; Soerjomataram, I.; Siegel, R.; Torre, L.; Jemal, A. Global Cancer Statistics 2018: GLOBOCAN Estimates of Incidence and Mortality Worldwide for 36 Cancers in 185 Countries. *CA Cancer J. Clin.* **2018**, *68*, 1–31. [CrossRef] [PubMed]
2. *Diet, Nutrition, Physical Activity and Colorectal Cancer*; World Cancer Research Fund: London, UK; American Institute for Cancer Research: Washington, DC, USA, 2018.
3. Thanikachalam, K.; Khan, G. Colorectal Cancer and Nutrition. *Nutrients* **2019**, *11*, 164. [CrossRef] [PubMed]
4. Kobaek-Larsen, M.; Thorup, I.; Diederichsen, A.; Fenger, C.; Hoitinga, M.R. Review of Colorectal Cancer and Its Metastases in Rodent Models: Comparative Aspects with Those in Humans. *Comp. Med.* **2000**, *50*, 16–26. [PubMed]
5. Tanaka, T. Colorectal Carcinogenesis: Review of Human and Experimental Animal Studies. *J. Carcinog.* **2009**, *8*, 5. [CrossRef]
6. Machado, V.F.; Feitosa, M.R.; da Rocha, J.J.R.; Féres, O. A Review of Experimental Models in Colorectal Carcinogenesis. *J. Coloproctology* **2016**, *36*, 53–57. [CrossRef]
7. Raskov, H.; Pommergaard, H.-C.; Burcharth, J.; Rosenberg, J. Colorectal Carcinogenesis-Update and Perspectives. *World J. Gastroenterol.* **2014**, *20*, 18151. [CrossRef]
8. Fagundes, D.J.; Taha, M.O. Modelo Animal de Doença: Critérios de Escolha e Espécies de Animais de Uso Corrente. *Acta Cirúrgica Bras.* **2004**, *19*, 59–65. [CrossRef]
9. Johnson, R.L.; Fleet, J.C. Animal Models of Colorectal Cancer. *Cancer Metastasis Rev.* **2013**, *32*, 39–61. [CrossRef]
10. Kwong, L.N.; Dove, W.F. APC and Its Modifiers in Colon Cancer. In *APC Proteins*; Näthke, I.S., McCartney, B.M., Eds.; Springer: New York, NY, 2009; Volume 656, pp. 85–106. ISBN 978-1-4419-1144-5.
11. Oliveira, R.C.; Abrantes, A.M.; Tralhão, J.G.; Botelho, M.F. The role of mouse models in colorectal cancer research—The need and the importance of the orthotopic models. *Anim. Model Exp. Med.* **2020**, *3*, 1–8. [CrossRef]
12. Treuting, P.M.; Dintzis, S.M.; Montine, K.S. *Comparative Anatomy and Histology: A Mouse, Rat, and Human Atlas*; Academic Press: Cambridge, MA, USA, 2017; ISBN 978-0-12-802919-0.
13. Vdoviaková, K.; Petrovova, E.; Maloveska, M.; Krešáková, L.; Teleky, J.; Elias, M.; Petrášová, D. Surgical Anatomy of the Gastrointestinal Tract and Its Vasculature in the Laboratory Rat. *Gastroenterol. Res. Pract.* **2016**, *2016*, 2632368. [CrossRef]
14. Evans, J.P.; Sutton, P.A.; Winiarski, B.K.; Fenwick, S.W.; Malik, H.Z.; Vimalachandran, D.; Tweedle, E.M.; Costello, E.; Palmer, D.H.; Park, B.K.; et al. From Mice to Men: Murine Models of Colorectal Cancer for Use in Translational Research. *Crit. Rev. Oncol. Hematol.* **2016**, *98*, 94–105. [CrossRef]
15. Suman, S.; Datta, K. Animal Models of Colorectal Cancer in Chemoprevention and Therapeutics Development. In *Colorectal Cancer: From Prevention to Patient Care*; Intech Open: London, UK, 2012; pp. 277–300. [CrossRef]
16. Newmark, H.; Yang, K.; Kurihara, N.; Fan, K.; Augenlicht, L. Western-Style Diet-Induced Colonic Tumors and Their Modulation by Calcium and Vitamin D in C57Bl/6 Mice: A Preclinical Model for Human Sporadic Colon Cancer. *Carcinogenesis* **2008**, *30*, 88–92. [CrossRef]
17. Miyamoto, M.; Tani, Y. A Study on Colon Cancer-Prone Rats of WF-Osaka Strain. *Med. J. Osaka Univ.* **1989**, *38*, 1–12.
18. Lorenz, E.; Stewart, H.L. Intestinal Carcinoma and Other Lesions in Mice Following Oral Administration of 1, 2, 5, 6-Dibenzanthracene and 20-Methylcholanthrene. *J. Natl. Cancer Inst.* **1940**, *1*, 17–40.
19. Rosenberg, D.W.; Giardina, C.; Tanaka, T. Mouse Models for the Study of Colon Carcinogenesis. *Carcinogenesis* **2009**, *30*, 183–196. [CrossRef]
20. Greene, F.L.; Lamb, L.S.; Barwick, M. Colorectal Cancer in Animal Models-A Review. *J. Surg. Res.* **1987**, *43*, 476–487. [CrossRef]
21. Reddy, B.S. Colon Carcinogenesis Models for Chemoprevention Studies. *Cancer Chemoprev.* **1998**, *12*, 963–973. [CrossRef]
22. Walpole, A.L.; Williams, M.H.C.; Roberts, D.C. The Carcinogenic Action of 4-Aminodiphenyl and 3:2′-Dimethyl-4-Amino-Diphenyl. *Br. J. Ind. Med.* **1952**, *9*, 255–263.
23. Reddy, B.S.; Mori, H. Effect of Dietary Wheat Bran and Dehydrated Citrus Fiber on 3,2′-Dimethyl-4-Aminobiphenyl-Induced Intestinal Carcinogenesis in F344 Rats. *Carcinogenesis* **1981**, *2*, 21–25. [CrossRef]
24. Davis, C.D.; Feng, Y. Dietary Copper, Manganese and Iron Affect the Formation of Aberrant Crypts in Colon of Rats Administered 3,2′-Dimethyl-4-Aminobiphenyl. *J. Nutri.* **1999**, *129*, 1060–1067. [CrossRef]
25. Feng, Y.; Finley, J.W.; Davis, C.D.; Becker, W.K.; Fretland, A.J.; Hein, D.W. Dietary Selenium Reduces the Formation of Aberrant Crypts in Rats Administered 3,2′-Dimethyl-4-Aminobiphenyl. *Toxicol. Appl. Pharmacol.* **1999**, *157*, 36–42. [CrossRef]
26. Ravoori, S.; Feng, Y.; Neale, J.R.; Jeyabalan, J.; Hein, D.W.; Gupta, R.C. Dose-Dependent Reduction of 3,2′-Dimethyl-4-Aminobiphenyl- Derived DNA Adducts in Colon and Liver of Rats Administered Celecoxib. *Mutat. Res.* **2008**, *638*, 103–109. [CrossRef]
27. Williams, G.M.; Iatropoulos, M.J.; Jeffrey, A.M.; Shirai, T. Protective Effect of Acetaminophen against Colon Cancer Initiation Effects of 3,2′-Dimethyl-4-Aminobiphenyl in Rats. *Eur. J. Cancer Prev.* **2002**, *11*, 39–48. [CrossRef]
28. Narisawa, T.; Weisburger, J.H. Colon Cancer Induction in Mice by Intrarectal Instillation of N-Methylnitrosourea (38498). *Proc. Soc. Exp. Biol. Med.* **1075**, *148*, 166–169. [CrossRef]
29. Narisawa, T.; Fukaura, Y.; Takeba, N.; Nakai, K. Chemoprevention of N-Methylnitrosourea-Induced Colon Carcinogenesis by Ursodeoxycholic Acid-5-Aminosalicylic Acid Conjugate in F344 Rats. *Jpn. J. Cancer Res.* **2002**, *93*, 143–150. [CrossRef]

30. Wang, Y.; Jin, H.-Y.; Fang, M.-Z.; Wang, X.-F.; Chen, H.; Huang, S.-L.; Kong, D.-S.; Li, M.; Zhang, X.; Sun, Y.; et al. Epigallocatechin Gallate Inhibits Dimethylhydrazine-Induced Colorectal Cancer in Rats. *World J. Gastroenterol.* **2020**, *26*, 2064–2081. [CrossRef]
31. Nakayama, Y.; Inoue, Y.; Minagawa, N.; Onitsuka, K.; Nagata, J.; Shibao, K.; Hirata, K.; Sako, T.; Nagata, N.; Yamaguchi, K. Chemopreventive Effect of 4-[3,5-Bis(Trimethylsilyl) Benzamido] Benzoic Acid (TAC-101) on MNU-Induced Colon Carcinogenesis in a Rat Model. *Anticancer Res.* **2009**, *29*, 2059–2065.
32. McSherry, C.K.; Cohen, B.I.; Bokkenheuser, V.D.; Mosbach, E.H.; Winter, J.; Matoba, N.; Scholes, J. Effects of Calcium and Bile Acid Feeding on Colon Tumors in the Rat. *Cancer Res.* **1989**, *49*, 6039–6043.
33. Hawks, A.; Magee, P.N. The Alkylation of Nucleic Acids of Rat and Mouse in Vivo by the Carcinogen 1,2-Dimethylhydrazine. *Br. J. Cancer* **1974**, *30*, 440–447. [CrossRef]
34. Druckrey, H.; Preussmann, R.; Matzkies, F.; Ivankovic, S. Selektive Erzeugung von Darmkrebs bei Ratten durch 1,2-Dimethyl-hydrazin. *Naturwissenschaften* **1967**, *54*, 285–286. [CrossRef]
35. Thorup, I.; Meyer, O.; Kristiansen, E. Effect of a Dietary Fiber (Beet Fiber) on Dimethylhydrazine-Induced Colon Cancer in Wistar Rats. *Nutr. Cancer* **1992**, *17*, 251–261. [CrossRef] [PubMed]
36. Markert, C.; Rogers, L.W.; Chiu, J.-F. Induction of Colonic Adenocarcinomas by 1, 2-Dimethylhiydrazine Intrarectally Administered in Rats. *Digestion* **1978**, *18*, 261–265. [CrossRef] [PubMed]
37. Bissahoyo, A.; Pearsall, R.S.; Hanlon, K.; Amann, V.; Hicks, D.; Godfrey, V.L.; Threadgill, D.W. Azoxymethane Is a Genetic Background-Dependent Colorectal Tumor Initiator and Promoter in Mice: Effects of Dose, Route, and Diet. *Toxicol. Sci.* **2005**, *88*, 340–345. [CrossRef] [PubMed]
38. Reddy, B.S. Studies with the Azoxymethane-Rat Preclinical Model for Assessing Colon Tumor Development and Chemoprevention. *Environ. Mol. Mutagenesis* **2004**, *44*, 26–35. [CrossRef]
39. Sohn, O.S.; Fiala, E.S.; Puz, C.; Hamilton, S.R.; Williams, G.M. Enhancement of Rat Liver Microsomal Metabolism of Azoxymethane to Methylazoxymethanol by Chronic Ethanol Administration: Similarity to the Microsomal Metabolism of N-Nitrosodimethylamine. *Cancer Res.* **1987**, *47*, 3123–3129.
40. Derry, M.; Raina, K.; Agarwal, R.; Agarwal, C. Characterization of Azoxymethane-Induced Colon Tumor Metastasis to Lung in a Mouse Model Relevant to Human Sporadic Colorectal Cancer and Evaluation of Grape Seed Extract Efficacy. *Exp. Toxic Pathol.* **2014**, *66*, 2166–2171. [CrossRef]
41. Gois, E.; Daniel, R.; Parra, R.; Almeida, A.; Rocha, J.; Garcia, S.; Féres, O. Hyperbaric Oxygen Therapy Reduces COX-2 Expression in a Dimethylhydrazine-Induced Rat Model of Colorectal Carcinogenesis. *Undersea Hyperb. Med.* **2012**, *39*, 693–698.
42. Prabhu, P.N.; Ashokkumar, P.; Sudhandiran, G. Antioxidative and Antiproliferative Effects of Astaxanthin during the Initiation Stages of 1,2-Dimethyl Hydrazine-Induced Experimental Colon Carcinogenesis. *Fundam. Clin. Pharmacol.* **2009**, *23*, 225–234. [CrossRef]
43. Davis, A.E.; Patterson, F. Aspirin Reduces the Incidence of Colonic Carcinoma in the Dimethylhydrazine Rat Animal Model. *Aust. N. Z. J. Med.* **1994**, *24*, 301–303. [CrossRef]
44. Silva, M.F.; Sivieri, K.; Rossi, E.A. Effects of a Probiotic Soy Product and Physical Exercise on Formation of Pre-Neoplastic Lesions in Rat Colons in a Short-Term Model of Carcinogenic. *J. Int. Soc. Sports Nutr.* **2009**, *6*, 17. [CrossRef]
45. Miliaras, S.; Miliaras, D.; Vrettou, E.; Zavitsanakis, A.; Kiskinis, D. The Effect of Aspirin and High Fibre Diet on Colorectal Carcinoma: A Comparative Experimental Study. *Tech. Coloproctology* **2004**, *8*, s59–s61. [CrossRef]
46. Jisha, N.; Vysakh, A.; Vijeesh, V.; Anand, P.S.; Latha, M.S. Methanolic Extract of Muntingia Calabura L. Mitigates 1,2-Dimethyl Hydrazine Induced Colon Carcinogenesis in Wistar Rats. *Nutr. Cancer* **2020**, 1–13. [CrossRef]
47. Ranhotra, G.S.; Gelroth, J.A.; Glaser, B.K.; Schoening, P.; Brown, S.E. Cellulose and Calcium Lower the Incidence of Chemically-Induced Colon Tumors in Rats. *Plant Foods Hum. Nutr.* **1999**, *54*, 295–303. [CrossRef]
48. Chung, C.-P.; Hsu, H.-Y.; Huang, D.-W.; Hsu, H.-H.; Lin, J.-T.; Shih, C.-K.; Chiang, W. Ethyl Acetate Fraction of Adlay Bran Ethanolic Extract Inhibits Oncogene Expression and Suppresses DMH-Induced Preneoplastic Lesions of the Colon in F344 Rats through an Anti-Inflammatory Pathway. *J. Agric. Food Chem.* **2010**, *58*, 7616–7623. [CrossRef]
49. Pence, B.C.; Buddingh, F. Inhibition of Dietary Fat-Promoted Colon Carcinogenesis in Rats by Supplemental Calcium or Vitamin D_3. *Carcinogenesis* **1988**, *9*, 187–190. [CrossRef]
50. Femia, A.P.; Salvadori, M.; Broekaert, W.F.; François, I.E.J.A.; Delcour, J.A.; Courtin, C.M.; Caderni, G. Arabinoxylan-Oligosaccharides (AXOS) Reduce Preneoplastic Lesions in the Colon of Rats Treated with 1,2-Dimethylhydrazine (DMH). *Eur. J. Nutr.* **2010**, *49*, 127–132. [CrossRef]
51. Craven, P.A.; DeRubertis, F.R. Effects of Aspirin on 1, 2-Dimethylhydrazine-Induced Colonic Carcinogenesis. *Carcinogenesis* **1992**, *13*, 541–546. [CrossRef]
52. Kaur Saini, M.; Nath Sanyal, S. Evaluation of Chemopreventive Response of Two Cycloxygenase-2 Inhibitors, Etoricoxib and Diclofenac in Rat Colon Cancer Using FTIR and NMR Spectroscopic Techniques. *Nutr. Hosp.* **2010**, *25*, 577–585.
53. Min, W.-K.; Sung, H.-Y.; Choi, Y.-S. Suppression of Colonic Aberrant Crypt Foci by Soy Isoflavones Is Dose-Independent in Dimethylhydrazine-Treated Rats. *J. Med. Food* **2010**, *13*, 495–502. [CrossRef]
54. Yamamoto, I.; Maruyama, H.; Moriguchi, M. Effect of P-Carotene, Sodium Ascorbate and Cellulose on 1,2-Dimethylhydrazine-Induced Intestinal Carcinogenesis in Rats. *Cancer Lett.* **1994**, *86*, 5–9. [CrossRef]
55. Jacobs, L.R. Enhancement of Rat Colon Carcinogenesis by Wheat Bran Consumption during the Stage of 1,2-Dimethylhydrazine Administration. *Cancer Res.* **1983**, *43*, 6.

56. Nelson, R.L.; Tanure, J.C.; Andrianopoulos, G. The Effect of Dietary Milk and Calcium on Experimental Colorectal Carcinogenesis. *Dis. Colon Rectum* **1987**, *30*, 947–949. [CrossRef]
57. Viñas-Salas, J.; Biendicho-Palau, P.; Piñol-Felis, C.; Miguelsanz-Garcia, S.; Perez-Holanda, S. Calcium Inhibits Colon Carcinogenesis in an Experimental Model in the Rat. *Eur. J. Cancer* **1998**, *34*, 1941–1945. [CrossRef]
58. Earnest, L.; Wall, K.; Jolley, S.; Bhattacharyya, K. Chemoprevention of Azoxymethane-Induced Colonic Carcinogenesis by Supplemental Dietary Ursodeoxycholic Acid. *Cancer Res.* **1994**, *54*, 5071–5074.
59. Reddy, B.S.; Hirose, Y.; Lubet, R.; Steele, V.; Kelloff, G.; Paulson, S.; Seibert, K.; Rao, C.V. Chemoprevention of Colon Cancer by Specific Cyclooxygenase-2 Inhibitor, Celecoxib, Administered during Different Stages of Carcinogenesis. *Cancer Res.* **2000**, *60*, 293–297.
60. Rao, C.V.; Indranie, C.; Simi, B.; Manning, P.T.; Connor, J.R.; Reddy, B.S. Chemopreventive Properties of a Selective Inducible Nitric Oxide Synthase Inhibitor in Colon Carcinogenesis, Administered Alone or in Combination with Celecoxib, a Selective Cyclooxygenase-2 Inhibitor. *Cancer Res.* **2002**, *62*, 165–170.
61. Kawamori, T.; Tanaka, T.; HarÃ, A.; Yamahara, J.; Mori, H. Modifying Effects of Naturally Occurring Products on the Development of Colonie Aberrant Crypt Foci Induced by Azoxymethane in F344 Rats. *Cancer Res.* **1995**, *55*, 1277–1282.
62. Reddy, B.S.; Nayini, J.; Tokumo, K.; Rigotty, J.; Zang, E.; Kelloff, G. Chemoprevention of Colon Carcinogenesis by Concurrent Administration of Piroxicam, a Nonsteroidal Antiinflammatory Drug with d,l-α-Difluoromethylornithine, an Ornithine Decarboxylase Inhibitor, in Diet. *Cancer Res.* **1990**, *50*, 2562–2568.
63. Chinthalapally, R.; Desai, D.; Rivenson, A.; Simi, B.; Amin, S.; Reddy, B. Chemoprevention of Colon Carcinogenesis by Phenylethyl-3-Methylcaffeate1. *Cancer Res.* **1995**, *55*, 2310–2315.
64. Kawamori, T.; Rao, V.; Seibert, K.; Reddy, B.S. Chemopreventive Activity of Celecoxib, a Specific Cyclooxygenase-2 Inhibitor, against Colon Carcinogenesis. *Cancer Res.* **1998**, *58*, 409–412.
65. Reddy, B.S. Chemopreventive Effect of S-Methylmethane Thiosulfonate and Sulindac Administered Together during the Promotion/Progression Stages of Colon Carcinogenesis. *Carcinogenesis* **1999**, *20*, 1645–1648. [CrossRef] [PubMed]
66. Steele, V.E.; Rao, C.V.; Zhang, Y.; Patlolla, J.; Boring, D.; Kopelovich, L.; Juliana, M.M.; Grubbs, C.J.; Lubet, R.A. Chemopreventive Efficacy of Naproxen and Nitric Oxide-Naproxen in Rodent Models of Colon, Urinary Bladder, and Mammary Cancers. *Cancer Prev. Res.* **2009**, *2*, 951–956. [CrossRef]
67. Kim, K.P.; Whitehead, C.; Piazza, G.; Wargovich, M.J. Combinatorial Chemoprevention: Efficacy of Lovostatin and Exisulind on the Formation and Progression of Aberrant Crypt Foci. *Anticancer Res.* **2004**, *24*, 1805–1812. [PubMed]
68. Rao, C.V.; Steele, V.E.; Swamy, M.V.; Patlolla, J.M.R.; Guruswamy, S.; Kopelovich, L. Inhibition of Azoxymethane-Induced Colorectal Cancer by CP-31398, a TP53 Modulator, Alone or in Combination with Low Doses of Celecoxib in Male F344 Rats. *Cancer Res.* **2009**, *69*, 8175–8182. [CrossRef] [PubMed]
69. Reddy, B.S.; Rao, C.V.; Rivenson, A.; Kelloff, G. Inhibitory Effect of Aspirin on Azoxymethane-Induced Colon Carcinogenesis in F344 Rats. *Carcinogenesis* **1993**, *14*, 1493–1497. [CrossRef] [PubMed]
70. Kanauchi, O.; Mitsuyama, K.; Andoh, A.; Iwanaga, T. Modulation of Intestinal Environment by Prebiotic Germinated Barley Foodstuff Prevents Chemo-Induced Colonic Carcinogenesis in Rats. *Oncol. Rep.* **2008**, *20*, 793–801. [CrossRef] [PubMed]
71. Li, H.; Schut, H.A.J.; Conran, P.; Kramer, P.M.; Lubet, R.A.; Steele, V.E.; Hawk, E.E.; Kelloff, G.J.; Pereira, M.A. Prevention by Aspirin and Its Combination with α-Difluoromethylornithine of Azoxymethane-Induced Tumors, Aberrant Crypt Foci and Prostaglandin E2 Levels in Rat Colon. *Carcinogenesis* **1999**, *20*, 425–430. [CrossRef]
72. Mølck, A.-M.; Poulsen, M.; Meyer, O. The Combination of 1a,25(OH)2–Vitamin D3, Calcium and Acetylsalicylic Acid Affects Azoxymethane-Induced Aberrant Crypt Foci and Colorectal Tumours in Rats. *Cancer Lett.* **2002**, *10*, 19–28. [CrossRef]
73. Weisburger, J.H.; Rivenson, A.; Aliaga, C.; Reinhardt, J.; Kelloff, G.J.; Boone, C.W.; Steele, V.E.; Balentine, D.A.; Pittman, B.; Zang, E. Effect of Tea Extracts, Polyphenols, and Epigallocatechin Gallate on Azoxymethane-Induced Colon Cancer. *Exp. Biol. Med.* **1998**, *217*, 104–108. [CrossRef]
74. Reddy, B.S.; Wang, C.X.; Kong, A.-N.; Khor, T.O.; Zheng, X.; Steele, V.E.; Kopelovich, L.; Rao, C.V. Prevention of Azoxymethane-Induced Colon Cancer by Combination of Low Doses of Atorvastatin, Aspirin, and Celecoxib in F 344 Rats. *Cancer Res.* **2006**, *66*, 4542–4546. [CrossRef]
75. Velmurugan, B.; Singh, R.P.; Agarwal, R.; Agarwal, C. Dietary-Feeding of Grape Seed Extract Prevents Azoxymethane-Induced Colonic Aberrant Crypt Foci Formation in Fischer 344 Rats. *Mol. Carcinog.* **2010**, *49*, 641–652. [CrossRef]
76. Reddy, B.S.; Patlolla, J.M.; Simi, B.; Wang, S.H.; Rao, C.V. Prevention of Colon Cancer by Low Doses of Celecoxib, a Cyclooxygenase Inhibitor, Administered in Diet Rich in ω-3 Polyunsaturated Fatty Acids. *Cancer Res.* **2005**, *65*, 8022–8027. [CrossRef]
77. Corpet, D.E.; Parnaud, G. Polyethylene-Glycol, a Potent Suppressor of Azoxymethane-Induced Colonic Aberrant Crypt Foci in Rats. *Carcinogenesis* **1999**, *20*, 915–918. [CrossRef]
78. Pierre, F.; Freeman, A.; Taché, S.; Van der Meer, R.; Corpet, D.E. Beef Meat and Blood Sausage Promote the Formation of Azoxymethane-Induced Mucin-Depleted Foci and Aberrant Crypt Foci in Rat Colons. *J. Nutr.* **2004**, *134*, 2711–2716. [CrossRef]
79. Watanabe, K.; Reddy, B.S.; Weisburger, J.H.; Kritchevsky, D. Effect of Dietary Alfalfa, Pectin, and Wheat Bran on Azoxymethane-or Methylnitrosourea-Induced Colon Carcinogenesis in F344 Rats. *J. Natl. Cancer Inst.* **1979**, *63*, 141–145.
80. Melo, A.; Mendonça, M.; Rosa-Castro, R. The Protective Effects of Fermented Kefir Milk on Azoxymethane-Induced Aberrant Crypt Formation in Mice Colon. *Tissue Cell* **2018**, *52*, 51–56. [CrossRef]

81. Clarke, J.M.; Topping, D.L.; Bird, A.R.; Young, G.P.; Cobiac, L. Effects of High-Amylose Maize Starch and Butyrylated High-Amylose Maize Starch on Azoxymethane-Induced Intestinal Cancer in Rats. *Carcinogenesis* **2008**, *29*, 2190–2194. [CrossRef]
82. Bonin, A.M.; Yáñez, J.A.; Fukuda, C.; Teng, X.W.; Dillon, C.T.; Hambley, T.W.; Lay, P.A.; Davies, N.M. Inhibition of Experimental Colorectal Cancer and Reduction in Renal and Gastrointestinal Toxicities by Copper–Indomethacin in Rats. *Cancer Chemother. Pharmacol.* **2010**, *66*, 755–764. [CrossRef]
83. Martin, J.E.; Le Leu, R.K.; Hu, Y.; Young, G.P. R-Flurbiprofen Suppresses Distal Nonmucin-Producing Colorectal Tumors in Azoxymethane-Treated Rats, without Suppressing Eicosanoid Production. *Am. J. Physiol. Gastrointest. Liver Physiol.* **2010**, *298*, G860–G864. [CrossRef]
84. Raju, J.; Bielecki, A.; Caldwell, D.; Lok, E.; Taylor, M.; Kapal, K.; Curran, I.; Cooke, G.M.; Bird, R.P.; Mehta, R. Soy Isoflavones Modulate Azoxymethane-Induced Rat Colon Carcinogenesis Exposed Pre- and Postnatally and Inhibit Growth of DLD-1 Human Colon Adenocarcinoma Cells by Increasing the Expression of Estrogen Receptor-β. *J. Nutr.* **2009**, *139*, 474–481. [CrossRef]
85. Le Leu, R.K.; Hu, Y.; Brown, I.L.; Woodman, R.J.; Young, G.P. Synbiotic Intervention of Bifidobacterium Lactis and Resistant Starch Protects against Colorectal Cancer Development in Rats. *Carcinogenesis* **2010**, *31*, 246–251. [CrossRef] [PubMed]
86. Liu, H.; Zhang, L.; Li, G.; Gao, Z. Xanthohumol Protects against Azoxymethane-Induced Colorectal Cancer in Sprague-Dawley Rats. *Environ. Toxicol.* **2020**, *35*, 136–144. [CrossRef] [PubMed]
87. Braga, V.N.L.; Juanes, C.D.C.; Peres Júnior, H.D.S.; Sousa, J.R.D.; Cavalcanti, B.C.; Jamacaru, F.V.F.; Lemos, T.L.G.D.; Dornelas, C.A. Gum Arabic and Red Propolis Protecteting Colorectal Preneoplastic Lesions in a Rat Model of Azoxymethane. *Acta Cir. Bras.* **2019**, *34*, e201900207. [CrossRef] [PubMed]
88. Tanaka, T.; Kohno, H.; Suzuki, R.; Yamada, Y.; Sugie, S.; Mori, H. A Novel Inflammation-Related Mouse Colon Carcinogenesis Model Induced by Azoxymethane and Dextran Sodium Sulfate. *Cancer Sci.* **2003**, *94*, 965–973. [CrossRef]
89. Tanaka, T.; Suzuki, R.; Kohno, H.; Sugie, S.; Takahashi, M.; Wakabayashi, K. Colonic Adenocarcinomas Rapidly Induced by the Combined Treatment with 2-Amino-1-Methyl-6-Phenylimidazo[4,5-b]Pyridine and Dextran Sodium Sulfate in Male ICR Mice Possess β-Catenin Gene Mutations and Increases Immunoreactivity for β-Catenin, Cyclooxygenas. *Carcinogenesis* **2005**, *26*, 229–238. [CrossRef]
90. Kohno, H.; Suzuki, R.; Sugie, S.; Tanaka, T. β-Catenin Mutations in a Mouse Model of Inflammation-Related Colon Carcinogenesis Induced by 1,2-Dimethylhydrazine and Dextran Sodium Sulfate. *Cancer Sci.* **2005**, *96*, 69–76. [CrossRef]
91. Suzuki, R.; Kohno, H.; Sugie, S.; Nakagama, H.; Tanaka, T. Strain Differences in the Susceptibility to Azoxymethane and Dextran Sodium Sulfate-Induced Colon Carcinogenesis in Mice. *Carcinogenesis* **2006**, *27*, 162–169. [CrossRef]
92. Robertis, M.; Massi, E.; Poeta, M.L.; Carotti, S.; Cecchetelli, L.; Signori, E.; Fazio, V.M. The AOM/DSS Murine Model for the Study of Colon Carcinogenesis: From Pathways to Diagnosis and Therapy Studies. *J. Carcinog.* **2011**, 10. [CrossRef]
93. Takayama, K.; Yamashita, K.; Wakabayashi, K.; Sugimura, T.; Nagao, M. DNA Modification by 2-Amino-1-Methyl-6-Phenylimidazo[4,5-b]Pyridine in Rats. *Jpn. J. Cancer Res.* **1989**, *80*, 1145–1148. [CrossRef]
94. Nakagama, H.; Nakanishi, M.; Ochiai, M. Modeling Human Colon Cancer in Rodents Using a Food-Borne Carcinogen, PhIP. *Cancer Sci.* **2005**, *96*, 627–636. [CrossRef]
95. Ito, N.; Hasegawa, R.; Sano, M.; Tamano, S.; Esumi, H.; Takayama, S.; Sugimura, T. A New Colon and Mammary Carinogen in Cooked Food, 1-Amino-1-Methyl-6-Phenylimidazo[4,5-b]Pyridine (PhIP). *Carcinogenesis* **1991**, *12*, 1503–1506. [CrossRef]
96. Guo, Y.; Liu, Y.; Zhang, C.; Su, Z.-Y.; Li, W.; Huang, M.-T.; Kong, A.-N. The Epigenetic Effects of Aspirin: The Modification of Histone H3 Lysine 27 Acetylation in the Prevention of Colon Carcinogenesis in Azoxymethane- and Dextran Sulfate Sodium-Treated CF-1 Mice. *Carcinogenesis* **2016**, *37*, 616–624. [CrossRef]
97. Chung, K.-S.; Cheon, S.-Y.; Roh, S.-S.; Lee, M.; An, H.-J. Chemopreventive Effect of Aster Glehni on Inflammation-Induced Colorectal Carcinogenesis in Mice. *Nutrients* **2018**, *10*, 202. [CrossRef]
98. Nam, S.Y.; Kim, J.S.; Kim, J.M.; Lee, J.Y.; Kim, N.; Jung, H.C.; Song, I.S. DA-6034, a Derivative of Flavonoid, Prevents and Ameliorates Dextran Sulfate Sodium-Induced Colitis and Inhibits Colon Carcinogenesis. *Exp. Biol. Med.* **2008**, *233*, 180–191. [CrossRef]
99. Zhao, Q.; Bi, Y.; Zhong, J.; Ren, Z.; Liu, Y.; Jia, J.; Yu, M.; Tan, Y.; Zhang, Q.; Yu, X. Pristimerin Suppresses Colorectal Cancer through Inhibiting Inflammatory Responses and Wnt/β-Catenin Signaling. *Toxicol. Appl. Pharmacol.* **2020**, *386*, 114813. [CrossRef]
100. Wu, M.; Li, J.; An, Y.; Li, P.; Xiong, W.; Li, J.; Yan, D.; Wang, M.; Zhong, G. Chitooligosaccharides Prevents the Development of Colitis-Associated Colorectal Cancer by Modulating the Intestinal Microbiota and Mycobiota. *Front. Microbiol.* **2019**, *10*, 2101. [CrossRef]
101. Moreira, T.G.; Horta, L.S.; Gomes-Santos, A.C.; Oliveira, R.P.; Queiroz, N.M.G.P.; Mangani, D.; Daniel, B.; Vieira, A.T.; Liu, S.; Rodrigues, A.M.; et al. CLA-Supplemented Diet Accelerates Experimental Colorectal Cancer by Inducing TGF-β-Producing Macrophages and T Cells. *Mucosal Immunol.* **2018**, *12*, 188–199. [CrossRef]
102. Huo, X.; Liu, D.; Gao, L.; Li, L.; Cao, L. Flavonoids Extracted from Licorice Prevents Colitis-Associated Carcinogenesis in AOM/DSS Mouse Model. *Int. J. Mol. Sci.* **2016**, *17*, 1343. [CrossRef]
103. Im, S.-A.; Kim, J.-W.; Kim, H.-S.; Park, C.-S.; Shin, E.; Do, S.-G.; Park, Y.I.; Lee, C.-K. Prevention of Azoxymethane/Dextran Sodium Sulfate-Induced Mouse Colon Carcinogenesis by Processed Aloe Vera Gel. *Int. Immunopharmacol.* **2016**, *40*, 428–435. [CrossRef] [PubMed]
104. Ki, H.-H.; Lee, J.-H.; Lee, H.-Y.; Lee, Y.-M.; Kim, D.-K. Triticum Aestivum Sprouts Extract Inhibits Azoymethane (AOM)/Dextran Sodium Sulfate (DSS)-Induced Colon Carcinogenesis in Mice. *Nutr. Cancer* **2018**, *70*, 928–937. [CrossRef]

105. Kim, M.; Miyamoto, S.; Yasui, Y.; Oyama, T.; Murakami, A.; Tanaka, T. Zerumbone, a Tropical Ginger Sesquiterpene, Inhibits Colon and Lung Carcinogenesis in Mice. *Int. J. Cancer* **2009**, *124*, 264–271. [CrossRef] [PubMed]
106. Kohno, H.; Suzuki, R.; Curini, M.; Epifano, F.; Maltese, F.; Gonzales, S.P.; Tanaka, T. Dietary Administration with Prenyloxy-coumarins, Auraptene and Collinin, Inhibits Colitis-Related Colon Carcinogenesis in Mice. *Int. J. Cancer* **2006**, *118*, 2936–2942. [CrossRef] [PubMed]
107. Kohno, H.; Suzuki, R.; Yasui, Y.; Miyamoto, S.; Wakabayashi, K.; Tanaka, T. Ursodeoxycholic Acid versus Sulfasalazine in Colitis-Related Colon Carcinogenesis in Mice. *Clin. Cancer Res.* **2007**, *13*, 2519–2525. [CrossRef] [PubMed]
108. Choi, Y.J.; Choi, Y.J.; Kim, N.; Nam, R.H.; Lee, S.; Lee, H.S.; Lee, H.-N.; Surh, Y.-J.; Lee, D.H. Açaí Berries Inhibit Colon Tumorigenesis in Azoxymethane/Dextran Sulfate Sodium-Treated Mice. *Gut Liver* **2017**, *11*, 243–252. [CrossRef]
109. Terasaki, M.; Masaka, S.; Fukada, C.; Houzaki, M.; Endo, T.; Tanaka, T.; Maeda, H.; Miyashita, K.; Mutoh, M. Salivary Glycine Is a Significant Predictor for the Attenuation of Polyp and Tumor Microenvironment Formation by Fucoxanthin in AOM/DSS Mice. *In Vivo* **2019**, *33*, 365–374. [CrossRef]
110. Kohno, H.; Suzuki, R.; Sugie, S.; Tanaka, T. Suppression of Colitis-Related Mouse Colon Carcinogenesis by a COX-2 Inhibitor and PPAR Ligands. *BMC Cancer* **2005**, *5*, 46. [CrossRef]
111. Predes, D.; Oliveira, L.F.S.; Ferreira, L.S.S.; Maia, L.A.; Delou, J.M.A.; Faletti, A.; Oliveira, I.; Amado, N.G.; Reis, A.H.; Fraga, C.A.M.; et al. The Chalcone Lonchocarpin Inhibits Wnt/β-Catenin Signaling and Suppresses Colorectal Cancer Proliferation. *Cancers* **2019**, *11*, 1968. [CrossRef]
112. Canene-Adams, K.; Sfanos, K.S.; Liang, C.-T.; Yegnasubramanian, S.; Nelson, W.G.; Brayton, C.; De Marzo, A.M. Dietary Chemoprevention of PhIP Induced Carcinogenesis in Male Fischer 344 Rats with Tomato and Broccoli. *PLoS ONE* **2013**, *8*, e79842. [CrossRef]
113. Narushima, S.; Sakata, T.; Hioki, K.; Itoh, T.; Nomura, T.; Itoh, K. Inhibitory Effect of Yogurt on Aberrant Crypt Foci Formation in the Rat Colon and Colorectal Tumorigenesis in RasH2 Mice. *Exp. Anim.* **2010**, *59*, 487–494. [CrossRef]
114. Carter, O.; Wang, R.; Dashwood, W.M.; Orner, G.A.; Fischer, K.A.; Löhr, C.V.; Pereira, C.B.; Bailey, G.S.; Williams, D.E.; Dashwood, R.H. Comparison of White Tea, Green Tea, Epigallocatechin-3-Gallate, and Caffeine as Inhibitors of PhIP-Induced Colonic Aberrant Crypts. *Nutr. Cancer* **2007**, *58*, 60–65. [CrossRef]
115. Tang, M.X.; Ogawa, K.; Asamoto, M.; Chewonarin, T.; Suzuki, S.; Tanaka, T.; Shirai, T. Effects of Nobiletin on PhIP-Induced Prostate and Colon Carcinogenesis in F344 Rats. *Nutr. Cancer* **2011**, *63*, 227–233. [CrossRef]
116. Santana-Rios, G.; Orner, G.A.; Xu, M.; Izquierdo-Pulido, M. Inhibition by White Tea of 2-Amino-1-Methyl-6-Phenylimidazo[4,5-b]Pyridine-Induced Colonic Aberrant Crypts in the F344 Rat. *Nutr. Cancer* **2001**, *41*. [CrossRef]
117. Hikosaka, A. Inhibitory Effects of Soy Isoflavones on Rat Prostate Carcinogenesis Induced by 2-Amino-1-Methyl-6-Phenylimidazo[4,5-b]Pyridine (PhIP). *Carcinogenesis* **2003**, *25*, 381–387. [CrossRef]
118. Guo, D.; Schut, H.A.; Davis, C.D.; Snyderwine, E.G.; Bailey, G.S.; Dashwood, R.H. Protection by Chlorophyllin and Indole-3-Carbinol against 2-Amino-1-Methyl-6-Phenylimidazo[4,5-b]Pyridine (PhIP)-Induced DNA Adducts and Colonic Aberrant Crypts in the F344 Rat. *Carcinogenesis* **1995**, *16*, 2931–2937. [CrossRef]
119. Hagiwara, A.; Boonyaphiphat, P.; Tanaka, H.; Kawabe, M.; Tamano, S.; Kaneko, H.; Matsui, M.; Hirose, M.; Ito, N.; Shirai, T. Organ-Dependent Modifying Effects of Caffeine, and Two Naturally Occurring Antioxidants Alpha-Tocopherol and n-Tritriacontane-16,18-Dione, on 2-Amino-1-Methyl-6-Phenylimidazo[4,5-b]Pyridine (PhIP)-Induced Mammary and Colonic Carcinogenesis in Female F344 Rats. *Jpn. J. Cancer Res.* **1999**, *90*, 399–405. [CrossRef]
120. Tan, W. Chemoprevention of 2-Amino-1-Methyl-6-Phenyli-Midazo [4,5-b] Pyridine-Induced Carcinogen-DNA Adducts by Chinese Cabbage in Rats. *World J. Gastroenterol.* **1999**, *5*, 138. [CrossRef]
121. Yamada, Y.; Mori, H. Multistep Carcinogenesis of the Colon in Apc Min/+ Mouse. *Cancer Sci.* **2007**, *98*, 6–10. [CrossRef]
122. Moser, A.; Gledhill, S.; Dove, W. Loss of Apc in Intestinal Adenomas from Min Mice. *Cancer Res.* **1994**, *54*, 5947–5952.
123. Floyd, R.A.; Towner, R.A.; Wu, D.; Abbott, A.; Cranford, R.; Branch, D.; Guo, W.-X.; Foster, S.B.; Jones, I.; Alam, R.; et al. Anti-Cancer Activity of Nitrones in the APCMin/+ Model of Colorectal Cancer. *Free Radic. Res.* **2010**, *44*, 108–117. [CrossRef]
124. Mutanen, M.; Pajari, A.-M.; Päivärinta, E.; Misikangas, M.; Rajakangas, J.; Marttinen, M.; Oikarinen, S. Berries as Chemopreventive Dietary Constituents—A Mechanistic Approach with the $Apc^{Min/+}$ Mouse. *Asia Pac. J. Clin. Nutr.* **2008**, *17* (Suppl. S1), 123–125.
125. Swamy, M.V.; Patlolla, J.M.R.; Steele, V.E.; Kopelovich, L.; Reddy, B.S.; Rao, C.V. Chemoprevention of Familial Adenomatous Polyposis by Low Doses of Atorvastatin and Celecoxib Given Individually and in Combination to APC^{Min} Mice. *Cancer Res.* **2006**, *66*, 7370–7377. [CrossRef]
126. Ritland, S.R.; Gendler, S.J. Chemoprevention of Intestinal Adenomas in the ApcMin Mouse by Piroxicam: Kinetics, Strain Effects and Resistance to Chemosuppression. *Carcinogenesis* **1999**, *20*, 51–58. [CrossRef]
127. Fini, L.; Piazzi, G.; Daoud, Y.; Selgrad, M.; Maegawa, S.; Garcia, M.; Fogliano, V.; Romano, M.; Graziani, G.; Vitaglione, P.; et al. Chemoprevention of Intestinal Polyps in $Apc^{Min/+}$ Mice Fed with Western or Balanced Diets by Drinking Annurca Apple Polyphenol Extract. *Cancer Prev. Res.* **2011**, *4*, 907–915. [CrossRef]
128. Phutthaphadoong, S.; Yamada, Y.; Hirata, A.; Tomita, H.; Hara, A.; Limtrakul, P.; Iwasaki, T.; Kobayashi, H.; Mori, H. Chemopreventive Effect of Fermented Brown Rice and Rice Bran (FBRA) on the Inflammation-Related Colorectal Carcinogenesis in $Apc^{Min/+}$ Mice. *Oncol. Rep.* **2010**, *23*, 53–59. [CrossRef]

129. Khor, T.O.; Hu, R.; Shen, G.; Jeong, W.-S.; Hebbar, V.; Chen, C.; Xu, C.; Nair, S.; Reddy, B.; Chada, K.; et al. Pharmacogenomics of Cancer Chemopreventive Isothiocyanate Compound Sulforaphane in the Intestinal Polyps of $Apc^{Min/+}$ Mice. *Biopharm. Drug Dispos.* **2006**, *27*, 407–420. [CrossRef]
130. Smith, D.L.H.; Keshavan, P.; Avissar, U.; Ahmed, K.; Zucker, S.D. Sodium Taurocholate Inhibits Intestinal Adenoma Formation in $APC^{Min/+}$ Mice, Potentially through Activation of the Farnesoid X Receptor. *Carcinogenesis* **2010**, *31*, 1100–1109. [CrossRef]
131. Tomimoto, A.; Endo, H.; Sugiyama, M.; Fujisawa, T.; Hosono, K.; Takahashi, H.; Nakajima, N.; Nagashima, Y.; Wada, K.; Nakagama, H.; et al. Metformin Suppresses Intestinal Polyp Growth in $Apc^{Min/+}$ Mice. *Cancer Sci.* **2008**, *99*, 2136–2141. [CrossRef]
132. Rajamanickam, S.; Velmurugan, B.; Kaur, M.; Singh, R.P.; Agarwal, R. Chemoprevention of Intestinal Tumorigenesis in APCmin/+ Mice by Silibinin. *Cancer Res.* **2010**, *70*, 2368–2378. [CrossRef]
133. Barnes, C.J.; Lee, M. Chemoprevention of Spontaneous Intestinal Adenomas in the Adenomatous Polyposis Coli Min Mouse Model with Aspirin. *Gastroenterology* **1998**, *114*, 873–877. [CrossRef]
134. Murphy, E.A.; Davis, J.M.; McClellan, J.L.; Gordon, B.T.; Carmichael, M.D. Curcumin's Effect on Intestinal Inflammation and Tumorigenesis in the $Apc^{Min/+}$ Mouse. *J. Interferon Cytokine Res.* **2011**, *31*, 219–226. [CrossRef]
135. Bobe, G.; Wang, B.; Seeram, N.P.; Nair, M.G.; Bourquin, L.D. Dietary Anthocyanin-Rich Tart Cherry Extract Inhibits Intestinal Tumorigenesis in APC^{Min} Mice Fed Suboptimal Levels of Sulindac. *J. Agric. Food Chem.* **2006**, *54*, 9322–9328. [CrossRef] [PubMed]
136. Mehl, K.A.; Davis, J.M.; Clements, J.M.; Berger, F.G.; Pena, M.M.; Carson, J.A. Decreased Intestinal Polyp Multiplicity Is Related to Exercise Mode and Gender in $Apc^{Min/+}$ Mice. *J. Appl. Physiol.* **2005**, *98*, 2219–2225. [CrossRef] [PubMed]
137. Sale, S.; Tunstall, R.G.; Ruparelia, K.C.; Butler, P.C.; Potter, G.A.; Steward, W.P.; Gescher, A.J. Effects of the Potential Chemopreventive Agent DMU-135 on Adenoma Development in the Apc^{Min+} Mouse. *Investig. New Drugs* **2006**, *24*, 459–464. [CrossRef] [PubMed]
138. Yamaguchi, K.; Cekanova, M.; McEntee, M.F.; Yoon, J.-H.; Fischer, S.M.; Renes, I.B.; Van Seuningen, I.; Baek, S.J. Peroxisome Proliferator-Activated Receptor Ligand MCC-555 Suppresses Intestinal Polyps in $Apc^{Min/+}$ Mice via Extracellular Signal-Regulated Kinase and Peroxisome Proliferator-Activated Receptor-Dependent Pathways. *Mol. Cancer Ther.* **2008**, *7*, 2779–2787. [CrossRef]
139. Sørensen, I.K.; Kristiansen, E.; Mortensen, A.; Nicolaisen, G.M.; Wijnands, J.A.H.; van Kranen, H.J.; van Kreijl, C.F. The Effect of Soy Isoflavones on the Development of Intestinal Neoplasia in ApcMin Mouse. *Cancer Lett.* **1998**, *130*, 217–225. [CrossRef]
140. Fan, K.; Kurihara, N.; Abe, S.; Ho, C.-T.; Ghai, G.; Yang, K. Chemopreventive Effects of Orange Peel Extract (OPE) I. OPE Inhibits Intestinal Tumor Growth in $Apc^{Min/+}$ Mice. *J. Med. Food* **2007**, *10*, 11–17. [CrossRef]
141. Baltgalvis, K.A.; Berger, F.G.; Peña, M.M.O.; Davis, J.M.; Carson, J.A. Effect of Exercise on Biological Pathways in $Apc^{Min/+}$ Mouse Intestinal Polyps. *J. Appl. Physiol.* **2008**, *104*, 1137–1143. [CrossRef]
142. Mabley, J.G.; Pacher, P.; Bai, P.; Wallace, R.; Goonesekera, S.; Virag, L.; Southan, G.J.; Szabó, C. Suppression of Intestinal Polyposis in $Apc^{min/+}$ Mice by Targeting the Nitric Oxide or Poly(ADP-Ribose) Pathways. *Mutat. Res. Fundam. Mol. Mech. Mutagenesis* **2004**, *548*, 107–116. [CrossRef]
143. Bose, M.; Hao, X.; Ju, J.; Husain, A.; Park, S.; Lambert, J.D.; Yang, C.S. Inhibition Of Tumorigenesis in $Apc^{Min/+}$ Mice by a Combination of (–)-Epigallocatechin-3-Gallate and Fish Oil. *J. Agric. Food Chem.* **2007**, *55*, 7695–7700. [CrossRef]
144. Williamson, S.L.H.; Kartheuser, A.; Coaker, J.; Kooshkghazi, M.D.; Fodde, R.; Burn, J.; Mathers, J.C. Intestinal Tumorigenesis in the Apc1638N Mouse Treated with Aspirin and Resistant Starch for up to 5 Months. *Carcinogenesis* **1999**, *20*, 805–810. [CrossRef]
145. Amos-Landgraf, J.M.; Kwong, L.N.; Kendziorski, C.M.; Reichelderfer, M.; Torrealba, J.; Weichert, J.; Haag, J.D.; Chen, K.-S.; Waller, J.L.; Gould, M.N.; et al. A Target-Selected Apc-Mutant Rat Kindred Enhances the Modeling of Familial Human Colon Cancer. *Proc. Natl. Acad. Sci. USA* **2007**, *104*, 4036–4041. [CrossRef]
146. Edelmann, L.; Edelmann, W. Loss of DNA Mismatch Repair Function and Cancer Predisposition in the Mouse: Animal Models for Human Hereditary Nonpolyposis Colorectal Cancer. *Am. J. Med Genet.* **2004**, *129C*, 91–99. [CrossRef]
147. Chang, Y.-J.; Hsu, W.-H.; Chang, C.-H.; Lan, K.-L.; Ting, G.; Lee, T.-W. Combined Therapeutic Efficacy of 188Re-Liposomes and Sorafenib in an Experimental Colorectal Cancer Liver Metastasis Model by Intrasplenic Injection of C26-Luc Murine Colon Cancer Cells. *Mol. Clin. Oncol.* **2014**, *2*, 380–384. [CrossRef]
148. Dunnington, D.J.; Buscarino, C.; Gennaro, D.; Greig, R.; Poste, G. Characterization of an Animal Model of Metastatic Colon Carcinoma. *Int. J. Cancer* **1987**, *39*, 248–254. [CrossRef]
149. Xie, X.; Briinnert, N.; Jensen, G.; Albrectsen, J.; Gotthardsen, B.; Rygaard, J. Comparative Studies between Nude and Scid Mice on the Growth and Metastatic Behavior of Xenografted Human Tumors. *Clin. Exp. Metastasis* **1992**, *10*, 201–210. [CrossRef]
150. Mittal, V.K.; Bhullar, J.S.; Jayant, K. Animal models of human colorectal cancer: Current status, uses and limitations. *World J. Gastroenterol.* **2015**, *21*, 11854–11861. [CrossRef]
151. Goldrosen, M.H. Murine Colon Adenocarcinoma. Immunobiology of Metastases. *Cancer* **1980**, *45*, 1223–1228. [CrossRef]
152. Donigan, M.; Loh, B.D.; Norcross, L.S.; Li, S.; Williamson, P.R.; DeJesus, S.; Ferrara, A.; Gallagher, J.T.; Baker, C.H. A Metastatic Colon Cancer Model Using Nonoperative Transanal Rectal Injection. *Surg. Endosc.* **2010**, *24*, 642–647. [CrossRef]
153. Tao, L.; Yang, J.-K.; Gu, Y.; Zhou, X.; Zhao, A.-G.; Zheng, J.; Zhu, Y.-J. Weichang'an and 5-Fluorouracil Suppresses Colorectal Cancer in a Mouse Model. *World J. Gastroenterol.* **2015**, *21*, 1125–1139. [CrossRef]

MDPI
St. Alban-Anlage 66
4052 Basel
Switzerland
Tel. +41 61 683 77 34
Fax +41 61 302 89 18
www.mdpi.com

Veterinary Sciences Editorial Office
E-mail: vetsci@mdpi.com
www.mdpi.com/journal/vetsci

www.ingramcontent.com/pod-product-compliance
Lightning Source LLC
Chambersburg PA
CBHW040815250926
51850CB00007BA/1171
9783036570839